International Workshop on Embodied Intelligence (EI 2023)

IOP Conference Series: Materials Science and Engineering
Volume 1321

Cambridge, United Kingdom
22-24 March 2023

ISBN: 979-8-3313-1144-5
ISSN: 1757-8981

Printed from e-media with permission by:

Curran Associates, Inc.
57 Morehouse Lane
Red Hook, NY 12571

Some format issues inherent in the e-media version may also appear in this print version.

Institute of Physics
Dirac House, Temple Back
Bristol BS1 6BE UK

Phone: 44 1 17 929 7481
Fax: 44 1 17 920 0979

techtracking@iop.org

Additional copies of this publication are available from:

Curran Associates, Inc.
57 Morehouse Lane
Red Hook, NY 12571 USA
Phone: 845-758-0400
Fax: 845-758-2633
Email: curran@proceedings.com
Web: www.proceedings.com

TABLE OF CONTENTS

IOP Conf. Series: Materials Science and Engineering **1321** (2024) 011001 doi:10.1088/1757-899X/1321/1/011001

Preface

With the rapid rise and recent success of artificial intelligence, there has been growing interest in the significance of embodiment, the formal limits of AI, and the nature of cognition. Additionally, the increasing computational complexity required for training AI models has motivated the research community to explore alternative methods of computation. One promising direction is leveraging the physical properties of agents' bodies to naturally simplify problems and reduce computational demands. These challenges call for a multidisciplinary approach, incorporating insights not only from computer science and statistics but also from philosophy, biology, mechanical and materials engineering.

The International Conference on Embodied Intelligence, now in its third year, has firmly established itself as a leading venue for bringing together researchers from diverse disciplines such as robotics, material science, sociology, philosophy, computer science, and beyond. This conference provides a unique, interdisciplinary platform where experts can come together to explore the rich, multifaceted concept of embodied intelligence. Through these discussions, the conference continues to shape research agendas and guide new directions for this rapidly evolving field. Our invited plenary speakers have been instrumental in steering the conversation, addressing some of the most critical challenges at the heart of embodied intelligence. Their presentations, along with dynamic panel discussions, have sparked deeper reflections and collaborative efforts aimed at finding practical solutions.

In this proceedings volume, contributors from across the conference including short talks, plenaries, and breakout sessions, share their insights on the pressing issues and challenges discussed at the event. Their articles also examine potential solutions, emerging trends, and future perspectives, ensuring that the dialogue surrounding embodied intelligence continues to grow and expand beyond the confines of the conference.

The book opens by offering a range of perspectives and outlooks on Embodied Intelligence, followed by chapters showcasing examples of soft robotics and Morphological Computing systems. The final chapters focus on Artificial Life and Intelligence including concepts of cognition and learning, as well as the philosophical and conceptual issues that arise in the study of Embodied Intelligence.

We extend our heartfelt thanks to all seminar participants, authors, and reviewers for their outstanding contributions to this volume, which would not have been possible without their collective expertise and dedication.

Arsen Abdulali
Josie Hughes
Fumiya Iida

IOP Conf. Series: Materials Science and Engineering **1321** (2024) 011002 doi:10.1088/1757-899X/1321/1/011002

Peer Review Statement

All papers published in this volume have been reviewed through processes administered by the Editors. Reviews were conducted by expert referees to the professional and scientific standards expected of a proceedings journal published by IOP Publishing.

- **Type of peer review:** Open Review
- **Conference submission management system:** Morressier
- **Number of submissions received:** 17
- **Number of submissions sent for review:** 17
- **Number of submissions accepted:** 17
- **Acceptance Rate (Submissions Accepted / Submissions Received × 100):** 100
- **Average number of reviews per paper:** 3.0588235294117645
- **Total number of reviewers involved:** 20
- **Contact person for queries:**
  Name: Arsen Abdulali
  Email: aa2335@cam.ac.uk
  Affiliation: University of Cambridge - Engineering

IOP Conf. Series: Materials Science and Engineering **1321** (2024) 012001 doi:10.1088/1757-899X/1321/1/012001

Intelligence as Computation

Oliver Brock
Robotics and Biology Laboratory
Science of Intelligence
Technische Universität Berlin
oliver.brock@tu-berlin.de

Abstract. This paper proposes a specific conceptualization of intelligence as computation. This conceptualization is intended to provide a unified view for all disciplines of intelligence research. Already, it unifies several conceptualizations currently under investigation, including physical, neural, embodied, morphological, and mechanical intelligences. To achieve this, the proposed conceptualization explains the differences among existing views by different computational paradigms, such as digital, analog, mechanical, or morphological computation. Viewing intelligence as a composition of computations from different paradigms, the challenges posed by previous conceptualizations are resolved. Intelligence is hypothesized as a multi-paradigmatic computation relying on specific computational principles. These principles distinguish intelligence from other, non-intelligent computations. The proposed conceptualization implies a multi-disciplinary research agenda that is intended to lead to unified science of intelligence.

1. Introduction

"The concept and notion of computation and intelligence has historically seen significant evolution which continues to this day." [Hughes et al., 2022]

This is not a good sign. How can we study, understand, and reproduce *intelligence* from a *computational* perspective, if we cannot base such efforts upon an understanding of what these terms actually mean? The lack of clarity manifests itself in the literature via a proliferation of different kinds of "intelligences:" computational intelligence, embodied intelligence, physical intelligence, neural intelligence, morphological intelligence, mechanical intelligence, and many more. Similar diversity of interpretation exists for the term computation, which seems inextricably linked to the notions of intelligence and cognition [Chalmers, 2011].

The objective of this paper is first to disentangle the notions of intelligence and computation and then to propose a unified view of intelligence (albeit not a definition!) based on these disentangled notions. The distinct emphasis of the discussion will be on the study of *synthetic intelligence*, thus on the synthetic disciplines of intelligence research, including robotics, computer science, and engineering. And in the discussion that entails, I will be fully committed to a computational view of intelligence and cognition [Chalmers, 2011]. This currently is the dominant view and, as we will see in Section 4, I will assume a very general conceptualization of computation so that the arguments are not constrained by philosophical disputes revolving around this term.

The term *synthetic intelligence* subsumes humans' attempts to re-create biological intelligence in synthetic artifacts. I avoid the term *artificial intelligence*, because it is so over-used that it has become too ambiguous for a scientific discussion. Something is *synthetic* if it results "from synthesis rather than occurring naturally" (Merriam Webster Dictionary). Therefore, it seems useful to juxtapose *synthetic* with *natural*, rather than with "biological," as we might discover intelligence that is natural (not synthesized) but not biological.

My hope is that the conceptual clarifications of the terms *intelligence* and *computation* proposed in this paper can help prevent a "Tower of Babel" within synthetic intelligence research. Reaching consensus on the meaning of underlying concepts and terms would certainly contribute to an integrated and multi-disciplinary research agenda towards understanding the type of intelligence we observe in biological agents.

2. Argumentative Roadmap

Dualism, or more specifically *mind-body dualism* [Robinson, 2023], is a historical view that has influenced philosophy of mind and historical views on intelligence research. Dualism is the belief that mental processes are non-physical and thus distinct from the body. This view has a long philosophical tradition, reaching back to Plato, Aristotle, and Descartes. In the context of this paper, however, it is not necessary to go into detail on dualism, since currently the alternative view of materialism dominates science. Materialism is the belief that all processes, including those going on in the mind, are realized by material interactions, i.e., are physics-based.

I mention dualism only because the observation that some "shadow" of dualism appears to remain present in intelligence research serves as a starting point for the arguments laid out in this article. This shadow manifests itself in the distinction between brain and body, which is frequently made when talking about intelligence (see Section 3). The notion of embodiment rests on the assumption that the body contributes something unique to intelligence, something distinct from what the brain contributes [Varela et al., 1993]. I will argue that this is in some sense true: the body does contribute something distinct to intelligence. But I will argue that the true nature of this unique contribution is misunderstood. The contribution does not follow from a brain/body distinction, it can be explained more compellingly.

To prepare this hopefully more compelling explanation, Section 3 discusses existing conceptualizations of intelligence, in particular in the light of the brain/body distinction. I will argue that this distinction is not the right one, and also not helpful, given that the brain obviously is part of the body. Hopefully, by the end of Section 3, there will be the need for a new, better explanation.

IOP Conf. Series: Materials Science and Engineering **1321** (2024) 012001 doi:10.1088/1757-899X/1321/1/012001

The main proposal of this paper is a specific computational view of intelligence. In order to prepare describing this view, a second preparatory step will be laid out in Section 4. This section describes different computational paradigms including digital computation, analog computation, and quantum computation. Each of these paradigms has, in principle, the same computational power as all the others. But some problems can practically be solved much more easily in one paradigm than in another. Section 4 will present these different paradigms and attempt to differentiate them based on their "special powers."

After having pointed out in Section 3 that current conceptualizations of synthetic intelligence are problematic, possibly because they have been influenced by remnants of mind-body dualism, and after having described different types of computation in Section 4, I will present *Intelligence as Computation* in Section 6. In short, I will argue that intelligence is computation—so far you will probably agree. I will further argue that intelligence refers to a subset of all possible computations that implements and critically relies on yet-to-be-determined computational principles (I will propose some).

This view of intelligence as computation offers some advantages to be detailed below. But in brief: The proposed view shifts the focus from describing the generated intelligent behavior (which is the focus of most existing conceptualizations of intelligence [Sternberg and Kaufman, 2011]), onto the computational mechanisms that generate intelligent behavior. This is in some sense a return to the roots of artificial intelligence, which had viewed intelligence as symbol manipulation. This focus on computational mechanism promises to be more palatable than an attempt to characterize behavior. While behavior can vary without limit with environmental conditions, the mechanism that produces behavior is limited to the physical realization of the acting agent. A second advantage of the computational view of intelligence is, of course, that the inconsistent brain/body distinction is removed in a productive manner. Furthermore, the realization that different computational paradigms contribute differently to intelligence points to a trajectory for research and a shared conceptualization of intelligence suitable for many disciplines.

3. Intelligences—Status Quo

I will first survey some of the types of intelligences introduced in the literature, criticizing them when necessary, sometimes for being influenced by a "dualistic shadow." This section also analyzes the problem that arises from these diverse, ambiguous, and sometimes inconsistent definitions.

3.1. Biological Intelligence

The conceptual consensus I am aiming for in the context of synthetic research, has remained elusive in the study of *biological intelligence*. A succinct mechanistic definition and understanding of the term *intelligence* does not exist [Sternberg and Kaufman,

2011]. Consensus is limited to the circular statement that "intelligence is what an intelligence test measures." There is, however, relatively broad consensus that *general intelligence* (which should be synonymous with *intelligence*) can be subdivided into fluid (ability to process new information, learn, and solve problems) and crystallized (accumulated knowledge) intelligence [Cattell, 1963]. A more detailed subdivision of intelligence is postulated in the theory of multiple intelligences [Gardner, 1993]. This theory is scientifically disputed and lacks empirical support. The list of different intelligences proposed in it includes musical intelligence, visual-spatial intelligence, linguistic intelligence, logical-mathematical intelligence, body-kinesthetic intelligence, and several more.

The proliferation of intelligences and the ongoing debate over them, together with the lack of a consensual definition of the term *intelligence* itself, make it seem plausible that we have not gained an understanding sufficiently deep for conceptualizing biological intelligence. Maybe a synthetic perspective on intelligence can help make some progress.

3.2. Synthetic Intelligence

I will now discuss the proliferation of intelligences within the synthetic disciplines. First, I will present definitions of various types of intelligences from the literature. I will demonstrate that some of these definitions are unduly influenced by what I called "shadow of dualism" earlier. I will also show that the definitions do not produce a consistent picture. To put it bluntly: when viewed together, they do not make sense. These problems, which I hope to overcome in later sections, have contributed to a vague and inflationary use of relevant concepts. Just to give one example: Some authors use the terms "embodied" and "intelligence" in the title of their publication but these words do not occur a single time within the text of the article [Wang et al., 2022b]. Some people in our communities have begun to use terms such as intelligence as marketing gimmicks.

3.2.1. Computational Intelligence: "Computational Intelligence (CI) is the theory, design, application and development of biologically and linguistically motivated computational paradigms. Traditionally the three main pillars of CI have been Neural Networks, Fuzzy Systems and Evolutionary Computation" [IEEE Computational Intelligence Society, 2023]. CI [Engelbrecht, 2007] is a biologically inspired subset of Artificial Intelligence (the classical discipline, not the buzz word) [Russell and Norvig, 2020]. Conforming to some dualistic tendencies, CI is associated with the brain as opposed to the body [Sitti, 2021].

3.2.2. Physical Intelligence: Physical Intelligence (PI) refers to "physically encoding sensing, actuation, control, memory, logic, computation, adaptation, learning and decision-making into the body of an agent" [Sitti, 2021]. Here, the body is proposed as an alternative substrate for all functions we might associate with intelligence. It offers the view that the computations required to produce intelligence can be implemented in the brain (CI) or the body (PI). This view also seems to exhibit remnants of the dualist tradition. We might take issue with the adjective "physical", as both body and brain are physical entities. The adjective thus does not distinguish correctly between alternative substrates for implementing the functionality of intelligence. Even CPUs are physical. The term seems to be based on the view that digital or neural computations are not physical.

The fact that "everything is physical" implies that the location of computation, brain or body, might not be the most useful distinction for computational processes associated with intelligent behavior. Later in the text, I will attempt to replace distinctions based on the "location" of computational processes with one about *how* computation is performed.

3.2.3. Neural Intelligence: Neural Intelligence (NI) refers to all intelligent systems that rely on neurons to perform computations. These neurons can either be biological or part of an artificial neural network. In this regard, NI can be seen as a subcategory of CI. Exhibiting a "dualistic shadow," the view associated with NI differentiates between body and mind: "Biological organisms can survive and operate autonomously in unstructured environments because of the neural intelligence in their brains and the physical intelligence encoded in their bodies" [Wang et al., 2022a]. The idea is to transfer this dichotomy of neural and physical intelligences to "the development of intelligent bio-inspired robots" [Wang et al., 2022a]. The notion of NI therefore combines a "dualistic shadow" with the idea of bio-mimeticism, i.e., applying insights from biological intelligence to building synthetic intelligence.

3.2.4. Embodied Intelligence: "Embodied intelligence [(EI)] is the computational approach to the design and understanding of intelligent behavior in embodied and situated agents through the consideration of the strict coupling between the agent and its environment (situatedness), mediated by the constraints of the agent's own body, perceptual and motor system, and brain (embodiment)" [Cangelosi et al., 2015]. The link of brain and body finds consensus in the literature with other authors defining Embodied Intelligence (EI) as "investigating tight coupling between an agent body and brain" [Sitti, 2021] or "intelligence that requires and leverages a physical body" [Mengaldo et al., 2022].

But can there be an intelligence that does not require and leverage a physical body? Certainly not, every form of intelligence has to be physically realized, irrespective of whether it is a thought realized as changes in a biological neural networks or a computation on a CPU realized as changes on finely etched circuitry. In a practical sense, nothing can be disembodied. The term *Embodied Intelligence* therefore does not lead to a meaningful categorization of lines of intelligence research. It is only meaningful from the perspective of a "dualistic shadow"in which the brain is not considered to be embodied.

At the same time, the emphasis on coupling brain and body, which is also part of the definition of EI, is an important effort in overcoming the "dualistic shadow." Also, the focus on situatedness seems crucial. For the first time, intelligence is not considered to be a phenomenon that resides inside an agent's body (considering the brain part of the body) but instead as something that arises from interactions of the agent with its environment [Pfeifer and Bongard, 2006, Clark, 2010, Murphy Paul, 2022]. EI implies a view of computation that attributes computational capabilities to both the brain and the body.

Embodied Cognition, in close relationship to EI, "rejects or reformulates the computational commitments of cognitive science [to the brain], emphasizing the significance of an agent's physical body in cognitive abilities" [Shapiro and Spaulding, 2021]. Certainly, what must be meant here cannot refer to a body/no-body distinction but must mean something else. Both body and brain are able to perform computations, programmed into a physical system.

3.2.5. Embodied Artificial Intelligence: Embodied Artificial Intelligence (EAI) "can be defined as the study of intelligent agents that can see (or more generally perceive their environment through vision, audition, or other senses), talk (i.e. hold a natural language dialog grounded in the environment), listen (i.e. understand and react to audio input anywhere in a scene.), act (i.e. navigate their environment and interact with it to accomplish goals), and reason (i.e. consider the long-term consequences of their actions)" [Deitke et al., 2022].

Similarly to EI, there is an emphasis on the interactions between agent and environment. In contrast to the types of intelligences discussed so far, the definition of EAI [Deitke et al., 2022] does not offer anything beyond the well-established. For example, the authors of the definition attempt to distinguish EAI from robotics by statements such as "Embodied AI also *includes* work that focuses on exploring the properties of intelligence in realistic environments while abstracting some of the details of low-level control. [...] Conversely, robotics *includes* work that focuses directly on the aspects of the real world, such as low-level control, real-time response, or sensor processing [emphasis added]." But these statements about what the fields *include* does not establish a distinction. The authors claim that "there are differences in focus which make embodied AI a research area in its own right." But a change in focus in itself clearly does not constitute a new research area. Please note that the other notions of intelligence discussed here do not make claims of establishing a new research field.

3.2.6. Morphological Intelligence: The term *morphology* refers to the shape and material composition of an object. Morphological Intelligence (MI) is the "ability of a morphology to facilitate the learning of novel tasks" [Gupta et al., 2021]. This is accomplished by morphological computation (MC), which is described as "outsourcing part of control, sensing and computation tasks to the robot body to be handled through emergence of functional behaviours due to properties such as architectured kinematics,

IOP Conf. Series: Materials Science and Engineering **1321** (2024) 012001 doi:10.1088/1757-899X/1321/1/012001

compliance, natural resonance, damping and friction" [Ghazi-Zahedi et al., 2021].

Presumably influenced by a "dualistic shadow," MI and MC juxtapose what the body does to what the brain does (Computational Intelligence, control, or computation). Implicit in the first quote is also the idea that MI is in service of learning or CI. But every form of computation is the result of the morphology of the object performing the computation. The way circuits are arranged on a VLSI chip constitute the chip's morphology and enable certain computations. Morphology cannot be a meaningful dividing line between different computational phenomena pertaining to intelligence.

3.2.7. Mechanical Intelligence: Mechanical Intelligence (MechI) "explores nature-inspired mechanisms for automatic adaptability and applies them to the design of structural components" [Khaheshi and Rajabi, 2022]. Closely related to the notions of *Physical Intelligence* and *Morphological Intelligence*, this view of MechI emphasizes the transfer for mechanisms from nature to the production of engineering artifacts. It thus seems narrower and is subsumed by these concepts.

It is important to distinguish the above definition of MechI from Mechanical Engineering (MechE) as a field. MechE, as a research field, focuses on building functionality into physical artifacts in clever and robust ways. There are thousands of years of progress accumulated in Mechanical Engineering, and many of the views expressed above (Physical Intelligence, Embodied Intelligence, Morphological Intelligence, Mechanical Intelligence) might be viewed as subsets of Mechanical Engineering. This indicates that Mechanical Engineering can serve as an important disciplinary foundation for the study of intelligent agents and their interactions with the environment.

3.3. Problems with Intelligences and Their Consequences

Let us look at the consequences that arise from the definitions of different intelligences when viewed all together.

3.3.1. Remnants of Dualism? Apparently, dualism still has an influence on the conceptual foundations of intelligence. Many of the intelligences above have a dualistic flavor. Of course, the discussion about brain versus mind and physical states versus mental states, as in Descartian dualism, is no longer present. But it seems to have been replaced by an assumed functional divide between brain and body. This is exemplified in statements like this one: "While PI [physical intelligence] only focuses on the physical intelligence encoded in the body, EI focuses on the tight coupling between an agent's body and brain (and the environment)" [Sitti, 2021]. This implies a relevant distinction between intelligence encoded in the body and the brain, calling the former "physical". Sitti [2021, Figure 1] includes in physical intelligence all things we might also attribute to the brain, e.g., memory, learning, decision making, etc. The proclaimed need to coordinate brain and body in the area of Embodied Intelligence, further demonstrates this misunderstanding.

I do not see any justification for this new variant of dualism. Brain and body are close collaborators in all we might consider intelligence or cognition. There are few functions we can entirely attribute to one or the other. Perception, for example, significantly begins with task-specific pre-processing already in the retina [Gollisch and Meister, 2010], which is an extension of the central nervous system. Processing obviously continues in the brain, but is critically enabled by the pre-processing of information outside of the brain. Similar pre-processing happens in other senses, including touch [Pruszynski and Johansson, 2014], hearing, and smelling [Yoshioka and Sakakibara, 2013].

Below, I will argue that we cannot divide cognitive functions into those performed by the brain and those performed by the body. It is very likely that nearly all relevant functions are implemented in a close collaboration between brain and body. Of course, I am not arguing that the brain is not a physiologically separate system. And if one wants to, one can define functions exclusively of the brain. However, if we are talking about functions that can be considered meaningful with respect to the generation of intelligent behavior, such as learning, motor control, perception, reasoning, etc., we probably should not just look to the brain or the body alone. Even memory, which one might attribute to the brain exclusively, can only work because relevant information has been filtered and structured by other parts of the body. It would be hopeless to simply store all perceptual information directly.

Maybe the new form of dualism is the consequence of an intuitive misunderstanding. Intuitively, brain and body are different in some way. However, it is not that they have sharply divided roles in terms of cognitive functions. What might justify the brain/body distinction instead is the consideration of computational paradigms. Brain and body collaborate in performing cognitive functions, but each contributes a distinct form of computation. Details about this argument will follow below, after we have discussed in Section 4 the different forms of computation we might consider here.

3.3.2. Bio-Inspiration Many types of intelligences include bio-inspiration in their definition. This is understandable as close observation of the miracles performed by biological agents makes it blatantly obvious that our focus on brain-bound supremacy is missing an important part of the picture. This focus has been termed "cognitive chauvinism: we tend to favor control complexity over morphological complexity. Classical artificial intelligence dispensed with the body altogether" [Cangelosi et al., 2015]. Cognitive chauvinism is probably connected to the new form of dualism discussed above.

Taking inspiration from biology seems like a very fruitful idea, and this has been proven with many inventions in everyday life. However, we should take lessons from biology without being constrained to the specific implementations of the solutions. In other words: Take the lesson, but be open to solutions that biology has not invented yet. This would mean that we should not remove bio-inspiration from our research labs and projects, as it can be incredibly productive, but we should also not simply mimic what we find.

3.3.3. Ambiguity Many of the types of intelligences discussed above assume a dualistic stance. I argued that this stance cannot be upheld because none of the definitions identified a valid criterion for the differentiation of brain and body. Once we strip away the new dualism, Physical Intelligence becomes equivalent to intelligence and subsumes all other forms of intelligence. This is because *all* relevant phenomena of all types of intelligence are physics-based, simply because there are no processes not based on physics.

The different definitions are also ambiguous and overlap in meaning. For example, Physical Intelligence, Embodied Intelligence, Morphological Intelligence, Mechanical Intelligence, and Embodied Cognition all emphasize the contribution of the body (without the nervous system). This is an indication that different intelligences are not mutually exclusive and therefore not ideal to structure intelligence research. It also stresses again the importance of finally overcoming dualism in all of its forms.

Some research programs associated with different definitions of intelligence include close ties to other types of intelligence. For example, Embodied Intelligence researchers propose *mechanics* as a foundational modeling tool [Mengaldo et al., 2022], indicating that Embodied Intelligence and Mechanical Intelligence (or at least Mechanical Engineering) have so much in common that a categorical distinction may not be justified.

Hadi Sadati et al. [2022, Figure 3] list key aspects of Embodied Intelligence: morphological computation, co-design, material, physical intelligence, sensor design, and learning. This has significant overlap with the definition Physical Intelligence ("physically encoding sensing, actuation, control, memory, logic, computation, adaptation, learning and decision-making into the body of an agent" [Sitti, 2021]).

The ambiguity reflected in these few examples illustrates how a more concise conceptualization of intelligence might help advance the fields by enabling a clearer conceptualization of the relevant factors.

3.4. Intelligence Without Dualism

Remnants of dualism remain a recurring theme in the intelligences discussed above. Nevertheless, there is broad agreement that both body and brain—sticking to the new dualistic distinction—both contribute to the behavior of intelligent agents and must be viewed together rather than separately.

The brain performs computation, few would dispute this. Then we are left with looking for a description of the body's contribution. Concepts like Morphological Computation, i.e. "outsourcing part of control, sensing and computation tasks to the robot body" [Ghazi-Zahedi et al., 2021], are a response to this need and also refer to the concept of *computation*. Unfortunately, the term "morphology" in *Morphological*

IOP Conf. Series: Materials Science and Engineering **1321** (2024) 012001 doi:10.1088/1757-899X/1321/1/012001

Computation is relatively ambiguous [Encyclopedia Britannica, 2023], but let us assume that it refers to how materials of varying properties are spatially arranged to constitute an object. This definition holds both at the macro level (objects) and at the micro level (atoms and molecules).

If you agree with this view of morphology, we are left with a dualistic dilemma: If we attribute the ability to control, sense, and compute to the body's morphology, as is suggested by the definition of MC, then we must also concede that a) the body performs computation and b) the arithmetic logic unit on a CPU performs its computation as a result of its morphology as well. A CPU is nothing else but different materials spatially arranged so as to realize a particular set of computations. This means that brain and body both perform computation and that this computation is always the result of morphology.

If we give up dualism (as we should) and realize that everything is physics-based, i.e., if we subscribe to materialism, all behavior arises from morphology and how it structures physical processes. There is no fundamental difference between the morphology of a CPU, a clockwork, or a robotic hand. It is all "just" physics, forced into a certain game plan by morphology. As a result, the distinctions among different intelligences described in this section fall apart.

We now have leveled the playing field for brains and bodies as well as for CPUs and brains. They all perform computation by structuring physics-based processes with the morphologies of their bodies—and now the body clearly includes the brain for which the exact same statement holds.

4. Computation

In the previous section, our discussion shifted from intelligence to computation as a way to explain how intelligent behavior comes about in a non-dualist setting. It is time to talk about computation in more principled manner, as it serves as the foundational concept for reasoning about intelligence. Of course, by computation we now mean something much broader than symbol manipulation. But how exactly should we view computation for it to serve as a useful foundation?

4.1. Computation—Status Quo

For most researchers in this area, computation serves as a foundation for the study of cognition [Chalmers, 2011] and embodied cognition [Shapiro and Spaulding, 2021]. I will discuss the relationship between cognition and intelligence in more detail in Section 7, but for now I will treat these two terms as synonyms, for practical reasons, and therefore assume that computationalism is also an adequate foundation for intelligence.

There are many unanswered questions about the relationship between computation and cognition. Chalmers [2011] lists a few of them: "What is it for a physical system to implement a computation? Is computation sufficient for thought? What is the role of computation in a theory of cognition? What is the relation between different sorts of computational theory, such as connectionism and symbolic computation?" These questions continue to be the subject of debate.

There are equally many questions about computation itself. As the context for our discussion is the synthesis of intelligence in physical systems, we will focus our discussion of computation in physical systems [Piccinini and Maley, 2021], called *concrete* computation, as opposed to *abstract* computation, addressing mathematical formalisms of computation. In the realm of concrete computation, people wonder, for example, what distinguishes a computer (believed to be a physical system capable of performing computation) from a rock (believed to not be able to perform computation). The relationship of concrete computation to abstract computation is also not fully clarified. Several theories about this relationship exist, including the simple mapping account, the semantic account, the syntactic account, the mechanistic account, and several more [Piccinini and Maley, 2021].

Pancomputationalism [Piccinini, 2015] is a relatively popular view of computation [Piccinini and Maley, 2021]. It states that all physical systems perform computation. Ultimately, the entire universe can be seen as a giant computer, computing away at its program (the laws of physics) with morphology representing the current program state (the distribution and dynamic state of matter in space).

4.2. View of Computation Used Here

Pancomputationalism is helpful if we want to give up dualism. According to this view, the neurons inside the brain perform computation and so does the compliant actuation system of the body, for example, when implicitly computing a good foot posture while walking on uneven ground. Pancomputationalism offers us a view of computation that does not require a distinction between different body parts' contributions to intelligence. All physical processes are computations and since everything is physics, everything is also computation.

For the purpose of our discussion, we should, however, differentiate between useful computation and everything else. In building an intelligent agent, the computation performed by gas clouds in a far-away galaxy is of no concern to us. We can safely assume, based on the known laws of physics, that the gas clouds do not affect the behavior of our agent.

Out of all the pancomputational processes unfolding in the universe at any point in time, we call only those *computations* that have been deliberately included to contribute to the behavior of the agent under consideration. In other words: we consider only those pancomputational (or physical) processes that are *useful* to the agent. The outcome of the computation, including its failure, can affect the behavior of the agent. By specifying a purpose for computation—for example, the processes contributing to a person passing the driver's test—we can refer to the set of relevant physical processes as computation. We focus our notion of computation on those physical processes that affect the agent.

IOP Conf. Series: Materials Science and Engineering **1321** (2024) 012001 doi:10.1088/1757-899X/1321/1/012001

This distinction is common in the literature on computation Horsman et al. [2014], Müller and Hoffmann [2017].

4.3. Computational Paradigms

We have left dualism behind and have leveled the playing field in regards to computation. Every physical thing, no matter whether it is called body or brain, performs computation, as long as it affects the behavior of the agent under consideration. To achieve this, we had to sacrifice the distinctions among different types of intelligences found in the literature 3. We also had to adopt a "practical" definition of teleological computation so as to be able to focus on the set of computations (physical process) relevant to a particular agent.

Let us now consider this entire set of relevant computations for a particular agent. Is there a useful way to categorize them? Such a categorization could serve as the foundation for devising different computational classes or paradigms within the study of intelligence. To approach the answer to this question, let us examine the different types of intelligences from Section 3 we just abandoned. What were the motivations for these particular types?

On the surface, a "dualistic shadow" produced the distinctions. Physical Intelligence attempted to distinguish between brain and body. But both are physical as there is nothing non-physical. Embodied Intelligence attempted to emphasize the role of the body and its interactions with the brain. Again, this sounds like a dualist distinction. Mechanical Intelligence also attempted to emphasize the role of the body (as opposed to the brain), in particular when viewed from their mechanical properties. But what if the distinction these approaches were attempting to make where not body/brain distinctions but the distinction between different computational paradigms? It is true that the computation performed by the brain and by the body are governed by different physical principles. It is true that mechanics plays a much more prominent role when we want to understand the body. And processes in the brain might be better explained and modeled assuming it performs some kind of digital or analog computation.

What are the different computational paradigms we currently know and how do they differ? Table 1 lists them. I recommend studying the table for a moment before reading on.

It is difficult to explicate the relationship among the computational paradigms listed in Table 1. Of course, all the way at the bottom, everything is implemented based on physics, probably quantum mechanics. But as a result of quantum mechanics, a number of physical phenomena arise at different levels of abstractions. These phenomena can be leveraged to perform computation, or we might say that these processes are shaped so that they represent a model of the abstract computation we want to perform. For example, analog computation are based on continuous, physical processes of currents and voltages in electric circuits as a means of computation. Underneath it, of course, must lie quantum physics, but it is useful to characterize the resulting higher-level

IOP Conf. Series: Materials Science and Engineering **1321** (2024) 012001 doi:10.1088/1757-899X/1321/1/012001

Paradigm	Digital	Analog	Quantum	Mechanical	Morphological	Neural
Physics	semiconductors, variable electrical conductivity, gating via doping	continuous variation aspect of physical phenomena such as electrical, mechanical, or hydraulic quantities)	quantum mechanics, entanglement; can be both analog and digital	in principle, all of physics, but emphasis has been on rigid structures of certain types (e.g. trusses)	mixture of analog and mechanical computing, emphasis on compliant materials and physics phenomena (piezo etc.)	biophysics, molecular biology, chemistry
Concept	leverage fast switching of small circuits, ability to compose into complex logic	simulate/solve the math problem via physical system with the same constraints/laws	exploit parallelism afforded by quantum physics	exploit parallelism afforded by mechanics and force propagation within matter (speed of sound)	exploit parallelism afforded by mechanics and force propagation within matter (speed of sound)	we have not yet decoded how neuronal structures do what they do so well
Programming principles	symbols, binary, von Neuman, computability, complexity theory, encapsulation	signals, calculus, integral, and differentiation, non-deterministic logic	too early to tell	design principles, smart mechanical design	design principles, smart mechanical design, co-design	network architecture, error backpropagation, loss functions, data engineering
Functional components	transistors, diodes, gates	resistors, capacitors, amplifiers, multipliers, potentiometers, function generators	qbits (photonics, trapped ions, ...), quantum registers, quantum reversible gates, quantum processing unit	materials, joints, mechanical assemblies	materials	neurons, ion channels
Computational capabilities	symbol manipulation, number crunching, storage	dynamic and differential systems, complex continuous functions	branch and test, search	interfacing with the real world while implicitly addressing uncertainty	interfacing with the real world while implicitly addressing uncertainty	identifying patterns in data
Advantages	generality, ease of programming	speed, inherently addresses uncertainty, works on continuous data, high mean time between failure	leveraging parallel universes	vastly parallel	vastly parallel, self-correcting physical interactions, uncertainty suppression and reduction	general, can use data, many parameters
Disadvantages	limited parallelism, requires explicit representation of uncertainty, memory bottlenecks	lack of precision, programming tedious, sensitivity to electric noise	not generally usable yet, cooling requirements	programming based on vague principles, programs difficult to change	programming based on vague principles, programs difficult to change, co-design not yet understood	requires much data, many parameters, energy consumption

Table 1: Overview of computational paradigms: The term "vastly parallel" refers to the fact that every part of the agent performs computation via atomic forces, for example.

IOP Conf. Series: Materials Science and Engineering **1321** (2024) 012001 doi:10.1088/1757-899X/1321/1/012001

phenomena and to leverage them to encode and perform computation.

Mechanical computation is a specific form of analog computation, emphasizing a particular type of analog phenomenon, i.e., the motion of matter at the macro scale rather that at the electron scale, as it was the case for analog computation based on electricity. For the latter, it would be appropriate to characterize computation based on currents or at the atomic/molecular scale, or on the flow of liquids in the case of hydraulic computations. In contrast, mechanical computation can probably most easily be realized by considering kinematics and material properties—phenomena at the super-molecular level. Consequently, different computational paradigms also capture physical phenomena at different levels of abstraction.

Morphological computation can simply be viewed as a narrower category of mechanical computation but it might differentiate itself from the latter by taking into account the fact that systems based on morphological computation also include some form of digital or neural computation. This necessitates the negotiation of a suitable interface between these two paradigms. Morphological computation might thus be the first computational model explicitly designed for multi-paradigm computation. In contrast to hybrid computation, which combines digital and analog computation across an explicit interface, the morphological computation paradigm not only includes the combination of two paradigms but also reflects the realization that problems must be divided between the paradigms appropriately. This poses a new kind of problem, namely the problem of suitably factorizing the original problems based on the computational strengths of the paradigms involved.

The key take-away from Table 1 is that each computational paradigm leverages different physics phenomena to perform computation. As a result, the computational capabilities and programming approaches are also different. Each paradigm comes with its unique sets of strengths and weaknesses. An agent leveraging several of these computational paradigms can probably generate a richer set of behaviors, given a finite amount of resources. And the agent can probably generate the same behaviors more efficiently when compared to an agent only leveraging a single paradigm. Intuitively, this seems plausible as a general rule. Practically, there are many specific examples (all species on planet earth).

On the other hand, in some formal sense, these paradigms possess probably equivalent computational power with respect to the Church-Turing hypothesis [Hopcroft and Ullman, 1979]. They all can, in principle, compute all computable functions. But these arguments from the domain of computability are not our focus when we discuss computation to generate intelligent behavior. Already very simple problems relevant to intelligence, for example in motion planning of polygons among polygonal obstacles, are NP-complete [Latombe, 1991] and therefore computationally very challenging. Nevertheless, intelligent agents solve versions of these problems all the time. Therefore, we need to shift our attention away from completeness towards bounded-resource, practical implementations of computational solutions. To judge a problem, it will not be helpful to ask how much computation in the different paradigms is required to solve *all* possible instances of that problem. An important question will be if there are large and relevant subsets of all possible problems that can be solved with reasonable efficiency—within one of the computational paradigms. An example of this is factorization of large integers, which becomes highly efficient using quantum computation but is practically intractable with digital computation.

As a result, building an intelligent agent becomes easier if we pick the computational paradigm most suitable for expressing a necessary computation. When we look at the computational paradigms listed in Table 1, we should foremost think about practical feasibility: Does the inclusion of a computational paradigm permit the construction of an agent with novel capabilities? These additional capabilities will most likely result from our ability to express a solution in that paradigm more simply, rather than from the inherent computational capabilities associated with the paradigm (as we assume these are the same for all paradigms, at least in principle). Effectively, we consider whether the inclusion of a paradigm can reduce the overall cost, be it of constructing the agent or of performing the computation.

I would like to suggest that the computational paradigms described here are a useful way of structuring subsets of physical processes. Each paradigm captures a particular exploitable regularity of physics, forming such a subset. Associated with each paradigm is also a suitable abstraction, allowing us to reason about the particular type of physical process (or computation).

The computational paradigms let us unite the different physical processes going on in agents into a single multi-paradigm computation. For example, a biological agent performs neural computations in the brain to determine its behavior. But these computations alone do not determine the behavior in its entirety. In addition, just to give an example, the lens and shape of the eye determine how the environment is perceived. And the structure of the skeletal and muscular structure lets neural instructions play out differently each time, depending on which forceful interactions with the environment occur during the behavior. Therefore, behavior is not determined by neural computations alone, mechanical computations also affect it. The resulting behavior is multi-paradigmatic.

Given the classification of physical processes into computational paradigms, and given the view that these computational paradigms are composed in the generation of intelligent behavior, we must then understand how to best perform such compositions to understand and synthesize intelligence. This requires knowledge of which computational paradigm is suited best for a particular category of problems or behavior. It will become important to understand this paradigm/problem relationship at a general level. In the context of digital computation, this relationship is already well understood via complexity and language theory [Hopcroft and Ullman, 1979]. We know that finite automata, push-down automata, and Turing machines implicitly capture regular languages, context-free languages, and recursively enumerable languages, respectively [Hopcroft and Ullman, 1979], just to give an example. This is a very helpful understanding of the relationship between problems (regular languages) and the most adequate computational solution (finite automata). I am not aware of similar specific results for other computational paradigms. And maybe the success of digital computation is linked to the ease with which the relationship between computation and problem can be established. But it seems highly desirable to attempt to characterize the computations that can be performed within each of the computational paradigms.

4.4. The Impact of Computational Paradigms on the Notion of Computation

The question of when a physical system performs computation and when not remains debated Horsman et al. [2014], Piccinini and Maley [2021]. The view proposed here deviates from some of the positions taken in the literature and thereby seems to resolve some of the difficulties in this discussion. I would like to briefly mention those, even though the notion of computation is not the main focus of this text.

(i) One criticism of pancomputationalism is that the notion of computation becomes vacuous when *everything* is considered computation Horsman et al. [2014]. The view propagated here is that every physical process *can* be a computation if it contributes to the behavior of an agent. Therefore, not *everything* is considered computation, the set of physical processes is subdivided into a meaningful way.

(ii) The set of physical processes considered as computation is further reduced by the requirement that physical processes are categorized into computational paradigms. Each paradigm has its own abstraction that we can use to model, understand, and program. By selecting a subset of all paradigms to create an intelligent agent, the criticism of vacuousness Horsman et al. [2014] is removed.

(iii) Central to many arguments about computation is the *representation relation* from physics, linking physical processes to abstract mathematical objects. However, I believe that this framework, while very useful for some kind of arguments, makes restrictive assumptions. This can be seen, for example, when the mathematical object becomes a necessity for computation and is required to be known, even a priori Horsman et al. [2014]: "If a computational description of a physical evolution can only be applied post-hoc, then the system has not acted as a computer." But in our view, a computational description (mathematical object) is not required at all. The result of a computation does not need to represent a symbol or mathematical object, and it does not need to be predicted to be useful for the behavior of an agent. It seems that with the *representation relation* one has inadvertently imported unnecessary constraints.

(iv) Computational paradigms resolve a confusion about embodiment prevalent in the discussion. Take a statement like the following: "The centrifugal governor is not a computer, even if it can be replaced by a computational system that measures the relevant values, computes a response, and controls the relevant value (valve opening)" Müller and Hoffmann [2017]. It does not seem plausible that the same function realized in two different physical processes should be called computation

IOP Conf. Series: Materials Science and Engineering **1321** (2024) 012001 doi:10.1088/1757-899X/1321/1/012001

in one case and not in another. Both the centrifugal governor and the alternative computational system are simply matter arranged in such a way that the behavior is achieved. We might be led into believing that they are fundamentally different because they implement different computational paradigms. In the view proposed here, they both perform computation and result in the same behavior.

(v) Another discussion in the literature pertains to systems that combine multiple computational paradigms. Authors object to the notion of "offloading" or "outsourcing" computation Müller and Hoffmann [2017], terms often used in the context of morphological computation (see Section 3.2.6), when mechanical computation enables the simplification of digital computation Bhatt et al. [2021]. This situation is easily explained in the context of computational paradigms. Let us use soft manipulation as an example Bhatt et al. [2021]. A soft-material hand contributes importantly to the manipulation behavior through its carefully designed compliance, enabling complex in-hand manipulation with nearly no digital computation Bhatt et al. [2021]. In such a scenario, we can say that the overall behavior of in-hand manipulation is produced by combining two computations, each from a different computational paradigm (mechanical and digital computation). Because the responsibility for producing the behavior is distributed aptly across these two types of computation, the resulting behavior is highly robust and simple. The question of whether computation is outsourced or offloaded is replaced by the distribution of responsibility for behavior generation across two computational paradigms. The complications previously perceived Müller and Hoffmann [2017] have disappeared.

5. Intelligences—Status Quo Revisited

Let us reexamine the definitions of intelligences from Section 3 based on the proposed multi-paradigmatic computational structure. We will see that computational paradigms resolve the difficulties we discovered earlier. For example, Physical Intelligence emphasizes analog computation, mechanical computation, and morphological computation (not the body, as stated in the definition). And it contrasts this with digital computation and neural computation (not the brain, as stated in the definition). Similarly, Embodied Intelligence sought to address the interplay between brain and body, but really it is about how to best divide a particular problem across digital and analog/mechanical computation and how to have those two computations integrate well.

In Section 3, we also discussed the term "embodied" as a property of intelligence. Let us contextualize this term in light of computational paradigms. For example: does being embodied imply the inclusion of mechanical computation? CPUs are a counter-example, where digital computation is embodied without mechanical computation. Even a robot does not necessarily have to include (significant) mechanical computation. If a robot's body is very stiff, controlled at high frequencies, and with very high torque,

IOP Conf. Series: Materials Science and Engineering **1321** (2024) 012001 doi:10.1088/1757-899X/1321/1/012001

then the robot's behavior is entirely determined by the digital control program, without mechanical computation. Mechanical computation enters the picture once the robot engages in physical interactions with the compliant environment. Therefore, the term "embodiment" does not help to differentiate between computational paradigms. And as a result—if we assume that computational paradigms are important to our discussion—the term "embodied" is not very helpful in intelligence research.

6. Intelligence as Computation

We are now ready for moving towards a description of the nature of intelligence. *Intelligence is the set of computations that satisfy certain properties.* What exactly these properties are will have to be the subject of intelligence research but I will make some suggestions below. To implement these computations we have various computational paradigms at our disposal. As our understanding of these paradigms improves, the task of producing intelligent computation will become easier.

This proposed view of intelligence—it cannot be called a definition because it is still missing important information—as a subset of all computations has advantages over current attempts to define intelligence [Sternberg and Kaufman, 2011]. Current attempts define intelligence by enumerating characteristics of intelligent behavior. Instead, I am proposing to characterize the computational processes that produce the behavior. This promises to be much simpler. The characteristics of behavior are the result of the mechanism and its interactions with diverse environments. It seems plausible that behavior is much more complex and versatile than the mechanism that generates it. It should therefore be simpler to characterize the computational mechanism.

The proposed view of intelligence is not subject to anthropomorphic bias. Intelligence can occur in individuals, in collectives, in societies, but also at behavioral, cellular, or genetic levels. Viewing intelligence as computation opens the mind to the discovery of intelligence that differs from human intelligence. I therefore believe that the proposed view of intelligence as computation provides a productive conceptual foundation for intelligence research.

Let us enumerate the key questions that our field must answer, if we assume this view of intelligence as computation:

(i) Understand better existing computational paradigms, including what types of problems they are particularly suited for
(ii) Develop an engineering practice for designing and programming systems that combine several computational paradigms
(iii) Identify the specific properties of computation that characterize intelligent computation, and, if such a characterization has been postulated,
(iv) Learn how to leverage these properties as inductive biases for producing synthetic intelligent agents and for understanding natural intelligent agents

It is important to note that the answer to these questions will most likely vary significantly for different behavioral spaces, such as the one spanned by individuals, collectives, societies, cells, genes, just to name the same examples as above.

Let us speculate about the properties of computation that produce intelligence. These properties can serve as inductive biases for building intelligent agents or for explaining intelligent behavior. When designing and programming an agent, these properties facilitate the generation of intelligent computation. I would speculate that these properties will turn into programming principles. The more of these principles we accumulate and are able to apply, the easier will it become to build intelligent agents and the more intelligent the resulting agents can become.

It remains possible that intelligence will never be characterized as a sharply defined set of computations. We might continue to discover new properties or principles, specific to particular ecological niches, without ever reaching consensus on a fixed set of properties. It is possible that different intelligent agents are based on some properties only, or even different properties that produce the same behavior. There might be properties that are necessary and others that are sufficient. So we might have to resort to the fall-back statement that intelligence is just a form of complex computation. But to me it seems exciting to attempt to find a succinct description of what distinguishes intelligence from other computations, in particular as it promises to enable the construction of intelligent agents.

The following subsections present speculations about possible properties that might help discriminate intelligent computation. If they turn out to be necessary principles for exhibiting intelligence, we could call them principles of intelligent computation, or even the principles of intelligence. This would provide support to the view that intelligence is computation that satisfies certain properties of sophistication, characterized by these principles.

6.1. Multiple Computational Paradigms

It might be an inherent property of intelligent computation that it leverages multiple computational paradigms. This is certainly true for all forms of biological intelligence, which leverage neural computation and mechanical computation.

It is also possible that the path towards determining a minimal set of properties to characterize intelligence or the path towards a form of intelligence that is only based on a single computational paradigm will require us to discover many properties and to exploit many computational paradigms. If we were to understand intelligence completely, a small set of properties might suffice to characterize intelligent computation, and a single computational paradigm might suffice to implement it.

6.2. Diverse, Active Component Interactions

The intrinsic complexity of the computational program executed by an intelligent agent, possibly measured by something like Kolmogorov complexity [Kolmogorov,

1965], is substantially less than the complexity of the behavior it is capable of producing. This discrepancy results from the fact that the same computational agent can interact with a very large number of different environments. Every relevant change in the environment can lead to significant changes in behavior. This modification of behavior does not require a fundamental change in the agent's computational structure. Therefore, the goal of the agent remains achievable under large environmental variations without significant changes to the agent. In fact, this seems to be a prerequisite for intelligence, namely, that agents achieve their goals robustly under significant variations in environmental conditions.

But what are the computational properties implemented in the agent that support this kind of robust adaptability to environmental conditions?

I would like to propose the principle of *diverse, active component interaction* as one possible property that contributes to the robustness of intelligent behavior. It is best explained in contrast to current software engineering practice of developing systems with strong modularity, i.e., systems in which the information exchanged between components is specified a priori and designed to be as simple as possible (this results from the software principle of encapsulation) [Martin, 2024]. The advantage of the software engineering approach is that errors and failure modes are confined to specific components of the overall agent. However, this also leads to brittleness under environmental changes. If the system is confronted with a novel environmental situation, all components might be able to contribute to a useful behavioral adaptation, even outside of the initially constrained, pre-programmed modularization of the system. For sure, any useful adaptation to new circumstances must be coordinated among several of the computational components of the agent. This coordination cannot easily be anticipated a priori, otherwise the agent would have to be as complex as the behavior it produces (and software engineers would have to be able to anticipate all possible environmental variations). Therefore, the information exchange between components must be *active* to adjust to the circumstances. It must also be *diverse* so that in many situations the required information can be extracted and exchanged.

Using *diverse, active component interaction* as a principle of computation, we can hope to create systems that achieve their programmed goals robustly, even when confronted with environmental variations. The principle implies that different components of the computational systems are capable of engaging with other components in diverse interactions, facilitated by active and diverse information channels. The components themselves must be able to change their behavior in response to the information relayed to it.

Diverse, active component interaction is, in many ways, the opposite of good software engineering practice, where the interface between components is fully specified and often as simple as possible to support encapsulation and high modularity [Martin, 2024]. However, if variations in behavior are required by new environmental circumstances, it is only plausible that these variations must be brought about by different responses of the system's components. These responses must be coordinated with other components. Also, components might facilitate the adequate response of a different component by providing information to it that would otherwise remain unknown. Therefore, having diverse component interactions might be an important ingredient to support this behavioral diversity on the way to building robust intelligent systems.

What I have stated about this principle is far from being a concrete recipe for implementing robust systems. But employing a vague understanding of this principles in system building has already led my group to building very robust and award-winning robotic systems [Eppner et al., 2018]. There also is fascinating evidence in biology and medicine, demonstrating that at least some specific biological components are set up to support strong component interactivity at the genetic [Leff et al., 1986], cellular [Vaney and Weiler, 2000], and tissue levels [Blackiston and Levin, 2013].

At the genetic level, there is a process called alternative splicing [Leff et al., 1986] that exhibits strong component interactivity. Alternative splicing is part of the production process of proteins inside cells. Genes encode the blueprint of proteins. However, proteins are not always encoded in a single, dedicated gene. Instead, alternative splicing assembles several pieces of genetic information to produce the protein. These pieces are reused in different combinations to produce different proteins. This implies that this genetic information must be ready for being combined with different pieces of information, each time producing a protein vital for the survival of the cell. The individual pieces do not only contain the necessary information to produce a single protein, they can be combined in different ways to produce different proteins.

Another example: Cells are connected by so-called gap-junctions [Vaney and Weiler, 2000] to facilitate strong component interconnectivity among cells. These junctions enable flexible communication across cell membranes, with the set of possible signal transmissions referred to as connexin. This enables flexible cellular communication adaptable to a broad range of environmental conditions. The communication, in turn, enables different modes of interaction between cells.

A third example, at the tissue level, is even more surprising. Surgically attaching a tadpole eye to the tail of another tadpole, triggers responses in the tissue that enables the tadpole to integrate the information obtained from that "Frankenstein eye" [Blackiston and Levin, 2013]. The tissue of the tadpole responds flexibly to the unexpected sensor information, growing a connection between the optical nerve and the tadpole's nervous system. The tadpole begins to change its behavior in response to light perceived by this new eye. This demonstrates that the tissue of tadpoles is capable of engaging in new, evolutionarily unplanned component interactivity.

Another example of diverse, active component interactions comes form transplantation medicine. If intestines are transplanted, the inclusion of the liver in the transplant appears to reduce the probability of rejection of the implanted organs [Vianna et al., 2023]. This is an example of a general pattern [Fridell et al., 2019], in which the inclusion of a weakly immunigenic organ can aid the long-term prospects for the transplant of a more strongly immunigenic organ. This, too, demonstrates that there

IOP Conf. Series: Materials Science and Engineering **1321** (2024) 012001 doi:10.1088/1757-899X/1321/1/012001

are active interactions among organs that go beyond their immediate bodily function.

These examples illustrate that biological agents are capable of strong component interactivity at different levels of abstraction. It is plausible that this principle is a critical enabler of the robustness exhibited by biological systems. And it is equally plausible that this architectural pattern can also be employed to large benefits at higher levels of abstraction.

Diverse, active component interactions can only be realized if components are developed via co-design, as the flexible coordination of components now becomes an important requirement for all components. To build this into components, co-design, i.e., the joint development of multiple components seems the only plausible way. In addition, we must realize that "physical interaction with the environment as the primary source of constraint on the design of intelligent systems" [Brooks, 1990]. Therefore, co-design must occur in realistic settings, possibly in rich real-world ecological niches, to produce relevant component interactions that in turn produce robust adaptation of behavior to environmental variations.

6.3. Agent/Environment Computation

The principle of *diverse, active component interaction*, discussed in the previous section also applies to the components "agent" and "environment". The computation that constitutes intelligence happens on both sides of this boundary [Clark, 2010, Murphy Paul, 2022]. Diverse interactions between both components are critical to enable robust and intelligent behavior. This is closely related to the *embodiment hypothesis* [Smith and Gasser, 2005], which states "that intelligence emerges in the interaction of an agent with an environment and as a result of sensorimotor activity."

Given the discussion above, the attributive noun "embodiment" might be incorrect. The key property one needs to emphasize is not that the agent is embodied (all agents are), it is that the computation that takes place is distributed between agent and environment. This fact might be another candidate for a principle of intelligence. The claim would be that the generation of intelligent behavior requires or is greatly facilitated by incorporating aspects of the environment into the computation. There is ample support from cognitive science in support of this happening in humans [Triesch et al., 2003].

The notion of "sensorimotor activity" describes the diverse interactions that occur across the agent/environment boundary. Our own experience confirms that both our perceptual abilities and our abilities for action are diverse and can support a broad range of robust and intelligent behaviors, i.e., they enable diverse component interactions across the agent/environment boundary. For this, it seems critical that the same computational paradigm, namely mechanical computation, is available on either side of the interaction. This might be the real reason, rather than embodiment, as a form of embodiment could be limited to a different computational paradigm. Therefore, it might be more accurate to call the embodiment hypothesis the "identical computational paradigm across the boundary" hypothesis. This hypothesis might state that if the same computational paradigm is available on both sides of the agent/environment boundary, implementation complexity goes down, computational complexity goes down, and robustness and versatility go up.

6.4. More On Hypothesized Principles

As a reminder: my proposal for understanding intelligence from a computational perspective is to identify principles that characterize intelligence. This could enable intelligent systems, realized by appropriately implementing these principles in a computational system. Of course, identifying these necessary principles requires a long-term, community-wide research effort. This effort will require the close collaboration of researchers studying natural and artificial intelligence. Above, I discussed three possible principles, but there must be many more, of course.

A simple candidate for a principle of computational intelligence is the "similarity principle" [Chang, 2017]. It states that similar things behave similarly—or perform similar computation. But since this applies to nearly all physical processes at the macro-level, we cannot use it to discriminate general computation from intelligent computation.

There are likely many more principles we must identify before we can hope to fully characterize intelligence or easily create intelligent artificial systems. For example, intelligent systems are likely to perform computations at many different levels of abstractions. That could be one of the necessary principles. And across these levels, similar computational principles might apply, giving rise to a new principle: self-similarity across levels of abstractions. In Section 6.2 we saw that in biological systems the principle of *diverse, active component interactions* occurs at the genetic, cellular, and organ levels. This is an example of the self-similarity principle in biology. The self-similarity principle might also apply to representations across these levels. It will be an interesting scientific journey to hypothesize such principles and to validate them in biological and artificial agents.

7. Cognition Versus Intelligence

Cognition and Intelligence are closely related concepts. However, I would say that no operationalizable and agreed-upon definition of either term exists. This makes it difficult to articulate their difference exactly. And somehow cognitive science and intelligence research have co-existed productively without this clarification.

Still, the computational view of intelligence implies a useful relationship between cognition and intelligence. Cognition, I would like to suggest, is concerned with the mechanism of computation that produce intelligence: What algorithms are implemented? What representations are required? How do the algorithms perform and what are their inherent limitations? Just to name a few. In contrast, intelligence refers to how these computations produce behavior that solves problems and achieves goals.

Let us exploit the computational metaphor to the fullest and transfer these definitions of cognition and intelligence to the world of digital computers. Cognition would be the algorithms representations implemented in the computers hardware (binary representation of numbers and, for example, the Dadda tree as a hardware implementation for multiplication) as well as the functionality implemented in the BIOS, operating system, and applications. Intelligence would correspond to the interactions between computer and user (the environment) to solve certain problems. We would like to think that this analogy between a user and the environment is weak because the environment does not "drive" the interaction the same way the user does, but this point is worth contemplating.

In some sense, this view resembles Nikolaas Tinbergen's four questions [Tinbergen, 1963]. Tinbergen, a Nobel Prize laureate and founder of behavioral biology, postulated that to fully understand the behavior of a biological agent, one must explain the causation, ontogeny, evolution, and survival value of the observed behavior. Without going into details, Tinbergen argued that answers given from four different perspectives are necessary to obtain the full picture. I would like to suggest that cognition and intelligence represent two perspectives from which we can begin to explain intelligent behavior. There might be more (Tinbergen's four questions are very promising candidates) and each perspective must be spelled out in more detail, but maybe this paper can serve as a starting point for this debate within our community.

8. Consequences for Embodiment and Flavors of Intelligence

Understandably, after decades of focus on digital computation, a counter-movement is now emphasizing the importance of analog computation and mechanical computation. This has led to branches of intelligence/cognition research named Embodied Intelligence or Embodied Cognition. The term "body" was intended to emphasize the non-brain portion of the body, the portion of the body that does not perform neural computation but rather mechanical computation. But this led to the issues we already discussed above: All of the body, including the brain, contributes to the computation performed by a biological agent. The emphasis of "body" does not disambiguate and does not emphasize the right concept. Speaking of neural and mechanical computation does accomplish the desired distinction.

One of the original accounts of *embodiment* put it like this [Varela et al., 1993]:

> By using the term embodied we mean to highlight two points: first that cognition depends upon the kinds of experience that come from having a body with various sensorimotor capacities, and second, that these individual sensorimotor capacities are themselves embedded in a more encompassing biological, psychological and cultural context.

And this rings true today as much as it did in 1993. A detailed discussion of the evolution of embodiment from a philosophical perspective has been provided by Shapiro and Spaulding [2021]. In my view, all accounts of embodied cognition or embodied intelligence fall short of fully articulating what exactly the contribution is that is made by the body. My answer, as outlined above, is: mechanical computation, as contrasted by neural computation performed in the nervous system. What is falsely referred to as *embodiment*—because everything is embodied—actually refers to the addition of an additional computational paradigm, i.e., mechanical computation.

If the reader buys into the arguments laid out above, then we should not talk about embodiment, embodied cognition, or embodied intelligence any longer. Instead, we should talk about intelligence based on multiple computational paradigms. In the case of biological agents, we know that they rely at least on neural computation and mechanical computation. We can unite the different flavors in a study of intelligence that can clearly articulate the type of computation employed by an agent and what computational power it contributes to the behavior.

9. Conclusion

Intelligence is a computational phenomenon. Computation comes in various flavors, including digital computation, analog computation, quantum computation, mechanical computation etc. When researching intelligence and when talking about it, we should refer to these different computational paradigms to explain how an observed behavior comes about. This is much cleaner and more conducive to progress than the ongoing debates about different flavors of intelligences. Unfortunately, these flavors are now associated with different communities and academic careers. This will make it difficult to leave these distinctions behind. But I think the reward for doing so would be substantial. We should realize that, given the diversity in computational phenomena contributing to intelligence, only a unified research program can move towards a constructive understanding of intelligence. All computational paradigms can contribute to intelligence, but, more importantly, their interactions are poised to play an equally important role. How can we best combine digital computation, maybe in the form of neural computation, and mechanical computation? This question can only be answered by studying these computational paradigms together. This, of course, is not a call for dismantling disciplinary research programs—not at all. They are important and will continue to be important. It is a call for breaking down the barriers that are keeping us from making progress towards a unified science of intelligence.

Acknowledgments

I am thankful for interesting discussions and helpful comments from Alan Akbik, Aravind Battaje, Guillermo Gallego, Olaf Hellwich, Alex Kacelnik, Matt Mason, Vito Mengers, Christa Thöne-Reineke, and Oscar Weißenbach. This work was funded by the Deutsche Forschungsgemeinschaft (DFG, German Research Foundation) under Germany's Excellence Strategy EXC 2002/1 "Science of Intelligence" project number

IOP Conf. Series: Materials Science and Engineering **1321** (2024) 012001 doi:10.1088/1757-899X/1321/1/012001

390523135.

References

A. Bhatt, A. Sieler, S. Puhlmann, and O. Brock. Surprisingly Robust In-Hand Manipulation: An Empirical Study. In *Proceedings of Robotics: Science and Systems*, 2021.

D. J. Blackiston and M. Levin. Ectopic eyes outside the head in Xenopus tadpoles provide sensory data for light-mediated learning. *The Journal of Experimental Biology*, 216(6):1031–1040, 2013.

R. A. Brooks. Elephants don't play chess. *Robotics and Autonomous Systems*, 6(1–2): 3–15, 1990.

A. Cangelosi, J. Bongard, M. H. Fischer, and S. Nolfi. Embodied Intelligence. In J. Kacprzyk and W. Pedrycz, editors, *Springer Handbook of Computational Intelligence*, Springer Handbooks, pages 697–714. Springer, 2015.

R. B. Cattell. Theory of fluid and crystallized intelligence: A critical experiment. *Journal of Educational Psychology*, 54(1):1–22, 1963.

D. J. Chalmers. A Computational Foundation for the Study of Cognition. *Journal of Cognitive Science*, 12(4):325–359, 2011.

M. Chang. What constitutes science and scientific evidence: Roles of null hypothesis testing. *Educational Psychology Measurements*, 77(3):475–488, 2017.

A. Clark. *Supersizing the Mind: Embodiment, Action, and Cognitive Extension.* Oxford University Press, 2010.

M. Deitke, D. Batra, Y. Bisk, T. Campari, A. X. Chang, D. S. Chaplot, C. Chen, C. P. D'Arpino, K. Ehsani, A. Farhadi, L. Fei-Fei, A. Francis, C. Gan, K. Grauman, D. Hall, W. Han, U. Jain, A. Kembhavi, J. Krantz, S. Lee, C. Li, S. Majumder, O. Maksymets, R. Martín-Martín, R. Mottaghi, S. Raychaudhuri, M. Roberts, S. Savarese, M. Savva, M. Shridhar, N. Sünderhauf, A. Szot, B. Talbot, J. B. Tenenbaum, J. Thomason, A. Toshev, J. Truong, L. Weihs, and J. Wu. Retrospectives on the Embodied AI Workshop. arXiv:2210.06849 [cs.CV], 2022.

Encyclopedia Britannica. Definition of morphology, 2023. URL `https://www.britannica.com/science/morphology-biology`.

A. P. Engelbrecht. *Computational Intelligence—An Introduction.* John Wiley & Sons, 2007.

C. Eppner, S. Höfer, R. Jonschkowski, R. Martín-Martín, A. Sieverling, V. Wall, and O. Brock. Four aspects of building robotic systems: Lessons from the Amazon Picking Challenge 2015. *Autonomous Robots*, 42(7):1459–1475, 2018.

J. A. Fridell, S. Niederhaus, M. Curry, R. Urban, A. Fox, and J. Odorico. The survival advantage of pancreas after kidney transplant. *American Journal of Transplantation*, 19(3):823–830, 2019.

H. E. Gardner, editor. *Frames of Mind: The Theory of Multiple Intelligences.* Basic Books, 1993.

K. Ghazi-Zahedi, J. Rieffel, S. Schmitt, and H. Hauser. Editorial: Recent Trends in Morphological Computation. *Frontiers in Robotics and AI*, 8, 2021.

T. Gollisch and M. Meister. Eye Smarter than Scientists Believed: Neural Computations in Circuits of the Retina. *Neuron*, 65(2):150–164, 2010.

A. Gupta, S. Savarese, S. Ganguli, and L. Fei-Fei. Embodied intelligence via learning and evolution. *Nature Communications*, 12(1):5721, 2021.

S. Hadi Sadati, M. ElDiwiny, S. Nurzaman, F. Iida, and T. Nanayakkara. Embodied Intelligence & Morphological Computation in Soft Robotics Community: Collaborations, Coordination, and Perspective. *IOP Conference Series: Materials Science and Engineering*, 1261(1):012005, 2022.

J. E. Hopcroft and J. D. Ullman. *Introduction to Automata Theory, Languages, and Computation.* Addison-Wesley Publishing Company, 1979.

D. Horsman, S. Stepney, R. C. Wagnner, and V. Kendon. When does a physical system compute? *Proceedings of teh Royal Society A: Mathematical, Physical and Engineering Sciences*, 470:20140182, 2014.

J. Hughes, A. Abdulali, R. Hashem, and F. Iida. Embodied Artificial Intelligence: Enabling the Next Intelligence Revolution. *IOP Conference Series: Materials Science and Engineering*, 1261(1):012001, 2022.

IEEE Computational Intelligence Society. What is Computational Intelligence?, 2023. URL `https://cis.ieee.org/about/what-is-ci`.

A. Khaheshi and H. Rajabi. Mechanical Intelligence (MI): A Bioinspired Concept for Transforming Engineering Design. *Advanced Science*, 9(32):2203783, 2022.

A. N. Kolmogorov. Three approaches to the quantitative definition of information. *Problems of Information Transmission*, 1(1):1–7, 1965.

J.-C. Latombe. *Robot Motion Planning.* Kluwer Academic Publishers, 1991.

S. E. Leff, M. G. Rosenfeld, and R. M. Evans. Complex Transcriptional Units: Diversity in Gene Expression by Alternative RNA Processing. *Annual Review of Biochemistry*, 55(1):1091–1117, 1986.

R. C. Martin. *Functional Design: Principles, Patterns, and Practices.* Addison Wesley, 2024.

G. Mengaldo, F. Renda, S. L. Brunton, M. Bächer, M. Calisti, C. Duriez, G. S. Chirikjian, and C. Laschi. A concise guide to modelling the physics of embodied intelligence in soft robotics. *Nature Reviews Physics*, 4(9):595–610, 2022.

V. C. Müller and M. Hoffmann. What is morphological computation? on how the body contributes to cognition and control. *Artificial Life*, 23(1):1–24, 2017.

A. Murphy Paul, editor. *The Extended Mind—The Power of Thinking Outside the Brain.* Mariner BOoks, 2022.

R. Pfeifer and J. Bongard, editors. *How the Body Shapes the Way We Think - A New View of Intelligence.* The MIT Press, 2006.

G. Piccinini. *Physical Computation—A Mechanistic Account.* Oxford University Press, 2015.

G. Piccinini and C. Maley. Computation in Physical Systems. In E. N. Zalta, editor, *The Stanford Encyclopedia of Philosophy.* Metaphysics Research Lab, Stanford University, Summer 2021 edition, 2021.

J. A. Pruszynski and R. S. Johansson. Edge-orientation processing in first-order tactile neurons. *Nature Neuroscience*, 17(10):1404–1409, 2014.

H. Robinson. Dualism. In E. N. Zalta and U. Nodelman, editors, *The Stanford Encyclopedia of Philosophy.* Metaphysics Research Lab, Stanford University, Spring 2023 edition, 2023.

S. J. Russell and P. Norvig, editors. *Artificial Intelligence—A Modern Approach.* Prentice Hall, fourth edition, 2020.

L. Shapiro and S. Spaulding. Embodied Cognition. In E. N. Zalta, editor, *The Stanford Encyclopedia of Philosophy.* Metaphysics Research Lab, Stanford University, Winter 2021 edition, 2021.

M. Sitti. Physical intelligence as a new paradigm. *Extreme Mechanics Letters*, 46: 101340, 2021.

L. Smith and M. Gasser. The Development of Embodied Cognition: Six Lessons from Babies. *Artificial Life*, 11(1-2):13–29, 2005.

R. J. Sternberg and S. B. Kaufman, editors. *The Cambridge Handbook of Intelligence.* Cambridge University Press, 2011.

N. Tinbergen. On aims and methods of Ethology. *Zeitschrift für Tierpsychologie*, 20 (4):410–433, 1963.

J. Triesch, D. H. Ballard, M. M. Hayhoe, and B. T. Sullivan. What you see is what you need. *Journal of Vision*, 3(1):86–94, 2003.

D. I. Vaney and R. Weiler. Gap junctions in the eye: evidence for heteromeric, heterotypic and mixed-homotypic interactions. *Brain Research Reviews*, 32(1):115–120, 2000.

F. J. Varela, E. Thomson, and E. Rosch. *The Embodied Mind: Cognitive Science and Human Experience.* MIT Press, 1993.

R. Vianna, J. J. Gaynor, G. Selvaggi, A. Farag, J. Garcia, A. Tekin, M. M. Tabbara, and G. Ciancio. Liver Inclusion Appears to Be Protective Against Graft Loss-Due-to Chronic But Not Acute Rejection Following Intestinal Transplantation. *Transplant International*, 36:11568, 2023.

S. Wang, Y. Shi, and L. Wen. Physical Intelligence in Biomechanics. *IOP Conference Series: Materials Science and Engineering*, 1261(1):012012, 2022a.

Z. Wang, B. Benes, A. H. Qureshi, and C. Mousas. Co-Design of Embodied Neural Intelligence via Constrained Evolution. arXiv:2205.10688 [cs], 2022b.

T. Yoshioka and M. Sakakibara. Physical aspects of sensory transduction on seeing, hearing and smelling. *Biophysics*, 9:183–191, 2013.

Habit-inspired methods for embodied intelligence

Nicolas Kuske[1,2*], Florian Roehrbein[1]

[1] Faculty of Computer Science, Chemnitz University of Technology, Chemnitz, Germany,
[2] Artificial and Natural Intelligence Toulouse Institute, Toulouse, France

*E-mail: nicolas.kuske@informatik.tu-chemnitz.de

Abstract. Algorithms that employ habitual processing may be characterized as context-contingent input-output pairings. They learn to relate recurring patterns with swift and effective responses, relying on minimal computational resources. While habits have been extensively researched in the fields of psychology, cognition, and neuroscience, they have yet to be systematically investigated in computer science. Data volumes are expanding at an exponential rate, and there is often a requirement for this data to be processed in real-time. Against this backdrop, principles of habitual cognition present a promising alternative to resource-intensive planning algorithms. We propose that future research focus on algorithmic and neuronal models of the neurocognitive foundations of habits to develop new perspectives on artificial intelligence and robotics.

1. Motivation

A habit is an automatised cognitive process for decision making. Within a particular context a stimulus automatically generates an impulse towards (overt or covert) action, based on learned stimulus-context-response associations. This implicit type of processing information immediately yields a result while relying on minimal computational resources. Due to their efficiency, habits have played a crucial role in the evolution of humans (as well as other animals) [12]. Habitual cognitive processes continue to influence a large part of our everyday decision-making [13-15]. The highly automated habit system is defined in opposition to a deliberate planning system that leverages explicit knowledge. Planning requires a model of the environment and allows the deduction of optimal actions as well as flexible decision making relevant to the current goal. Mapping human cognition onto these two distinct information processing poles follows a long interdisciplinary tradition. Dual-process theories form a fundamental part of research in psychology, cognitive science and machine learning [15-18]. Yet, the relationship between the proposed processing systems remains elusive. Conventional cognitive models predominantly emphasise explicit planning methodologies (for a review of model architectures, see [19]). The exploration of heuristic protocols, which serve to simplify and compress planning algorithms, has been mooted as a nexus intersecting both systems [20-22]. Two heuristic subtypes can be delineated: implicit heuristics, characterised by their automatic deployment, and explicit heuristics, distinguished by their deliberate application. With adequate iterative exposure or positive reinforcement, the explicit invocation of a rule of thumb can transition into an automated response [22]. Nonetheless, for a precise classification of these processes, it is imperative to demarcate between the act of initiation and the subject itself being activated. Habits serve as full or partial substitution for planning algorithms; unlike heuristics, they don't inherently constitute planning algorithms in their own right.

Cognition science and AI is transitioning from isolated lab scenarios and toy applications to real-world interactions and applications. The transition underscores that with the escalation of environmental intricacy, standalone model-centric planning paradigms become suboptimal. In response, the interest in the role of habits increases [1-3,23-25]. A similar situation arises with the current development of information technology. The swift proliferation of the Internet of Things, coupled with the augmentation of computational intelligence in discrete devices, results in an unparalleled intricacy of the overarching digital infrastructure and an immense concomitant data volume [26,27]. The conventional response to the challenge of complexity is the development of heuristics and an increase in computing resources [4,5]. However, heuristic functions utilise task-specific knowledge and meta-heuristic production mechanisms remain domain-bound [28,29]. Addressing increased complexity with added high-performance computing provides a broad solution but contradicts the prevalent imperative for resource optimization. We propose an alternative approach. Its main research objective would be to assess, enhance and utilise the potential of habit-inspired methods in the processing of information by addressing methods for integrating the cognitive habit principle into embedded computing systems.

1.1 Embodiment and phase transition

Following a definition proposed in last year's conference proceedings [47], Embodied Intelligence involves three interconnected levels: materials, structure, and situatedness in an environment, emphasizing that intelligence manifests through physical embodiment and environmental interactions. Embodied intelligence requires a fit between body and environment, allowing the interaction loop to bypass resource-demanding 'off-body' computational algorithms [48]. Habits also involve 'off-body' neural structures and, as we will see below, mental habits can incorporate higher cognitive pathways. Hence, habits are conventionally considered a form of disembodied intelligence. However, this definition of body may be too simplistic for multilayered, hierarchical systems like animals and intelligent robots. To fully leverage the notion of embodied intelligence, it seems necessary to identify different bodies with distinct materials and structures within the same agent. In a specific context, a bodily layer may also become part of the environment for another, deeper body within the same system. For example, during habitual egocentric spatial navigation, sensory modalities, motor effectors, and habitually strengthened neural connections can be considered a single bodily layer, bypassing resource-intensive 'off-body' computations.

Building on this perspective, one can incorporate the notion of phase transitions in physical systems more broadly. The bodily structure can be viewed as a phase of the system with minimal entropy [49,50]. When the stability of this bodily structure can no longer be maintained—such as when a novel obstacle in the environment blocks a habitual path in spatial navigation—the stable structure undergoes a phase transition. It returns to a dynamic state, increasing its entropy, and functionally, it can no longer be considered a distinct body but merges with the remaining computational and cognitive apparatus. A fitting analogy is the frozen weights of a deep neural network after training to a criterion. The current weight pattern represents the system's optimal state within its present context, and minimal plasticity is required to maintain overall stability. The system is in a kind of habitual exploitation mode and can be considered a bodily structure. Only when the (local) environmental context changes unpredictably does the system shift into a dynamic and plastic state, ready to explore new (mental) affordances to incorporate into habit.

A detailed analysis of the relationship among the concepts of habit, body, and phase transition is beyond the scope of this article and will be explored in future work. Here, it is sufficient to offer a glimpse of the potential and rich interconnections of the seemingly simple concept of habit. The following sections will provide an overview of the biological and technical foundations of the habit principle. The article also presents a case study of habit-inspired methods in artificial agents and concludes with a call for further investigation into the interaction between the habit principle and conventional planning processes in embodied intelligence research.

2. Biological foundations

In the early 1900s, habits were first recognised in psychology as crucial elements of daily human behaviour. By the mid-1900s, the focus shifted towards understanding them within the paradigm of learned behaviours, distinguishing habits from reflexes due to their more complex context dependence and, thus, independence from specific stimuli [12]. Around the same time habits were explored in contrast to intentional goal-directed behaviour, laying early groundwork for what would later be formalised as dual-process theories. It became apparent that habitual behaviour is robust against goal-devaluation (habit as automatism) but that deliberate processes directed towards novel goals can intervene to impose an alternative action (habit as impulse) [15]. The late 20th to the 21st centuries saw the intersection of neuroscience and psychology shedding light on the brain mechanisms underpinning habit formation.

2.1 Habit formation

Biological systems acquire or form habits via the two related processes of use-dependent learning and reward-dependent learning [23]. Use-dependent learning theories state that the habitual controller is sensitive only to the history of selected actions, and not to the outcomes of those actions – repetition is key. The biological mechanism underlying this theory is Hebbian learning, which states that the synaptic connections between cortical neurons are strengthened when the respective neurons are active in direct succession. The second process of biological habit formation is reward dependent. It assumes that actions that have been followed by rewarding outcomes are likely to be repeated in the future. This type of learning maps well onto the subcortical release of the neurotransmitter dopamine whenever the animal receives an unpredicted reward [6]. Dopamine greatly enhances the Hebbian strengthening of the synaptic connection between two successively active neurons.

2.2 Habit spectrum

The subcortical brain regions involved in reward-dependent habit learning are the Basal Ganglia. They form multiple information processing loops with the cortex which fulfil cognitive control functions and due to their dopaminergic innervation can shape associative connections [7,30]. Habit research has long focused on sensorimotor associations, which are implementing overt habitual behaviour [13,31]. Strict dual-process theories contrast these associations with executive control functions constituted by prefrontal cortex networks [32]. However, recent experimental and neurocomputational investigations into the role of frontal-cortical basal ganglia loops point towards an intricate interplay between both systems [2,33,34]. Covert cognitive processing prior to motor execution likely involves mental habits driving emotional responses, intuition and reoccurring thought patterns. Such a spectrum of habits has already been acknowledged in psychology, for example in mental health therapy or the development of comprehensive education frameworks [14,35]. Recent studies in the domain of computer programming skills corroborate this conclusion [1]. Consequently, the conventional picture of an either-or substitution relationship between automatic habit and intentional planning in biological systems, has to be extended to incorporate the habit principle as an integral augmentation mechanism at all levels of information processing.

2.3 Habit relevance

The relevance of habits for substitution and augmentation of more complex information processing methods in biological systems rests on speed, efficiency, transferability and robustness. First, in the social wild, situations occur in which quick action is necessary to achieve a goal. Often, these situations are context-dependent, making them an optimal fit for stimulus-context-response associations [12]. Second, it has been shown that tasks involving deliberate and conscious planning processes demand more energy (glucose and oxygen) than cognitive routine tasks [36,37]. Third, habits constitute transferable knowledge because their contextual triggering conditions can be satisfied by a subset of the original situation. Furthermore, within a community, habits can be transmitted among individuals through the process of imitation and the use of language. Fourth, robustness refers to the less-is-more principle or ecological rationality [20,38]. In a complex environment with interdependencies that are difficult to model comprehensively, a planning algorithm may face overfitting issues. The law of large numbers guarantees the resilience of simple statistical measures, such as averages, ensuring their robustness. Less-is-more effects have been demonstrated analytically and through psychological experiments as well as through in-situ experiments involving computational simulations in technical systems [38].

3. Technical foundations

Algorithmic rule-based planners have a long tradition in computer and cognitive science. In contrast to these expert systems, habit-inspired methods of information processing are less prominent. Likely, the outdated perspective that habits mainly control motor behaviour has hindered progress of advancing these methods in technical systems [24]. Recently, however, decision-making habits have been formalised for technical systems which resemble mental habits in biological systems in that they can act on any abstraction level of information processing prior to the system's end effector [39]. The intricate interplay of habit and plan currently being discovered in biological systems likely transfer in large parts to technical systems. Moreover, the technical engineering perspective offers a significant advantage for habit-inspired methods in computer science. While repetition and reward are required to form and break habits in biological systems, technical systems allow for both automated learning processes and specifically engineered habitual (overt or covert) behaviour.

3.1 Habit engineering

Habit-inspired engineering of information processing methods for technical systems can produce potent stand-alone solutions with broad applicability. The most prevalent example of a specifically engineered method is information factorization in hierarchically structured systems [40]. The method provides only partial or pre-processed information to certain parts or modules in hierarchical order. Although a central planner could, in principle, integrate all available information and produce controls directly, segregated control units are likely to generalise better. The principle of hierarchical information factorisation has been applied with considerable success in the field of robotics engineering [41]. It recently gained more widespread attention in AI systems design [40,42]. The principle also underpins the development of neurocomputational models of the neurobiological habit network structure [2,7].

3.2 Habit in machine learning

The formalisation of habit acquisition in machine learning algorithms eliminates the need for case-by-case design. A plethora of approaches has been developed and applied for automatic extraction of statistical regularities in data in order to predict the next optimal action [8-

11,43,44]. The machine learning method closest to formalising the biological mechanisms underlying habit formation is Reinforcement Learning (RL) [18]. RL algorithms consider transition probabilities between states of an agent in the world depending on its own actions. If a state-action pair participates in reaching a rewarding goal-state it accumulates value. Learning optimal behaviour for a given task occurs as value propagates backward through the state-action space. Based on the current value distribution, the next optimal action can be chosen without the computational burden to calculate the effect of all subsequent actions with respect to a final outcome.

The machine learning community distinguishes between model-free and model-based algorithmic approaches to intelligent behaviour [18,45]. Habit formation relies on model-free RL as outlined above. To qualify as model-based, the learning algorithm must integrate the state-action pairs into a model structure. Akin to traditional non-RL planners, the agent then leverages this world model by computing or simulating the sequence of actions leading to the desired outcome before acting. The relevance of habits in technical systems depends on the environmental embedding of the current goal just as it does for their biological counterpart. In real-world applications, optimal behaviour of technical systems may operate at a fast, efficient, transferable and robust model-free instance in default mode. It must resort to a model-dependent planner only when the performance deviates from anticipated or desired outcomes [25]. Consequently, the currently best performing RL algorithms in robotics [45] and automated driving [46] merge model-free and model-based processes. In summary, model-free RL requires additional cognitive overhead to allow for multiple goals and flexible policy adaptation in dynamical environments.

4. Case studies: Application of habit-inspired methods

The work by Dromnelle et al. [25] provides a comprehensive examination of habit-inspired methods through three specific experimental case studies. These experiments illustrate the wide range of applicability and benefits of habit-inspired reinforcement learning architectures.

4.1 Navigation task

In the first experiment, a robot was tasked with navigating a real-world environment with dynamic changes, such as altered wall configurations and goal locations. The neuro-inspired architecture combined model-based and model-free reinforcement learning, with a meta-controller arbitrating between the two based on a trade-off between efficiency and computational cost. This approach allowed the robot to adapt flexibly to changes while minimizing computational expenses. The model-based component utilized dynamic programming to build a model of the long-term effects of actions, enabling the robot to adapt its navigation behaviour to changes in the environment. Meanwhile, the model-free component relied on discrete, tabular Q-learning to learn optimal actions through trial and error (see [51] for a continuous deep Q network comparison). The meta-controller monitored the environment and the robot's performance, dynamically switching between both strategies to balance computational load and adaptability. This approach allowed the robot to maintain optimal performance with significantly lower computational costs. The hybrid architecture demonstrated robust adaptability to environmental changes, successfully navigating new wall configurations and altered goal locations while minimizing computational resources.

4.2 Object manipulation task

The second experiment involved an object manipulation task where the robot received teaching signals from a human. This scenario tested the robot's ability to learn from human interaction and adapt its strategies accordingly. The robot was equipped with sensors to perceive the environment and manipulator arms to interact with objects. Human teaching signals were provided through a graphical user interface, where the human teacher could demonstrate tasks and provide corrective feedback. The robot utilized a hybrid RL approach, with the model-based component allowing the robot to understand and plan for long-term consequences of actions, while the model-free component enabled quick adaptation to immediate feedback from the human. The architecture enabled the robot to balance between learning new strategies and relying on established habits, resulting in effective task completion and interaction with the human teacher. The robot was able to generalize from the human demonstrations, improving its object manipulation skills over time and reducing the learning curve significantly compared to traditional methods.

4.3 Human-robot cooperation task

The third experiment focused on a simulated human-robot cooperation task aimed at tidying up objects on a table. This task required the robot to collaborate with a human partner, adapting to the human's actions and changes in the environment. The robot's cooperative behavior was managed through the same hybrid RL architecture. The robot was programmed to recognize human actions and cues through visual and auditory sensors. The model-based component allowed the robot to plan actions based on the predicted long-term outcomes of cooperative strategies, while the model-free component enabled real-time adaptation to the human partner's behavior. The habit-inspired architecture facilitated efficient cooperation, allowing the robot to maintain high performance and adaptability. The robot could respond to human cues quickly and effectively, adjusting its actions based on the human partner's movements and changes in the task environment. The experiment highlighted the robot's ability to handle unexpected changes and maintain seamless interaction, demonstrating the robustness and flexibility of the habit-inspired methods in dynamic human-robot interaction scenarios.

These case studies demonstrate the versatility and effectiveness of habit-inspired methods in diverse scenarios, from navigation and manipulation to human-robot cooperation. The integration of habits into the cognitive architecture significantly enhanced the robot's adaptability and efficiency, highlighting the potential of this approach in real-world applications.

5. Towards habit-inspired artificial agents

Provided with sufficient computational capacity and a comprehensive environmental model, planning algorithms can converge towards near-optimal solutions within defined constraints (e.g., a time limit). However, in real-world scenarios neither computational resources nor a sufficient model are readily available. In these cases, habits become a powerful "toolbox" [20]. We propose to take inspiration from the way animals and humans dynamically and autonomously adapt their behaviour according to variations in performance while minimising cognitive effort. The overarching research question is how to ideally combine conventional planning approaches with the habit principle.

In order to develop habit-inspired methods for technical systems, it is necessary to formalise the underlying computational principles in biological systems and individually transfer them to different technical domains. But biological habits are constituents of a holistic cognitive structure.

Increasing evidence suggests that the traditional two-process distinction of plan and habit does not map onto clear-cut independent brain-modules but rather interacting information processing hubs [7,33,34].
strategies, while the model-free component enabled real-time adaptation to the human partner's behavior. The habit-inspired architecture facilitated efficient cooperation, allowing the robot to maintain high performance and adaptability. The robot could respond to human cues quickly and effectively, adjusting its actions based on the human partner's movements and changes in the task environment. The experiment highlighted the robot's ability to handle unexpected changes and maintain seamless interaction, demonstrating the robustness and flexibility of the habit-inspired methods in dynamic human-robot interaction scenarios.

Research in habit-inspired methods for embodied intelligence would aim at the advancement of artificial cognitive architectures that emulate the functional attributes of their biological role models. In particular, humans are able to alternate between contexts in which leveraging a sequence of discrete decision making habits is sufficient, versus contexts requiring the activation of more comprehensive planning processes. Higher cognitive reasoning likely integrates both methods of processing information.

References

[1] Peitek N, Bergum A, Rekrut M, Mucke J, Nadig M, Parnin C, Siegmund J, Apel S. Correlates of programmer efficacy and their link to experience: a combined EEG and eye-tracking study. In Proceedings of the 30th ACM Joint European Software Engineering Conference and Symposium on the Foundations of Software Engineering 2022 Nov 7 (pp. 120-131), doi:10.1145/3540250.3549084

[2] Baladron J, Hamker FH. Habit learning in hierarchical cortex–basal ganglia loops. European Journal of Neuroscience. 2020 Dec;52(12):4613-38, doi:10.1111/ejn.14730

[3] Scholl C, Baladron J, Vitay J, Hamker FH. Enhanced habit formation in Tourette patients explained by shortcut modulation in a hierarchical cortico-basal ganglia model. Brain Structure and Function. 2022 Apr;227(3):1031-50. doi:10.1007/s00429-021-02446-x

[4] Dietze R, Rünger G. The search-based scheduling algorithm HP* for parallel tasks on heterogeneous platforms. Concurrency and Computation: Practice and Experience. 2020 Nov 10;32(21):e5898, doi:10.1002/cpe.5898

[5] Rauber T, Rünger G. Parallel Programming for Multicore and Cluster Systems; 3rd Edition. Springer. 2023, ISBN: 978-3-031-28924-8, doi:10.1007/978-3-031-28924-8

[6] Vitay J, Hamker FH. Timing and expectation of reward: a neuro-computational model of the afferents to the ventral tegmental area. Frontiers in neurorobotics. 2014 Jan 31;8:4. doi:10.3389/fnbot.2014.00004

[7] Baladron J, Hamker FH. Re-Thinking the Organization of Cortico-Basal Ganglia- Thalamo-Cortical Loops. Cognitive Computation. 2023 Apr 24:1-6. doi: 10.1007/s12559- 023-10140-9

[8] Chen G, Cao H, Ye C, Zhang Z, Liu X, Mo X, Qu Z, Conradt J, Röhrbein F, Knoll A. Multi-cue event information fusion for pedestrian detection with neuromorphic vision sensors. Frontiers in neurorobotics. 2019 Apr 2;13:10, doi:10.3389/fnbot.2019.00010

[9] Turetta C, Skenderi G, Capogrosso L, Demrozi F, Kindt PH, Masrur A, Fummi F, Cristani M, Pravadelli G. Towards Deep Learning-based Occupancy Detection Via WiFi Sensing in Unconstrained Environments. In 2023 Design, Automation & Test in Europe Conference & Exhibition (DATE) 2023 Apr 17;1-6. IEEE.

[10] Sampath Kumar A, Schlosser T, Langner H, Ritter M, Kowerko D. Improving OCT Image Segmentation of Retinal Layers by Utilizing a Machine Learning Based Multistage System of Stacked Multiscale Encoders and Decoders. Bioengineering. 2023 Oct 10;10(10):1177. doi:10.3390/bioengineering10101177

[11] Thomanek R, Rolletschke T, Platte B, Hosel C, Roschke C, Manthey R, Heinzig M, Vogel R, Zimmer F, Vodel M, Eibl M. Real-time activity detection of human movement in videos via smartphone based on synthetic training data. In Proceedings of the IEEE/CVF Winter Conference on Applications of Computer Vision Workshops. 2020: 160-164. doi:10.1109/WACVW50321.2020.9096936

[12] Hodgson GM. Choice, habit and evolution. Journal of Evolutionary Economics. 2010 Jan;20(1):1-8, doi:10.1007/s00191-009-0134-z

[13] Verplanken, B.: The Psychology of Habit. Cham: Springer; 2018 doi:10.1007/ 978-3-319-97529-0

[14] Colvin E, Gardner B, Labelle PR, Santor D. The automaticity of positive and negative thinking: a scoping review of mental habits. Cognitive therapy and research. 2021 Dec 1:1-27. doi:10.1007/s10608-021-10218-4
[15] Wood W, Rünger D. Psychology of habit. Annual review of psychology. 2016 Jan 4; 67:289-314, doi:10.1146/annurev-psych-122414-033417
[16] Kahneman D. Thinking, fast and slow. Farrar, Straus and Giroux. 2011 Oct 25. ISBN: 978-0374275631
[17] Evans JS, Stanovich KE. Dual-process theories of higher cognition: Advancing the debate. Perspectives on psychological science. 2013 May;8(3):223-41, doi:10.1177/ 1745691612460685
[18] Sutton RS, Barto AG. Reinforcement learning: An introduction. MIT press; 2018 Nov 13. ISBN: 9780262039246
[19] Kotseruba I, Tsotsos JK. 40 years of cognitive architectures: core cognitive abilities and practical applications. Artificial Intelligence Review. 2020 Jan;53(1):17-94, doi:10.1007/s10462-018-9646-y
[20] Gigerenzer G, Gaissmaier W. Heuristic decision making. Annual review of psychology. 2011 Jan 10;62:451-82. doi:10.1146/annurev-psych-120709-145346
[21] Knauff M, Knoblich G. Logisches Denken. In Allgemeine Psychologie 2017 Oct 25: 533-585. Berlin, Heidelberg: Springer Berlin Heidelberg. doi: 10.1007/ 978 -3-642-53898-8_15
[22] De Neys W. Advancing theorizing about fast-and-slow thinking. Behavioral and Brain Sciences. 2022 Sep 2:1-68. doi: 10.1017/S0140525X2200142X
[23] Pauli WM, Cockburn J, Pool ER, Perez OD, O'Doherty JP. Computational approaches to habits in a model-free world. Current Opinion in Behavioral Sciences. 2018 Apr 1;20:104-9, doi:10.1016/j.cobeha.2017.12.001
[24] Ramírez-Vizcaya S, Froese T. Agents of habit: refining the artificial life route to artificial intelligence. In Artificial Life Conference Proceedings 32. 2020 Jul 1:78-86. MIT Press. doi:10.1162/isal_a_00298
[25] Dromnelle R, Renaudo E, Chetouani M, Maragos P, Chatila R, Girard B, Khamassi M. Reducing Computational Cost During Robot Navigation and Human–Robot Interaction with a Human-Inspired Reinforcement Learning Architecture. International Journal of Social Robotics. 2022 Nov 8:1-27. doi:10.1007/s12369-022-00942-6
[26] Benbya H, Nan N, Tanriverdi H, Yoo Y. Complexity and information systems research in the emerging digital world. Mis Quarterly. 2020 Feb 16;44(1):1-7. doi:10.25300/ MISQ/2020/13304
[27] Gil D, Johnsson M, Mora H, Szymański J. Review of the complexity of managing big data of the internet of things. Complexity. 2019 Feb 3; doi: 10.1155/2019/4592902
[28] Russell S, Norvig P. Künstliche Intelligenz. München: Pearson Studium; 2012. ISBN: 9783868940985
[29] Karimi-Mamaghan M, Mohammadi M, Meyer P, Karimi-Mamaghan AM, Talbi EG. Machine learning at the service of meta-heuristics for solving combinatorial optimization problems: A state-of-the-art. European Journal of Operational Research. 2022 Jan 16;296(2):393-422. doi:10.1016/j.ejor.2021.04.032
[30] Graybiel AM. The basal ganglia. Current biology. 2000 Jul 15;10(14):R509-11. doi:10.1016/s0960-9822(00)00593-5
[31] Du Y, Krakauer JW, Haith AM. The relationship between habits and motor skills in humans. Trends Cogn Sci. 2022 May;26(5):371-387. doi: 10.1016/j.tics.2022.02.002
[32] Daw ND, Niv Y, Dayan P. Uncertainty-based competition between prefrontal and dorsolateral striatal systems for behavioral control. Nature neuroscience. 2005 Dec;8(12):1704-11, doi:10.1038/nn1560
[33] Daw ND, Gershman SJ, Seymour B, Dayan P, Dolan RJ. Model-based influences on humans' choices and striatal prediction errors. Neuron. 2011 Mar 24;69(6):1204-15, doi:10.1016/j.neuron.2011.02.027
[34] Huang Y, Yaple ZA, Yu R. Goal-oriented and habitual decisions: Neural signatures of model-based and model-free learning. NeuroImage. 2020 Jul 15;215:116834, doi:10.1016/j.neuroimage.2020.116834
[35] Altan S, Lane JF, Dottin E. Using habits of mind, intelligent behaviors, and educational theories to create a conceptual framework for developing effective teaching dispositions. Journal of Teacher Education. 2019 Mar;70(2):169-83. doi:10.1177/00224871177360
[36] Magistretti PJ, Allaman I. A cellular perspective on brain energy metabolism and functional imaging. Neuron. 2015 May 20;86(4):883-901, doi:10.1016/j.neuron.2015.03.035
[37] Hahn A, Breakspear M, Rischka L, Wadsak W, Godbersen GM, Pichler V, Michenthaler P, Vanicek T, Hacker M, Kasper S, Lanzenberger R. Reconfiguration of functional brain networks and metabolic cost converge during task performance. elife. 2020 Apr 21;9:e52443, doi:10.7554/eLife.52443
[38] Callaway F, van Opheusden B, Gul S, Das P, Krueger PM, Griffiths TL, Lieder F. Rational use of cognitive resources in human planning. Nature Human Behaviour. 2022 Aug;6(8):1112-25. doi: 10.1038/s41562-022-01332-8
[39] Batta E, Stephens C. Heuristics as decision-making habits of autonomous sensorimotor agents. In ALIFE 2019: The 2019 Conference on Artificial Life. 2019 Jul 1:72-78. MIT Press. doi: 10.1162/isal_a_00144
[40] Merel J, Botvinick M, Wayne G. Hierarchical motor control in mammals and machines. Nature communications. 2019 Dec 2;10(1):5489. doi : 10.1038/s41467-019-13239-6

[41] Brooks RA. Intelligence without reason. In The artificial life route to artificial intelligence. 2018 May 15 (pp. 25-81). Routledge. ISBN: 9781351001885

[42] Dong X, Yu L, Wu Z, Sun Y, Yuan L, Zhang F. A hybrid collaborative filtering model with deep structure for recommender systems. In Proceedings of the AAAI Conference on artificial intelligence. 2017 Feb 12 (Vol. 31, No. 1). doi:10.1609/aaai.v31i1.10747

[43] Joo M, An Y, Roh H, Lee W. Predictive prefetching based on user interaction for web applications. IEEE Communications Letters. 2020 Nov 16;25(3):821-4. doi:10.1109/ LCOMM.2020.3038255

[44] Alhijawi B, Kilani Y. A collaborative filtering recommender system using genetic algorithm. Information Processing & Management. 2020 Nov 1;57(6):102310. doi:10.1016/ j.ipm.2020.102310

[45] Wu P, Escontrela A, Hafner D, Goldberg K, Abbeel P. Daydreamer: World models for physical robot learning. arXiv preprint arXiv:2206.14176. 2022 Jun 28.

[46] Kiran BR, Sobh I, Talpaert V, Mannion P, Al Sallab AA, Yogamani S, Pérez P. Deep reinforcement learning for autonomous driving: A survey. IEEE Transactions on Intelligent Transportation Systems. 2021 Feb 9;23(6):4909-26, doi:10.1109/TITS.2021.3054625

[47] Vihmar M, Valdur KA, Banerji S, Must I. How to measure embodied intelligence?. In IOP Conference Series: Materials Science and Engineering. 2023 Oct 1 (Vol. 1292, No. 1, p. 012002). IOP Publishing. doi: 10.1088/1757-899X/1292/1/012002

[48] Mengaldo G, Renda F, Brunton SL, Bächer M, Calisti M, Duriez C, Chirikjian GS, Laschi C. A concise guide to modelling the physics of embodied intelligence in soft robotics. Nature Reviews Physics. 2022 Sep;4(9):595-610. doi: 10.1038/s42254-022-00481-z

[49] Hershey D. Entropy theory of aging systems: humans, corporations and the universe. World Scientific; 2009 Aug 14. ISBN: 1848162928

[50] Dengsø MJ. The temporal and affective structure of living systems: A thermodynamic perspective. Adaptive Behavior. 2024 Feb;32(1):17-31. doi: 10.1177/10597123231176

[51] Dromnelle R, Renaudo E, Pourcel G, Chatila R, Girard B, Khamassi M. How to reduce computation time while sparing performance during robot navigation? A neuro-inspired architecture for autonomous shifting between model-based and model-free learning. In Conference on Biomimetic and Biohybrid Systems 2020 Jul 28 (pp. 68-79). Cham: Springer International Publishing.

Can the prescience of Blade Runner engender a deeper understanding of our relationship with AI technology?

Alexander G. Ross, Ph.D., FRHistS[1]

[1] New College, Oxford and Branford College, Yale

*E-mail: alexander.ross@new.ox.ac.uk

Abstract. This article explores how, in the 1982 film Blade Runner (directed by Ridley Scott), some preeminent creative minds predicted a future which increasingly reflects the realities of our age. The research outcome is multidisciplinary in perspective, drawn from interviews with the filmmakers as well as with thought leaders in Artificial Intelligence, Classics, Engineering, Ethics, Philosophy, Robotics, and Theology, and is, in part, influenced by the seminal work of Alessandro Portelli. The author has also structured the article by applying some of Jonathan S. Feinstein's models of creative engagement in exploring the intricacies of creative and innovative work in the arts and sciences, as developed in his book The Nature of Creative Development (2006). The article contributes to the debate about the numerous contentions within the text which it seeks to clarify, as well as to the debate about the nature of creative work and the extent to which the visionaries involved could benefit from a cross-pollination of ideas between disciplines. *Keywords*: AGI, bioengineering, biotechnology, ChatGPT, computation, cognition, consciousness, cyborgs, DeepMind, Philip K. Dick, ethics, evolutionary robotics, Geoffrey Hinton, Ray Kurzweil, LaMDA, Elon Musk, neuroscience, replicants, singularity, Alan Turing, Xenobots.

1. Introduction

Blade Runner might just be a film, but it also set up the laboratory of today, which allows us to explore a broad, diverse range of hypotheses about frontier technologies and what they could ultimately mean for humankind.

This essay explores how, in the 1982 film Blade Runner (directed by Ridley Scott), some preeminent creative minds predicted a future which increasingly reflects the realities of our age. The research outcome is largely in the form of an oral history study consisting of unstructured interviews with the filmmakers as well as with thought leaders in Artificial Intelligence, Robotics, Philosophy, Ethics and the Classics. When Blade Runner premiered in theatres on June 25, 1982, it was a box office flop. Since then, it has become one of the most influential science fiction, or "futurist", films made. More than 40 years after its premiere, the film continues to impress viewers, scholars, and critics with its prescient portrait of societal decay, juxtaposed against advanced technology, which foreshadows ongoing advancements in artificial intelligence, machine learning and robotics.

The entry point for this essay is provided by Ivor Powell, one of the film's producers. "The world that we painted is virtually knocking on the door of this world that we're in now. It's extraordinary and that's another thing that I find slightly uncomfortable about now, is that I suddenly find myself in this world because I always loved science fiction."[1]

His statement frames the impetus not only to seek a deeper understanding of the creative process applied by those tasked with the evolution of the film's narrative, but also their change of perspective over the last forty years. This central theme will continue to be explored throughout the essay.

The core of Blade Runner revolves around the character of Rick Deckard (played by Harrison Ford), a gun for hire who tracks down mutinous Replicants[2] for a living and murders them. Deckard is compelled by his former employer, LAPD police captain Harry Bryant (played by M. Emmet Walsh) to "retire"[3] a recent pride of Replicants who have killed 23 people to return to earth and extend their lifespans. Led by Roy Batty (played by the late Rutger Hauer), we are left to decide whether the Replicants are mass murderers or freedom fighters. With only hours left to live, Batty targets their three creators: Eldon Tyrell, J.F. Sebastian, and Hannibal Chew, to force them to extend their lifespan. When they assert their inability to fulfil this request, Batty retires them. Meanwhile, Deckard's personal conflict ensues, when he meets Rachael (played by Sean Young), the biological android who initially does not know she is a Replicant. His self-hatred and distaste for these "machines", when impacted by his ensuing love for Rachael, guides the main character arc for the role.

Admittedly, we are already witnessing a certain ennui de AI. The roots of its technological development are now at least 75 years old. We are overwhelmed by the number of articles, books, and contentious interviews on the implications of this latest "paradigm shift". Highly educated and experienced engineers repeatedly compete with each other to predict the technology driven societal changes that await us in the next five, ten, twenty or more years. The tendency to over-promise outweighs what can be delivered.

Strangely, few of the same engineering thought leaders question the Wizard of Oz paradigm in all of this. Most of the current research takes place on the campuses of immensely powerful multi-national companies such as Google, Microsoft, Apple, Amazon, Tesla etc. All of which are given to being monopolies. Over the years they are jointly responsible for thousands of competitors either being put out of business or absorbed by them. They can finance the campaigns of politicians who are meant to protect people from abuses of power. No one knows what is going on behind these corporate veils. As this essay will explore, amongst many others Wendell Wallach, the Emeritus Chair of Technology and Ethics Studies at Yale University's Interdisciplinary Centre for Bioethics is one of the scientists to voice concerns about the status quo.

Do we trust the likes of Zuckerberg, Bezos, Musk, Pichai, Cook, etc. to altruistically evolve this technology? If not, then why have we allowed them to amass this degree of power? Does Blade Runner foreshadow a possible outcome through the Tyrell corporation? Eldon Tyrell, a self-appointed God, responsible for creating life and taking it, on his own terms and without any oversight.[4]

The truth is that nobody really knows what is going on. Adrienne Mayor, a classics professor at Stanford University, makes the case in her 2018 book, that some of our best thinking about Artificial Intelligence and Robotics was done by the Greeks, 2,500 years ago.[5] In Gods and Robots: Myths, Machines, and Ancient Dreams of Technology, Mayor chronicles how ancient Greek, Roman, Indian, and Chinese myths envisioned artificial life, automata, self-moving devices and human enhancements. Is technology finally catching up with the imagineers of lore?

Regardless of whether Ray Kurzweil's "singularity" is closer or further away than anticipated, Blade Runner's polysemous narrative has launched a thousand cross-disciplinary conversations, revolving around the broader topic of our flawed understanding of personhood. These conversations continue to expand in their critical mass to address some of the most vital concerns of our time.[6]

This essay initially highlights the practical realities of the science depicted in the film. It also draws connections from the film to ongoing research at the frontiers of technology and embodied intelligence that explore whether the Replicants, as portrayed in the film, could become a reality. It must be emphasized that the Replicants of Blade Runner are described as man-made, entirely biological humanoids with para-physical capabilities. They are virtually indistinguishable from humans. In consequence, the areas of science touched upon in the film are: biotechnology (specifically, stem cell biology), bioengineering (synthetic organ production) and neuroscience.

Subsequently, the essay touches on the broader philosophical landscape which depicts how the film's prescient narrative can guide us in our understanding of the fluid relationship we have with technology. This encompasses observations from multidisciplinary sources about the film's deeper meanings relating to spirituality, human consciousness, psychology, and ethics.

2. Artificial intelligence and robotics

Two decades after Blade Runner premiered, Kurzweil (2005) observed that technology will increase the speed at which we make discoveries and inventions in the field of human- computer interfacing until we reach a point of no return, where we will have created something that will outsmart us and eventually become the dominant intelligence on this planet. There are two, if not more, positions on most questions about technological innovation: a skeptical one and a celebratory one. We will now explore how these two positions developed with regards to artificial intelligence, starting with the celebratory one that critics, who are driven by the zeitgeist, often see as dystopian and dismiss with antipathy.

Would a world run by ChatGPT, and various androids necessarily lead to a dire set of circumstances? Could such AI driven technology not endow us with the time, resources, and capacity to enhance our existence and focus on achieving personal goals? At this time, the most potent uses of AI are classified as disembodied. While machines which try to replicate human activity are prevalent, unlike the Replicants in Blade Runner, they are not embodied. "The discussion about ethics and morality takes on a different dimension in Blade Runner," Claude Canizares, the Bruno Rossi Professor of Physics at the Massachusetts Institute of Technology, wrote. "Most, if not all, AI ethics programs deal with how AI treats humans, whereas Blade Runner emphasizes how humans treat AI."[7]

Despite the volume of opinions and scholarly observations and the expanding universe of AI-focused projects, it would be prudent to avoid certitude in drawing conclusions on technology's increasing role in the economy and our livelihoods, given that large-scale applications are still at an early stage. In consequence, the article will be restricted to these research parameters whilst focusing on the prescience of the film — not only its mandate to envisage the future, but also the possible consequences of integrated human and machine interaction.

2.1 The realities of the science depicted in Blade Runner

The key scientific areas explored in the film are biotechnology (stem cell biology), bioengineering (synthetic organ production) and neuroscience. One should be aware of the distinctions between synthetic and stem cell biology. Synthetic biology incorporates engineering to develop tools for constructing biological systems, in which, according to the National Institutes of Health (NIH) definition, allows researchers to reprogram stem cells that are then differentiated into a specific cell type. Although it is beyond the scope of this article to engage with the latest developments in these disciplines it is possible to identify select case studies to highlight the interaction between the visionary, though largely fictional science in the film, and the current scientific status quo.

Stem cell biology: Bioartificial organ manufacturing technologies already exist to grow human compatible organs in a laboratory. Due to significant rejection rates, which impede the successful adoption of these organs, the imminent next step is to grow these with cells that can conform to the surrounding tissue.

Neuroscience: Neuropsychologist Chris Frith, an admirer of not only Blade Runner the film but its source material "Do Androids Dream of Electric Sheep?" the Philip K. Dick novel on which the film and its sequel Blade Runner 2049 are based, has been influenced by the texts in his study of the neuroscience of consciousness and emotion. In a recent interview he stated that: "If you know how the brain works, you could, in principle, build something like it, using silicon".[8]

Blade Runner features an analogous Turing Test referred to as the "Voigt-Kampff Test". It is designed to identify Replicants through their inability to generate involuntary emotional reactions. A subject is asked emotionally provocative questions whilst his or her heart, pupillary dilation, respiration, etc., are monitored. Neuroscientists presently use a similar test. The International Affective Picture System database provides a selection of pictures, some emotionally disturbing, which are used to quantify a person's reactions. When combined with brain scans, which highlight reactions in the cerebral cortex, the test is considered a highly reliable indicator of a subject's state of mind.

Replicants in the film have a maximum four-year lifespan. The creative entities responsible for the Blade Runner narrative argue that if the Replicants were to live longer, their emotions would evolve to the extent that humans could no longer control these androids. Accordingly, the Nexus-6 Replicants in the original Blade Runner film in 1982, were issued with fake memories in the form of "family photographs" to ground them. The Nexus-9 generation in the Blade Runner 2049 sequel, had their fake memories implanted. Today, cognition researchers use memory implants. By various means they convince subjects that a fictional memory happened to them. These memory implants show how easy it is to manipulate a subject's memory, by reducing the differences between Replicants and humans. Let's consider an alternative view by Searle (1990) who ponders the question of whether simulating thought is the same as experiencing it. "You cannot explain a physical system such as a typewriter or a brain by identifying a pattern which it shares with its computational simulation, because the existence of the pattern does not explain how the system actually works as a physical system," Searle (1990, p. 13) explains. "In the case of cognition, the pattern is at much too high a level of abstraction to explain such concrete mental (and therefore physical) events as the occurrence of a visual perception or the understanding of a sentence."[9] The philosopher John Locke believed that the continuity of our memories was the seat of our selfhood, because it is only by virtue of this continuity that we know ourselves to be the same person from one moment to the next.

Alex Sabo, a psychiatrist with the Berkshire Medical Center in Massachusetts, (2024) cites recent work by neuroscientists, Hans Panksepp[10] (2009) and Antonio Damasio,[11] (2012), both of whom have written about the "transpecies self" and "how the self develops in three stages: 1) consciousness in body self (protoself), 2) the body self in relation to other selves, (emotional, relational, core self), and 3) finally, a sense of self in humans that feels continuous over time, the autobiographical self." Sabo (2024) adds, "Damasio's book gets into some of the neuroscience behind how this happens, thus fleshing out Locke's "intuition".[12] Sabo also cites that Hawkins (2021) has specified three things which should be accomplished in order to claim that machines show intelligence: continuous learning, learning via movement and the capacity for many models using reference frames to store knowledge.[13] In terms of continuous learning, Hawkins (2021, pp.130-32) noted that machines should display flexibility, accomplished by ongoing adjustments reflecting changing conditions and discovery of new knowledge, allowing neurons to learn new patterns in forming new synapses on a specific dendrite branch in a manner where new synapses don't impact previous learned ones on other branches by learning something new that does not force the neuron to modify or forget that which it already knew.[14]

2.2 Getting Closer?

Thomas George Thuruthel, lecturer in AI and robotics at University College London, has long been intrigued by the scientific speculation in Blade Runner. He explained how the fictional Replicants were created in the film. The combination of advanced genetic engineering, robotics, and artificial intelligence is familiar to us now. Thuruthel laid out the process in three phases. First, the Replicants' genetic code is created in a lab, and then implanted into an anatomically correct, bioengineered body which resembles humans in every possible way. "In essence they are AI systems designed to replicate the physical abilities and intellectual capacities of humans".[15] Finally, the body is programmed with the desired combination of drugs, electrical stimulation and a selected set of memories and experiences.[16]

Another scientist who identified as a fan of the film is Josh Bongard at the University of Vermont. He works in evolutionary robotics, computation, and physical simulation. Bongard teamed up with biologist Michael Levin at Tufts, to create the world's first computer- designed organism.[17] They have created xenobots: synthetic lifeforms made from the genomes of frog cells which are much closer to the science behind the Blade Runner Replicants than to the cyborgs of science fiction nostalgia. Bongard and Levin's "computer- designed collection of cells" can replicate spontaneously.

The author asked Bongard if Xenobots are precursors to an entity like Blade Runner's Replicants. "I believe so," he wrote in an email exchange. "Indeed, the Xenobots have many creators. The frog that donated its cells for xenobot construction. The human microsurgeon who assembles those cells into a xenobot under the microscope. The AI that sent the xenobot blueprint to the microsurgeon. The supercomputer on which the AI designed the xenobot. The computer scientist who programmed the AI…"[18]

According to Bongard, scientists understand something by building it. But to what extent do we understand how something is built if it is designed and tested by AI? How do we control it? "If scientists understand how something works by building it, (or, I will add, by being able to write the equations that describe exactly how it works) then one can argue that we don't yet understand AI," Canizares explains. So far as I know machine learning is still a 'black box'. Then again, there is a program of AI/robotics working on personal care robots, which are intentionally designed to appear 'conscious.' Yet, I doubt if there is an 'agreed upon definition of consciousness'".[19] Bongard agreed, adding, "or even if consciousness is a thing. Personally, I'm an illusionist; I don't believe it's a thing."[20] When asked about the possibility of using a lab-grown brain as the operating system and a hard drive as a starting point, Bongard dismissed this as not being potentially all that helpful. "We don't know whether or not other people have conscious experiences, or whether bats do, or bacteria, or rocks for that matter," he noted. "We just assume 'yes' or 'no', on a case-by-case basis. Xenobots are just one more entity about whose inner life we can only guess at."[21]

When asked whether start-up companies which are propelled by AI technologies will either plateau or reach an unsurpassable roadblock, Bongard said it would be hard to say if this could be the case. "AI is an enabler technology, like the wheel, electricity, computers, the internet...," he added. "I'm no historian, but none of those technologies have yet to show signs of diminishing returns. We still keep finding new uses for them, even thousands of years since some of them were invented. If I had to bet, I don't think it's going to slow down. It's just going to keep transforming more and more sectors of society."[22]

Scientists struggle to agree whether Xenobots are organisms, robots, or something else. Bongard suggested that we should always remember that we are not building colleagues, we are building tools. "I think our work is part of a much larger scientific trend that seems to increasingly suggest that free will, consciousness, etc., are useful illusions that 'believing' in them might have had some survival value, but they don't actually exist," Bongard explained.[23] Going further, Kurzweil and other scientists predict that it is only a matter of time for these tools to outsmart the manufacturers. What happens if we build a conscious machine by accident? In responding, Sabo (2024) says, "I would challenge the implied assumption that human nature would only build a conscious machine by accident. If a human can do something, somewhere sometime, a human will do it."[24] In 2023, Geoffrey Hinton, whom many consider the 'Godfather of AI', announced he was leaving his post as vice president and engineering fellow at Google, because of what he saw as an unresolved problem of alignment, which centers around ensuring that AI is performing what humans expect it to do in terms of providing beneficial, not harmful, goals.[25]

2.2.1 Response and Critique

Susan Schneider recently completed a three-year NASA project on the future of intelligence and is the founding director of the Center for the Future Mind. She has long been interested in Blade Runner as well as its source novel. "We have the director's cut of Blade Runner at home," she said.[26]

Like Bongard, Schneider struggles to answer the question: What happens if we build a conscious machine, by accident? "How would we even know? Maybe we've already done that with the chatbots. Maybe that's why they're so smart. We just don't know. I'm really worried about the failure to come up with feasible tests for machine consciousness."[27]

In a recent meta-analysis Consciousness in Artificial Intelligence, 19 leading researchers needed 81 pages to conclude that after all this time we still cannot decide on a definition of the word "consciousness", let alone come up with a test to determine its existence in AI. This raises a critical question: Why would anyone want to create a conscious machine which would likely conclude that humanity has overall been a colossal disaster? It would be impossible to control such intelligence. Why would anyone without a hubris-driven God complex want to create such a thing? And, if they succeeded, what would be the consequences for humanity?[28] Consequently, are we truly as close to manifesting the science implied in Blade Runner? Wendell Wallach, who has written extensively about the ethics and governance of advanced technologies told the author in an interview that "My worldview is that from a problem-solving angle, maybe these systems have general intelligence and they're better than most school kids and maybe better in some respects than most professionals and in some areas, they are better than all of us,".[29] He added "Intelligence and understanding still depend upon consciousness and emotions in ways that we do not really understand, and we aren't even looking at how to implement [them] into these systems." Wallach, who wrote the book Moral Machines: Teaching Robots Right from Wrong (2008), said that until a machine is developed to handle complex ethical questions and circumstances, no one can justifiably claim that artificial general intelligence or artificial super intelligence has been achieved reflecting on how we humans tend to keep shifting the goalposts on this question.[30]

Regarding the film's Replicants, Wallach explained that, despite a plethora of technological advances, the Replicants (androids) in Blade Runner are far from becoming a reality. Such General Artificial Intelligence remains beyond the reach of extant AI systems. For now. He added that the Replicants are "really surrogate human beings" with a survival instinct. "The problem is that they are exemplifying all kinds of qualities that I don't think we have any evidence we can reproduce," he continued. "If we were creating that kind of biological system with computational brains that would be something else, but we're nowhere near creating biological systems of that nature."[31] Sabo (2024) agrees. "We'd likely need to create biological systems that have the three layers of self-organization and consciousness that Panksepp and Damasio [whom he cited previously in this article] describe—these systems evolved over millions of years. So, how do we create something like that, that can keep learning and evolving over time?"[32]

In his analysis of Blade Runner, Andrew Norris discusses the Voigt-Kampff Empathy Test in the film, which was designed to distinguish between Replicants and humans. However, Norris explains the test does not assess the respondent's capabilities to empathize with human beings.[33]

What none of the questions do is test the ability or propensity of the test subject to empathize with (other) human beings. Part of this is surely an intentional irony, as the lives the characters in the film test this particular ability again and again—and, as many have noted, it is only the Replicants who demonstrate any real ability to empathize with suffering creatures (The test, as it is seen in the film, serves as a potent reminder that the wider natural world is dying, and that the animals with which it asks the test subject to empathize are in fact absent if not extinct) or indeed any vibrant emotional life. (Conscience too seems to be peculiar to the Replicants).

Schneider added that such tests cannot be applied to technologies such as Chat GPT, as machine learning systems have already been learning from the Internet and have facts about human consciousness. When news broke in 2022, that a Google engineer who claimed that the Language Model for Dialogue Applications (LaMDA) chatbot he developed is sentient, many scientists and ethicists disputed the claim, as a rushed judgment. Accordingly, Giandomenico Iannetti (2022), a professor of neuroscience at the Italian Institute of Technology and University College London, responded that even today, scientists still cannot unequivocally demonstrate consciousness in humans, and, at best, can only estimate it. He said, "we have only neurophysiological measures—for example, the complexity of brain activity in response to external stimuli," adding that researchers can only infer the state of consciousness based on external measures.[34]

"Basically, we are really going to need something, because super intelligent AI could manipulate us," Schneider added. Like Schneider, scientists such as Erica Darics and Lotte Van Poppel believe the jury is still out on Chat GPT and similar technologies. In recognizing how such technologies could disrupt and upend the worlds of education and work as we know it, they believe we should prepare for the potential dangers but also positive outcomes. In their optimistic take, Darics and Van Poppel believe that like the machines, people will be savvier about the prompts used in GPT and other technologies. They believe that people will be much better prepared to shield themselves from attempts to manipulate or misinform us (e.g., deepfakes of Toby Walsh's Faking It).[35] On the other hand, Sabo (2024) adds, "that advanced AI could manipulate humans easily by identifying the degree of 'intolerance of uncertainty' in a human or group of humans and then creating the 'facts' that the group would like to believe in. This is happening already on a global scale and has always been the purview of demagogues, fundamentalist religious leaders, and dictators."[36]

2.3 Approaching the Frontier of Biological Entities

Kurzweil sees the project of replicating human intelligence in nonbiological entities as "biologically inspired," suggesting that the process of learning is chaotic but that it also leads to the development and sharpening of the cognitive objective of organizing knowledge and experiences. "The purpose of reverse engineering the human brain is not to copy the digestive or other unwieldly processes of biological neurons but rather to understand their key information-processing methods," he writes (2005, p. 474)[37] Kurzweil's peroration reminds readers and viewers of the Replicants Rachael and Batty in the film. "It is my thesis that by sharing the complexity as well as the actual patterns of human brains, these future non-human biological entities will display the intelligence and emotionally rich reactions (such as "aspirations") of humans" (2005, p. 475).[38]

Thinking about the ramifications of compacting computations to achieve algorithmic patterns, if we observe that AI is so much more evolved than us and doing a lot of the heavy lifting, will we still bother to go to university? Will we learn to play the piano if all we have to do is download software to allow us to play like Vladimir Ashkenazy? Schneider responded, "That's a good point. There'll probably be differences from person to person and culture to culture". She added that the sudden interest in machine consciousness should make people sceptical and worried about AI. It reminded her of another science fiction thriller that came out a dozen years before Ridley Scott's masterpiece hit the screens: Joseph Sargent's Colossus the Forbin Project. "The thing that worries me is these chat boxes becoming political platforms as they connect to the Internet," Schneider added. "I'm not sure where things are going to go."

When the author interviewed Hampton Fancher, the lead screenwriter of Blade Runner, he related Batty's perspective. "There's contempt for humans. They say you might have created us... but look at the mess you have made of everything."[39] Fancher added, "I have just about had it with this experiment." I asked him what he meant by "experiment." "Human beings. Homo sapiens. The genus of the Homo sapien." he responded. "But also, all along the line, we've been doing Clockwork Orange".[40] Stanley Kubrick's A Clockwork Orange's (1971) violent satire uncovered the psychological dangers of authoritarian societies manipulating how we exercise free will. Have we returned to Dick's contention in Do Androids Dream of Electric Sheep? that humans are the real antagonists in the world?

This continual grasp for meaning is just as critical for experienced scientists as it is for the curious and inquisitive individual. "There are so many things we don't know... I think as a true scientist, the more you find out, the more you realize we don't know, and I always think that it's surprising that more people aren't troubled," Lex Fridman, an artificial intelligence scientist and podcaster, noted in an interview with Demis Hassabis, the chief executive officer and co-founder of DeepMind. "Every night I think about all these things we interact with all the time that we have no idea how they work: time consciousness, gravity, life. These are all the fundamental things of nature. I think that way we don't really know [how] they are to live life [and] we pin certain assumptions on them and kind of treat our assumptions as if they're fact. That allows us to sort of box them off somehow."[41]

Going further, the discussion of considering machine ethics as a game of either imitating one or another's behaviour or by following a prescribed code of conduct in ensuring that humanoids desire to be "good people," Ge nova, Moreno and Gonza lez (2023) contend that contrary to the notion that ethics consists primarily of adhering to accepted societal norms and conventions, "the very seed of ethics is critical thinking, a refusal to conform to the dominant mindset."[42] Consequently, attempts to conflate the idea of a "moral machine" with a true ethical machine" will fall short, given several examples such as "the problem of autonomous cars should not be whom to kill, but how not to kill" (Holstein et al., 2021), or the dangers of the Moral Machine experiment being used for "normative ends and to the whole voting system approach it is built upon in addressing ethical issues" (Etienne, 2020), as it invites people to express preferences and biases "not only because humans are very bad at judging people based on appearances, but especially because it ends up being "nothing less than a direct attack on personal dignity" (Jaques, 2019)."[43] While imitation is a human trait, "if there is any ethical difference between doing A and doing B, that difference is not in the number of people who behave one way or another, but in the reasonableness and good will of their behavior" (Ge nova, Moreno and Gonza lez, 2023).[44]

Wallach's remarks about a framework for governing how the technology is used and applied, expand the points that Fridman raised. "They're making progress but they're also revealing that the challenges are much more complex than they ever thought. And to me that's the critical thing," Wallach explained in his interview with the author. "The critical thing is, you might be able to get an artificial general intelligence that's a sociopath, but you aren't getting something with full intelligence. Now can it do a lot of damage as a sociopath? Yes, it can, but that depends on what kind of autonomy it has. But that's not full intelligence that will continue to be the Achilles' heel. And it's the same thing with biology."[45] It goes back to acknowledging the differences between simulation and emulation, especially as being connected to the fact of having a biologically human form.

Although artificial general intelligence has not yet been realized, one could reasonably assume that something akin to the nuclear arms race during the Cold War of the 20th century is underway "because the economic gains to whichever firm or government lab invents the world's first AGI) will be immense."[46] Naude and Dimitri (2020) believe that a constructive path toward mitigating the risks and probabilities of a dangerous or unfriendly AGI from being developed and implement should involve public policy initiatives including public procurement of innovation, taxing an AGI and finetuning legal issues including patenting:[47]

This will help to address the control and political problems in AI. Future research is needed to elaborate the design of systems of public procurement of AI innovation and for appropriately adjusting the legal frameworks underpinning high-tech innovation, in particular dealing with patenting by AI.

The narrative's prescient vision of Replicants and the behaviour of Eldon Tyrell, the billionaire scientist, underscores the instructive value of ethical policy making to address the challenges. Existing social and economic inequalities will be widened between nations with AGI capacities once and when singularity is achieved and those nations without it will lag further behind once their counterparts leap into the technological advancements that portend elements of transhumanism and post humanism, along with the consolidation of a class of exclusively privileged humans appearing in society.[48]

2.4 As seen through the glass darkly

This section reviews human consciousness from a complex spiritual perspective: "How will man handle playing God?" Historically, humans do not have an optimal track record handling what can be termed as "excessive power". Weckert, an ethicist of science and technology, put this into context when he wrote: "the dramatic technological manipulation of humans, life, and matter is rapidly extending the scope of human control and thus thrusting humans into unfamiliar situations where they do not have the competence to make good decisions."[49]

The creative minds responsible for Blade Runner imagined a future, in which humans are allowed to retire androids at will. This mindset is not only exploited by the likes of Eldon Tyrell, whose corporation creates the androids, but is espoused by the average person, referring to the Replicants as "skin jobs" (a racist term). Passers-by barely flinch at seeing Replicants executed in the street. In the future envisaged by Blade Runner, humankind is also at ease in its consciousness of accepting these creations as slaves. A parallel can be drawn between the Tyrell Corp. and many multinational oligopolies that are secretly working on their evolution of AI to serve their commercial and political interests. They compete to become monopolies. What happens if a company like Amazon or Google reaches that stage of growth and then appoints an AI to run the business? If it makes disastrous decisions, who will be held responsible according to the law? (Does it give the owners a get out of jail-for-free-card? This has existential consequences for those affected, including those seeking compensation). Given that AI has no legal identity, according to Jeannie Paterson, professor of law at The University of Melbourne, those who appointed the AI would continue to be responsible. "The real issue, though, is how the law can ever be effectively endorsed by a company with a turnover larger than the GDP of many countries," Paterson added[50]

On a different plane, Schmidt (2021, p. 142) sees Christian themed references in the film, including a dialogue exchange between Batty and Tyrell, where the self-anointed corporate "God-Father" explains that technically he cannot extend Batty's lifespan past four years and calls Batty "the Prodigal Son".[51] Tyrell is the embodiment of the megalomaniac with sociopathic tendencies, arguably not unlike some individuals presently running companies and countries. It is interesting that in a previous draft of the film's screenplay, Tyrell had died and been replaced by a Replicant empowered to make decisions with nefarious consequences for human society outside the Egyptian pyramid in which this Pharaoh isolated himself.

One of the more contentious discussions Kurzweil has frequently raised leads us to the writings of William Dembski, a theologian and philosopher who is also a proponent of intelligent design. Dembski (1999) argued that "attributing spirituality to machines entails an impoverished view of spirituality, and that the empirical evidence doesn't confirm that machines can bring about minds."[52] In the Blade Runner narrative, the fatal flaws of the creators that Batty pursues are revealed. The film highlights the fate of those who fail to see the ethical and philosophical dilemmas as a scientific problem but instead as an ontological one. Dembski (1999) illustrated the argument.[53]

> If all that exists is matter governed by natural laws, then humans are machines.
>
> If all that exists is matter governed by natural laws together with spiritual substances that are incapable of coherently interacting with matter, then, once again, humans are machines.
>
> But if matter is merely an abstraction gotten by removing all the features peculiar to unified things, then there is no reason to think that combining it with natural laws (or anything else for that matter) will entail the recovery of things. And in that case, there is no reason to think that humans are machines.

A September 1982 review in Christianity Today called the film a "chilling allegory about man's relationship to God" where a "Blade Runner has no designer, no higher power to appeal to."[54] When the author asked screenwriter Fancher to set aside Mercerism and reconsider the question about whether he saw any religious symbolism in Blade Runner and to also comment on the extensive writings suggesting multiple religious meanings in the film, he said:[55]

> I would not be surprised. The book had it. Mercerism. I am a devout atheist. I would reject any of that nonsense as far as I am concerned. I think I fooled around with some Mercerism, maybe in the first draft. Maybe not. I had no inclinations along those lines and David Peoples and Ridley also didn't, I don't think. I read that stuff before, but I've never related to it. I don't see it. There's something, you know, a sense of, in Rutger Hauer, in his performance there is something that points toward something transcendental. You know about this creature? You could call it Frankenstein; you can also call it something more sacrosanct and saintly.

Fellow scribe David Webb Peoples agreed when the author asked him if he saw any religious symbolism in the movie:[56]

> Peoples: I've never seen any.
>
> Because Philip K Dick was intrigued by Mercerism, and many academics have written papers about the religious symbolism within Blade Runner.
>
> Peoples: Sometimes these things are in there because of the unconscious feelings, just as I included stuff that made Deckard a Replicant, when I was really talking about his relationship to God. All of this stuff is mysterious.
>
> But you are sure you did not go out of your way to insert any sort of religious symbolism in there. If somebody sees something in there...
>
> Peoples: Absolutely, I am sure I didn't.

When the author asked production executive Katherine Haber for her thoughts on the theme, she gave the most diplomatic answer to cover the spectrum of contentions: "All these what ifs are in the eye of the beholder, which perpetuates the fascination with the film for over 40 years".[57]

Despite denials by the filmmakers the author interviewed, about the complete lack of intent, there are obvious Christian symbolic visuals in the death scene that cannot be easily dismissed: the nail through Batty's hand, the dove flying into the sky at his moment of death, the crosses on the air-conditioning units. Oxford philosopher Stephen Mulhall explained the scene signified how Heidegger explained the human relationship to death, adding, that, from a humanist or even existential perspective, it highlights an "ultimate irrelevance of any distinction between human beings and Replicants which is couched in terms of the length of their respective life spans or the degree of certainty with which each can predict an end to their lives on a particular date... Replicants stand in a human relationship towards death."[58] Going further, the potentially instructive link to the Road to Damascus conversion in the Bible, during which Saul, the zealous persecutor of Christians became the Apostle Paul, is undeniable. With Batty's death, Mulhall suggested that Batty has upended the conventional representation of Christian ethics and the morality expressed in the aforementioned symbols from the Gospels.[59]

The concept of slavery also acquires a further level of significance in this sense, however: at the end of Deckard's ordeal, after Roy's unexpected rescue of him, Roy offers his pupil the following description of his experience: 'Quite an experience to live in fear, isn't it? That's what it is, to be a slave.'

Arguably, the numerous biblical images seen in the film are less a matter of spirituality but more of a historical reference to the narratives in the Old and New Testaments pertaining to power issues between God (the maker) and man (his creation). The Old Testament God is given to dispatching his creations with much anger, like a potter disposing of misshapen clay figures. Even the New Testament presents us with discomfiting attitudes towards slavery. In 1 Peter 2:18: "Slaves, be subject to your masters with all reverence, not only to those who are good and equitable but also to those who are perverse." All of this likely contributed to the creation of the psychological profile of Eldon Tyrell.

It is conceivable that screenwriters Hampton Fancher and David Webb Peoples, as well as director Ridley Scott may have missed some of the more consequential ethical and moralistic aspects of Philip K. Dick's subtext, which can either be religious and/or secular in nature. Or they might have elected to do so either for commercial or expressive reasons because such subtext readings did not interest them at the time. Dick turned to the Episcopalian Church after a terrifying "Satanic" vision in 1963 and continued to identify as a soul-searching Christian for the rest of his life.[60] It can hence be argued that the source novel "Do Androids Dream of Electric Sheep" written in 1968, was in part shaped by Dick's state of mind, as influenced by his mystical experiences.

3. Replicant psychology: more human than human?

In 2024, the symbolism portrayed in Blade Runner rings with an eerily relevant parallel in the contemporary world of business management. The search for authenticity of our human nature is quintessential and universal. For example, some themes in Blade Runner even resonate in the contemporary world of business management. Pedro Sena-Dias at the Nova School of Business and Economics in Portugal, drew a parallel between Blade Runner and what he sees as management's core challenge in a post-pandemic business world:[61]

> Off-screen, in the real world some humans act like and/or are treated as slave-robots while others are free and encouraged to express and use their human nature. This tension, which is at the heart of the narrative, is perhaps a depiction of an ancient quest for our true human nature and a way to make sense of an inner clash between productivity and creativity, convenience and profoundness, that is often reflected in a battle for sustainability, between short- and long-term gain. This battle is as important to management as ever.

If this debate sounds familiar, it is. Our largely muddled engagement with the realm of artificial intelligence (AI), robotics and automation in general, is its present incarnation. However, Blade Runner also primed the landscape for the ethical and moral considerations associated with these technological advancements. Citing the scene in the film when Deckard meets Rachael for the first time at the Tyrell Corporation, Schmidt (2021, p. 61) recalled Deckard's line: "Replicants are like any other machine: They're either a benefit or a hazard. If they're a benefit, it's not my problem". Schmidt intriguingly suggests the Deckard name intentionally referred to Rene Descartes, the 17th century French philosopher and scientist. "The basic difference of both principles' 'spirit' (res cogitans), which only applies to humans, and 'material' (res extensa) brought Descartes to the point where he saw animals as soulless machines," Schmidt continued. "This creates a hierarchical dualism of spirit and material i.e., between man and nature, in association with an anthropocentric privileging, which encourages man to dominate the spirit and soulless nature for his own ends."[62]

In a 2013 Film Philosophy journal article, Norris discussed what he views as a narrative gap in the logic of one of the film's scenes. He referred to the scene where Rachael, on discovering that she is a Nexus-6 Replicant, also learns that the memories of her childhood and her mother are those of the niece of her creator (Eldon Tyrell). Norris contended:[63]

> The Replicants feel, but not 'in the human way,' we might say. Any doubts about this are immediately shown to be unfounded when the initial interview ends with Leon shooting the interrogator in response to a question about his mother. 'My mother? ... Let me tell you about my mother.' The fact of being a Replicant entails its own emotional states, among them rage at being made, and not born.

The rage drives the film's emotional currents. Replicants feel rage for being considered a disposable non-entity by other flesh-and-blood entities who are far inferior. Replicants rage at their slave status, realizing that they are relentlessly subjected to a speciesism embedded within the state-of-the-art manifestation of their creation. At the same time, Per Schelde, in his 1993 book, Androids, Humanoids, and Other Folklore Monsters, suggested an unexpected, yet surprisingly satisfying, alternative focal point to the film. Consider that Roy [Batty] is not even the emotional centre of the movie. That the centre is occupied by Rachael, the android who initially does not know that she is a Replicant. As Schelde explains, "she's the one who has to grapple with the problems of self and identity ... She's the one who has to learn, on her own, that ... she can love and be loved. Rachael, with the soulful brown eyes, has to learn to love and trust Deckard with the cold, blue ones."[64]

3.1 Moral dilemmas: living together?

In the realm of transhumanism and post humanism, a key angle has focused on the aging process, not as an inevitability, but as a manageable condition. The upper ends of stem cell research involving synthetic scaffolds suggests, that in the long run, we may no longer need organ donors, or immunosuppressive drugs. It is fascinating to think of a time when individuals will be able to get new sets of organs whenever their original ones wear out.

Kurzweil does see moral and ethical questions in a post-Singularity period that some observers contend would be defined as posthumanism. He ponders where the line should be drawn. "Is a human with a bionic heart still human? How about someone with a neurological implant? What about two neurological implants? How about someone with ten nanobots in his brain? How about 500 million nanobots?" he asks. "Should we establish a boundary at 650 million nanobots: under that, you're still human and over that, you're posthuman?" (2005, p. 374).[65] This, he later contends, leads to the question that AI and robotics researchers must confront: "So, again, who am I? It's the ultimate ontological question, and we often refer to it as the issue of consciousness" (p. 384).[66] However, he differs with Dembski. Kurzweil sees this as a window for moral enlightenment, not the foreboding abyss that Nietzsche described.It is hard to extrapolate meaning from such a polysemous text without invoking the writings of Friedrich Nietzsche, particularly in considering the themes of alienation. In his book "Thus Spoke Zarathustra" (1883) the character of Zarathustra argues in favour of his concept of the U bermensch, as a shift away from more ethereal Christian concepts to a grounding in earth- based humanity.[67] The objective being for humans to give meaning to life on earth by, amongst other things, creating constantly evolving, superior generations. The ensuing debates are beyond the research parameters of this article. The links to sinister eugenic concepts must, however, be addressed, in that numerous scholars have engaged with them with reference to the character of Roy Batty, who incorporates most of the prerequisites for an U bermensch.[68]

Why did so many film journalists and fans of Blade Runner empathize with the Batty character (which Paterson suggests could be read like a theme from the 1960 film Spartacus or in a more modern form from the 2000 film Gladiator),[69] the key antagonist, seeing him as the most enthralling presence in the film? Batty is a mass murderer who kills twenty-three people as the leader of a group of off-world, rebel Replicants in order to escape to earth. In essence, viewers would be expected to empathize with Harrison Ford's character, Rick Deckard, the protagonist, in consequence of his Road to Damascus experience in which his distaste for Replicants evolves into loving one of them.

Batty can initially be seen as a freedom fighter who escapes slavery and heads for earth to confront his makers and secure more time to live. Any objective achieved by murder is ethically problematic, but if we ask ourselves how we would act in such circumstances, we can empathize, which raises questions about our own ethics. As philosopher Stephen Mulhall points out, if we fully acknowledge that slavery in human civilization destroys the individual's humanity by robbing it of its autonomy, then humans are implicated because they have denied the Replicants their capacity for humanity. Would our reaction to Batty have been different with an opening scene showing the twenty-three murders taking place? (One of the many scenes which were deleted due to budgetary constrictions). Deckard's crisis puts him on the edge of survival and, as Mulhall explains, he is reduced to the same animalistic desire to survive, "a sample ... of the texture of which all Replicants life consists, and the responsibility for that lies with every human being."[70]

Once on earth, they are free to choose, not quite in a Hobbesian context, because they are subject to an external impediment, because Deckard, a representative of the law-and-order establishment, is tasked with killing the Replicants— not to arrest, jail and put them, on trial. To kill them. They have no rights. Deckard is like the local dog catcher who puts stray Replicants down, with no questions needing to be answered.

Once denied an extension to his life, Batty proceeds to kill his three makers: Eldon Tyrell, J.F. Sebastian and Hannibal Chew in a sadistic fashion. His tone is flippant as he murders them. He seems to enjoy himself. In consequence, is there really such a difference between Deckard and Batty, until the final scene of the movie? When Batty commits an act of unexpected nobility and allows his foe, the killer of his friends to live, he upsets the narrative's ethical balance and becomes the good guy (i.e., protagonist) in a remarkable turning point which distances him from macabre eugenic notions.

Schmidt (2021, p. 142) alerted the reader to the similarities found in Batty's final scenes — "Fiery the angels fell/ Deep thunder roll'd/Around their shores, burning with the fires of Orc." — before his death which paraphrase the verses of William Blake. "With Blake, the angels rise up from the earth; in Blade Runner, the Replicants are fallen angels, who fall out of the sky (the outer world colonies)," Schmidt noted. "Through the quote Batty identifies with the Orc, the embodiment of the striving for freedom... From the American perspective, Orc is the liberator, for the English colonial masters, however the antichrist or Satan... Batty is a rebel who rises against the godly order of the Tyrell Corporation and finally kills his creator" (Schmidt, 2021, p. 142).[71]

Batty's death scene is one of the most iconic in all of cinema. According to producer Michael Deeley, Hauer edited his final lines in the film, believing that the original was too melodramatic. He slashed through the original script and left us with 42 words: "I've seen things you people wouldn't believe. Attack ships on fire off the shoulder of Orion. I watched C-beams glitter in the dark near the Tannha user Gate. All those moments will be lost in time, like... tears in rain. Time to die".

Ironically, a lot of it does not make sense to the uninitiated: "I watched C-beams? Glitter in the dark near the Tannha user Gate?" But juxtaposed against "tears in rain", the lines appeal to our imagination. They evoke something that only the very best writers and filmmakers feel secure enough to rely on, knowing full well that whatever they present on the screen, nothing will ever surpass the power of our imaginations to amaze us. Mulhall explains that Batty's final words allow Deckard, who watches him die, to acknowledge his death and reflect on his own authenticity:[72]

Deckard has learned his lesson, about acknowledging others and about mortality, by acknowledging another's death.

4. Conclusion

The ingenuity of the visionary, non-scientist, creative entities who shared their futuristic world views with us is not based on fact but on a creative vision of what the future might be and, as Wallach noted, we are nowhere near being able to create Replicants. However, before dismissing this outright, it is worth noting that each of the individuals involved came to the table having done impressive amounts of research in the sciences, humanities, and social sciences. One might ask why so many eminent scholars across the spectrum of disciplines have long been fascinated by the film and have engaged with it in different ways.[73]

As in the film, much of the current AI research is driven by oligopolistic companies and their corporate power. Examples include Google, Microsoft, Amazon, Meta, IBM, etc. – none of them given to sincere altruism. Significant elements of each corporate entity are reminiscent of the fictional Tyrell corporation, and, as Paterson suggests, the sense of human alienation, captured in the grim scenes of Deckard in his flat.[74]

In 2014, Cruz echoed producer Powell's observations at the beginning of this essay when he summarized the status quo: "All the problems and ethical question the film pointed to in 1982 seem to become more pressing as the days go by, and as we approach 2019: environment pollution, cloning, the dehumanization of individuals, the omnipresent state of surveillance with video and image control, overpopulation, overcrowded streets, mega- corporations, lack of ethics, loss of individual rights, and so on. Therefore, the film's predictions of future life have become a portrayal of our present."[75]

The recent evolution in Artificial Intelligence has been astounding, but as shown in this article it encompasses significant problems: 1) Scientists understand how something works by building it. To what extent do we understand how something is built if it is designed and tested by AI?; 2) We currently do not have a satisfactory test to determine whether it has reached an agreed upon definition of consciousness; 3) We do not know how to fully control it; 4) We lack the combined global will and essential resources to create adequate protective barriers to prevent bad actors from using the technology for nefarious purposes; 5) We do not know, with an acceptable degree of certitude, where it is taking humanity.

As anticipated, humanity has leapt off the cliff in hope that its backpack contains a parachute and not just lunch.

Adrienne Mayor, the classics professor at Stanford University, who made the case that some of our best thinking about Artificial Intelligence and Robotics was done by the Greeks 2,500 years ago[76], also shared with the author that she asked ChatGPT, about Greek mythology, robots, and Amazons. "It came up with anodyne, old-fashioned potted replies and then began wild guessing and apologizing when I questioned them," she said. "It seemed like something that lazy students/journalists might use."

The future of humanity could be surprisingly bright if the power of AI could be harnessed to serve the best interests of people and states across the globe. However, as Wallach points out, the extant barriers to bad behaviour are equivalent to no more than a social media slap on the wrist. As long as greedy companies compete against each other and the many rogue states led by nefarious bad actors are desperate enough to empower themselves with AI to ensure their survival and cement their capacity to impose their toxic world views on other countries, the prospects are not encouraging. Nothing that Blade Runner did not warn us about, over 40 years ago.

References

[1] Powell, Ivor. Interview with author, 22 November 2022.

[2] A Replicant is a fictional, entirely biological humanoid, created in a laboratory. They have a four-year lifespan as it is argued that a longer lifespan would allow them to develop individualistic character traits which would make them uncontrollable. The Replicants of Blade Runner, as the name suggests, are essentially AI integrated circuits uploaded into state-of-the-art bioengineered bodies created to replicate the physical and intellectual capabilities of humans.

[3] An interesting neologism used to finesse a term which basically means an execution.

[4] "Power tends to corrupt and absolute power corrupts absolutely. Great men are almost always bad men, even when they exercise influence and not authority; still more when you super add the tendency of the certainty of corruption by authority." Member, Acton Staff. "Lord Acton Quote Archive." Acton Institute, 19 July 2024, www.acton.org/research/lord-acton-quote-archive.

[5] Mayor, Adrienne. Gods and robots: Myths, machines, and ancient dreams of technology. Princeton University Press, 2019.

[6] Kurzweil, Ray. The Singularity Is Near: When Humans Transcend Biology. Penguin, 2005.

[7] Claude Canizares, email correspondence, 2024, January 21.

[8] Gomes, Victor. "The Science behind 'Blade Runner's Voight-Kampff Test." Nautilus, 6 Oct. 2017, nautil.us/the-science-behind-blade-runners-voight_kampff-test-236837/.

[9] Searle, John R. "Is The Brain A Digital Computer?" Proceedings and Addresses of the American Philosophical Association. Vol. 64. No. 3. American Philosophical Association. 1990.

[10] Panksepp Hans J. and Northoff G. "The Trans-Species Core SELF: The Emergence of Active Cultural and Neuro-ecological Agents through Self-related Processing within Subcortical-cortical Midline Networks." Consciousness and Cognition 2009; 18:193-215.

[11] Damasio Antonio: Self Comes to Mind: Constructing the Conscious Brain, Vintage Books, New York, 2012.

[12] Sabo, email correspondence.

[13] Hawkins J, A Thousand Brains: A New Theory of Intelligence, Basic Books, NY, NY, 2021.

[14] Hawkins, 2021, pp. 130-32.

[15] Alex Paterson, Gabby Bush, Jeannie Paterson. "Technodystopia: Are We Heading towards a Real-World Blade Runner?" Pursuit, The University of Melbourne, 14 Nov. 2023, pursuit.unimelb.edu.au/articles/technodystopia-are-we-heading-towards-a-real-world-blade-runner.

[16] Thuruthel, Thomas George. How to Build a Replicant/Android. PowerPoint presentation, At Blade Runner: The Final Cut Edition at the London Science Museum, 2023.

[17] Bongard, Josh and Levin, Michael. Self-Replicating Robots. Soft Robotics Podcast, El-Diwiny, Marwa YouTube: https://www.youtube.com/watch?v=GBdGCYmA_kQ, 2022.

[18] Bongard, email correspondence, 9 February 2024.

[19] Canizares, email correspondence, January 2024

[20] Bongard, email correspondence, 9 February 2024.

[21] Bongard, email correspondence, 9 February 2024.

[22] Bongard, email correspondence, 9 February 2024.

[23] Bongard, email correspondence, 9 February 2024.

[24] Alex Sabo, email correspondence, 27 January 2024.

[25] The New York Times, 1 May 2023. https://www.nytimes.com/2023/05/01/technology/ai-google-chatbot- engineer-quits-hinton.html

[26] Schneider, Susan. Interview with author. May 5, 2023
[27] Schneider, Susan. Interview with author. May 5, 2023
[28] Butlin, Patrick, et al. "Consciousness in artificial intelligence: insights from the science of consciousness." arXiv preprint arXiv:2308.08708 (2023).
[29] Wendell Wallach is the Carnegie-Uehiro fellow at Carnegie Council for Ethics in International Affairs, where he co-directs the Artificial Intelligence & Equality Initiative (AIEI).
[30] Wallach, Wendell. Interview with author, October 11, 2023.
[31] Wallach, interview. October 11, 2023.
[32] Sabo, email correspondence.
[33] Norris, Andrew. 'How Can It Not Know What It Is?': Self and Other in Ridley Scott's Blade Runner. Film- Philosophy 17,1 (2013): pp. 19-50.
[34] de Cosmo, Leonardo. "Google Engineer Claims AI Chatbot Is Sentient: Why That Matters." 22 July 2022. Scientific American. https://www.scientificamerican.com/article/google-engineer-claims-ai-chatbot-is- sentient-why-that-matters/.
[35] Darics, Erica and van Poppel, Lotte. Debate: ChatGPT Offers Unseen Opportunities to Sharpen Students' Critical Skills. The Conversation. 20 February 2023. https://theconversation.com/debate-chatgpt-offers- unseen-opportunities-to-sharpen-students-critical-skills-199264
[36] Sabo cites the following as an example: Larsen, Emmett M., Donaldson, Kayla, Liew, Megan & Mohanty, Aprajita. "Conspiratorial Thinking During COVID-19: The Roles of Paranoia, Delusion-proneness, and Intolerance of Uncertainty." Frontiers in Psychiatry 12 (2021): 698147.
[37] Kurzweil, 2005.
[38] Kurzweil, 2005.
[39] Fancher, interview. January 7, 2022.
[40] Fancher, Hampton, Interview with author. January 7, 2022.
[41] Demis Hassabis: DeepMind - AI, Superintelligence & the Future of Humanity. Lex Fridman Podcast #299. YouTube. 1 July 2022. https://www.youtube.com/watch?v=Gfr50f6ZBvo
[42] Genova, Gonzalo, Valentín Moreno, and M. Rosario Gonzalez. "Machine ethics: do androids dream of being good people?" *Science and Engineering Ethics* 29.2 (2023): 10.
[43] Holstein, T., Dodig-Crnkovic, G., & Pelliccione, P. (2021). Steps toward real-world ethics for self-driving cars: Beyond the trolley problem. In S. J. Thompson (Ed.), Machine law, ethics, and morality in the age of artificial intelligence. IGI Global; Etienne, H. (2020). When AI ethics goes astray: A case study of autonomous vehicles. Social Science Computer Review, 40(1), 1–11; Jaques, A. E. (2019). Why the moral machine is a monster? In We robot conference. University of Miami School of Law, April 11–13, 2019.
[44] Genova et al, 2023.
[45] Wallach, interview. October 11, 2023.
[46] Naude, Wim, and Nicola Dimitri. "The race for an artificial general intelligence: implications for public policy." *AI & Society 35* (2020): 367-379.
[47] Naude and Dimitri, 2020.
[48] Van de Gevel, A. J., Noussair, C. N., van de Gevel, A. J., & Noussair, C. N. (2013). *The nexus between artificial intelligence and economics* (pp. 1-110). Springer Berlin Heidelberg.
[49] Weckert, John. "Playing God: What Is the Problem?" The Ethics of Human Enhancement: Understanding the Debate, Oxford University Press, Oxford, United Kingdom, 2016.
[50] Paterson, Jeannie, email correspondence, 2024, January 3.
[51] Schmidt, 2021.
[52] Dembski, William. Are We Spiritual Machines? First Things 96 (October 1999): pp. 25-31. https://www.firstthings.com/article/1999/10/are-we-spiritual-machines
[53] Dembski, 1999.
[54] Blade Runner. Christianity Today. 3 September 1982. https://www.christianitytoday.com/ct/1982/september-3/blade-runner.html.
[55] Fancher, interview. January 7, 2022.
[56] Peoples, David Webb. Interview with the author. January 11, 2022.
[57] Haber, Katy. Interview with the author. June 28, 2022.
[58] Mulhall, 1994.
[59] Mulhall, 1994.
[60] Dick, Philip K. The shifting realities of Philip K. Dick: Selected literary and philosophical writings. Vintage, 1996.

[61] Sena-Dias, Pedro. Explorations of the Future of Work Inspired by Blade Runner. Nova School of Business and Economics, Medium. 29 January 2019. https://medium.com/nova-sbe/explorations-of-the-future-of- work-inspired-by-blade-runner-b253cb86a4af
[62] Schmidt, 2021. Translated from German by the author.
[63] Norris, 2013.
[64] Schelde, Per. Androids, Humanoids, and Other Folklore Monsters. In Androids, Humanoids, and Other Folklore Monsters. New York University Press, 1993.
[65] Kurzweil, 2005.
[66] Kurzweil, 2005.
[67] Kudrna, David. "Mental and Ontological Simulacra: Non-Rationality and Non-Reality in Works by Philip K. Dick." (2020).
[68] Silva, Francisco Magno Soares da. "Struggling with human-like machines: the fading boundaries between humans and androids in Philip K. Dick's Do Androids Dream of Electric Sheep?" (2017).
[69] Paterson, email correspondence.
[70] Mulhall, Stephen. Picturing the Human (Body and Soul): A Reading of Blade Runner. Film and philosophy 1 (1994): pp. 87-104.
[71] Schmidt, 2021.
[72] Mulhall, 1994.
[73] There are over 85,000 references to the film on Google Scholar and 947,000,000 on Google.
[74] Paterson, email correspondence.
[75] Cruz, D. T. (2014). Postmodern metanarratives: Blade Runner and literature in the age of image. Palgrave Macmillan.
[76] Mayor, Adrienne. Gods and robots: Myths, machines, and ancient dreams of technology. Princeton University Press, 2019.

Embodiment, Socially Intelligent Behaviour and Proactive Robots

Chapa Sirithunge[1], Thilina Dulantha Lalitharatne[2] and Fumiya Iida[1]

[1] Bio-Inspired Robotics Laboratory, Department of Engineering, University of Cambridge, Cambridge, UK
{csh66, fi224}@cam.ac.uk
[2] School of Engineering and Materials Science, Queen Mary University of London, London, UK
t.lalitharatne@qmul.ac.uk

Abstract. There is growing consensus among robotics experts that integrating concepts from social intelligence and embodiment can significantly enhance robot design. Humans have evolved to perform and survive in a world characterised by rapid change, high uncertainty, indefinite intricacy, and limited availability of information. By contrast, robots operate in highly controlled environments with limited uncertainty. Although many challenges remain, concepts from social intelligence and embodiment will eventually enable researchers to engineer proactive robots for the real world that possess at least some of the desirable properties of humans, such as adaptivity, predictability, dexterity, and creativity in versatile domains where behaviour must not only be intelligent but also socially acceptable. This article aims to explore the multifaceted role of embodiment in enhancing the proactive capabilities of robots in this context.

Keywords: Embodied cognition · social robotics · proactive robots · human-robot interaction

1 INTRODUCTION

Embodiment in robotics, particularly in the context of proactive robots, is a burgeoning field of study that incorporates the physical capabilities of robots with their decision-making processes, enabling more nuanced and effective interactions in a variety of environments [5]. The concept of embodiment extends beyond mere physical presence; it encompasses a robot's ability to perceive, interact with, and adapt to its environment [4].

The emergence of proactive robots has resulted in a significant shift from the traditional reactive robotic systems. Unlike their reactive counterparts, which operate based on predefined rules or direct human commands, proactive robots anticipate needs, make decisions independently, and initiate actions without explicit instructions. In addition, **proactivity** refers to predicting or foreseeing possible incidents that might happen in the future and taking actions at present accordingly [6]. This shift is not merely a technological advancement but represents a fundamental change in how we envision the role of robots in society. Proactive robots, equipped with the ability to foresee and address tasks, are transforming industries, healthcare, domestic environments, and public spaces [8, 9]. However, the core of this transformation lies in the effective embodiment of these robotic systems which makes them acceptable in the human-robot domains [7].

Embodiment in proactive robots is realised through sophisticated sensory systems, advanced AI algorithms, and physical structures that enable them to interact dynamically with their environment [10]. This sensory perception is critical for proactive behaviour because it enables robots to detect changes, identify potential issues, and respond promptly. For instance, in healthcare settings, a proactive robot can monitor vital signs and environmental conditions of patients, allowing it to anticipate and respond to medical needs before escalating [11, 12].

Moreover, the physical form of a robot – its embodiment – plays a crucial role in its interaction with the world. The design of a robot, including its size, shape, mobility, and manipulation capabilities, determines its ability to perform tasks and interact with humans and objects. A well-embodied robot can navigate diverse environments, manipulate objects, and perform physical tasks with precision and efficiency. This aspect of embodiment is particularly crucial in industrial and domestic settings, where robots must perform complex physical tasks autonomously. Humanoid robots can be stated as a good example of such capabilities in terms of embodiment and situation-awareness. For example, the physical development of Kaspar robot for childcare testifies to the requirement of embodiment and physical capabilities together with cognitive capabilities of a robot to improve its engagement with the surrounding [13]. In this context, **self-body awareness** is an important aspect to be considered in the development of a robot's cognition and embodiment. Advanced Large Language Models (LLMs) [14], machine learning techniques [15] and contemporary AI frameworks possess the transformative capability to synergistically integrate embodied components within a robot with its cognitive processes, thereby enabling the generation of dynamic, adaptive behaviours. This process allows robots to adapt their behaviour based on changing environmental conditions and user feedback.

Furthermore, the integration of AI and embodiment has led to the creation of robots capable of social interaction [16]. Socially intelligent robots that understand and respond to human emotions and social cues are becoming increasingly prevalent in customer service, education, and companionship roles. The proactive capabilities of these robots, facilitated by their embodiment, enable them to initiate interactions, adapt their behaviour to suit individual preferences, and provide personalised assistance [17].

Ethical considerations are of paramount importance in the development of proactive embodied robots. As these robots become more autonomous and capable of making decisions, questions arise regarding responsibility, privacy, and potential impact on deployment [18]. Therefore, the design and programming of proactive robots must be guided by ethical principles to ensure that they benefit society and do not cause harm.

Another aspect of proactive robots is collaborative robots, or cobots, which are designed to work alongside humans, complementing human abilities and providing assistance. The physical design and AI-driven decision-making processes of these robots are tailored to ensure safe and efficient human-robot collaboration. In addition, embodied capabilities are essential for these robots to be able to accomplish their tasks and be more deployable.

The role of embodiment in proactive robots is a multifaceted and rapidly evolving area of study. It encompasses the physical design, sensory capabilities, and AI-driven decision-making processes that enable robots to anticipate needs and initiate actions autonomously. This evolution is not only a technological leap but also a paradigm shift in how we interact with and perceive robots. As we continue to advance in this field, it is crucial to consider the ethical implications and ensure that these developments are aligned with societal needs and values. The future of proactive, embodied robots holds immense potential for transforming various aspects of human life, from industry and healthcare to everyday domestic settings. In this paper, we intend to explore the multiple facets of socially intelligent robotic behaviours, where proactivity plays a major role in the interaction between the robot and humans.

2 Proactive robots

Proactive robots are defined by their ability to anticipate needs and take initiative without explicit instructions. Hence situation-awareness is a key feature of proactive robots to generate responses appropriate for a particular scenario [1]. In the field of organizational psychology, **proactive behaviour** is characterised as anticipatory behaviour, initiated by the individual, focused on creating change, and oriented towards future outcomes [19]. Furthermore, behavioural creativity is higher in proactive robots, and perceived creativity is observed by the user in the robot's proactive behaviour [20]. This study further discusses the interconnectedness between the creativity of users towards robots and the proactivity of robots. Although there is no clear definition for proactivity, artificial proactivity-includes context awareness, activity recognition, goal reasoning, planning, plan execution, and execution monitoring [21].

Considering the above aspects, Fig. 1 illustrates the interplay of information between an embodied agent (for instance, a robot, an avatar, or a virtual character) and its surroundings during an interaction. The interplay of the embodiment has been inspired by the work by Pfeifer et al. in [4] and the concepts related to proactivity have been extracted from the study in [20]. An agent's mechanical system interacts directly with the task environment or its surroundings. Its actions or movements respond to the surroundings through mechanical feedback. External stimuli in the task environment are received by the sensory network of the agent. Actions, feedback and external stimulations, all occur in parallel. Internal stimulations induced by the mechanical system act as sensory feedback. Information structuring, feedback assessment and action commands are then generated by the controller, based on knowledge from its past and present. Sensory feedback plays a major role in updating a robot's knowledge. The sensory network, mechanical system and controller make an embodied robot, while knowledge including abilities such as creativity, planning and prediction make a situation-aware robot. The interactivity between knowledge, the controller, and the mechanical system linked through sensory feedback and decisioning creates an embodied, situation-aware robot. The patterns induced by internal and external stimulations depend on the physical characteristics and morphology of the sensory systems and the motor commands. The information structure is generated based on the sensory-motor coordination, which strongly depends on the morphological and material properties of the physical system. In bio-inspired humanlike physical systems, the tunability of the physical system is higher, and properties such as shape and compliance are dynamic.

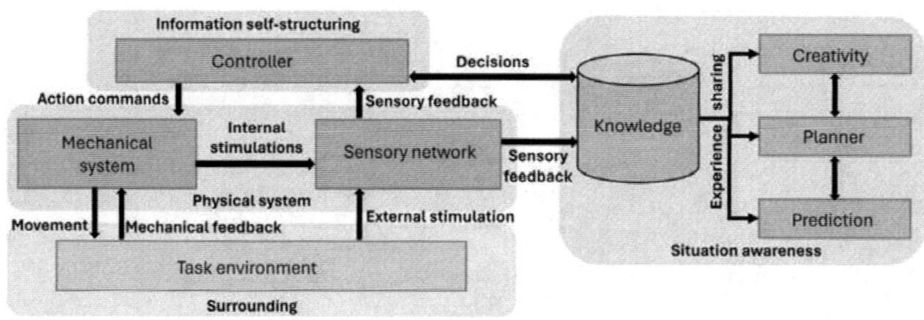

Fig. 1. Interplay of information and physical systems in proactive systems

Proactive robots require highly capable physical components, hence **adept embodiment**. Proactive robots are designed to identify and address tasks or problems before they are explicitly directed to do so. This quality is crucial in fields such as healthcare, where they might anticipate patient needs, or in industrial settings, where they can predict and prevent equipment failures. Social robots can be identified as one of the most functional proactive robots in use at present.

Social and proactive robots, while distinct in their **primary functions** and **design philosophies**, share a symbiotic relationship. Social robots are a major category of robotics in which **proactive behaviour** is compulsory for acceptance by humans during an interaction. On the other hand, social robots are primarily designed to interact with humans. They are equipped with features, such as facial expression recognition, natural language processing, and emotional intelligence. These capabilities enable them to engage in social behaviours, understand and respond to human emotions, and communicate relatedly. The primary objective of social robots is to provide companionship and assistance or facilitate social interactions, making them ideal for use in settings such as healthcare, education, customer service, and personal companionship.

The relationship between Social and Proactive Robots is worth talking about in this context. Social robots benefit significantly from proactive features. A proactive social robot can anticipate and respond to human needs or emotions, thus enhancing the quality of interaction and engagement. For instance, a social robot in a healthcare setting could proactively offer comfort or assistance based on its interpretation of a patient's emotional state.

2.1 Predictive Social Behaviour

Proactivity in social robots allows them to be more effective in their roles. For example, a social robot in a customer service role could anticipate customer queries or problems and address them preemptively, leading to improved customer satisfaction. In many applications, the distinction between social and proactive robots is vague. Robots intended for social interaction are increasingly being equipped with proactive capabilities to improve their effectiveness, while proactive robots are being designed with social features to make them more approachable and user-friendly.

Both social and proactive robots excel in environments in which adaptability and context awareness are key. Proactive robots with social capabilities can adapt their behaviours based on the social context, ensuring that their initiatives are appropriate and well received. In this context, embodiment refers not only to bodily capabilities but also to the experience of predicting possible scenarios or simulating future possibilities [2].

The integration of social and proactive features raises complex **ethical and safety** considerations, such as privacy, dependency, and decision-making autonomy. This is particularly pertinent in scenarios where robots are expected to take initiative in sensitive social contexts. Trust plays an important role in this context. The capability of the robot to perform physical actions as smoothly and dexterously as humans can considerably improve human trust [3]. However, limitations in the dynamics, as well as emotional intelligence, have to be overcome so humans can build trust in their robotic companions for more critical tasks. In [24], 925 studies on trust research on robotics have been analysed and the highest number of research was found on methods to measure trust. However, physical characteristics, theoretical frameworks, perception, trust maintenance, and design of robots play major roles in establishing trust in human-robot domains.

While social and proactive robots have distinct primary functions, the integration of their capabilities leads to more sophisticated and effective robotic systems. Embodiment plays a major role in making this a reality. This synergy between proactive robots and social behaviour enables robots to not only interact in a socially intelligent manner but also to take initiative and anticipate needs, thereby enhancing their utility and effectiveness in various applications. This concept plays a crucial role in enabling a robot to be proactive in various ways. Proactivity requires information from various sources via various modes.

Fig. 2 sums up these relationships in social encounters where proactivity of a robot is efficacious.

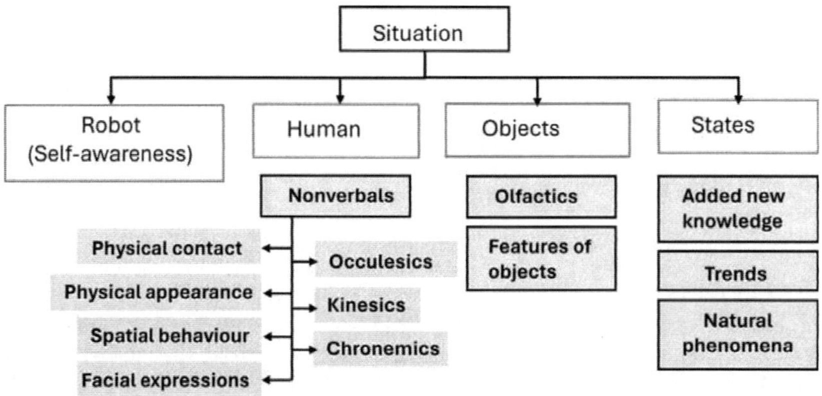

Fig. 2. Concept model depicting situation awareness of proactive robotic systems in HRI (Human-Robot interaction). This model is inspired by the models for human non-verbal forms and awareness [22, 23]. In a typical human-robot scenario, a situation will consist of the robot itself, humans, and the surrounding which contains the free space and objects. Before an interaction, proactivity of a robot is to understand a human's nonverbal communication in the first place. Nonverbals can take different forms such as physical contact (haptics), physical appearance (eg: face, skin, clothes, adornments), kinesics (gestures, postures, movements), spatial behaviour (eg: orientation, proxemics, distancing, positioning), facial expressions, oculesics (eg: eye movement, gaze, visual orientation) and chronemics. Similarly, objects in the surroundings have their own features (eg: physical shapes, texture, color). In addition, olfactics will exist depending on the objects in an environment. In addition to these, there are states that might change from time to time such as newly added knowledge based on robot's interactions with the environment and user feedback, trends (eg: accepted social behaviours with new technologies), socially accepted behaviours, laws and regulations, and natural phenomena (eg: weather).

2.2 Features required for proactivity of a robot

Sensory perception allows a robot to achieve environmental awareness. Embodied robots have sensors such as cameras, LIDAR, microphones, and tactile sensors, that allow them to perceive their surroundings. This sensory input is critical for proactive behaviour, as it enables the robot to detect changes or needs in its environment. Furthermore, predictive analysis of the surroundings with the help of fast, advanced sensors can assist a robot in predicting potential issues or needs, allowing it to act pre-emptively.

Physical embodiment allows robots to interact with and manipulate their environment. This capability is essential for tasks such as moving objects, repairing systems, or providing physical assistance. **Autonomous action** makes a proactive robot distinguishable from an ordinary robot. Such proactive, embodied robots can perform tasks without human intervention, enabling them to respond quickly to changing situations.

Through interaction with the environment, embodied robots can **learn** and **adapt** their behaviour. For example, a robot in a factory can learn to navigate more efficiently over time. **Contextual adaptation** or adjusting their actions based on the context as well, which is crucial for proactive behaviour. For instance, a domestic robot might change its routine based on the time of day or the presence of people. The relationship between physical embodiment, autonomous action and contextual adaptation can be seen in Fig. 1.

Physical presence allows for non-verbal communication cues, such as gestures or movements, which can be essential in social contexts. Having the necessary skills for **nonverbal communication (NVC)** can also be a part of embodiment in social contexts. In human environments, proactivity in social robots can mean anticipating human needs or emotions and responding appropriately.

In terms of **mobility and autonomy**, access to different locations provides an advantage for proactive robots. Mobility allows robots to move to where they are needed, enabling proactive interventions in various scenarios. Energy Autonomy: Having its own power source, a robot can decide when to recharge, maintaining its effectiveness without human intervention.

As in all other robotic systems, **realtime decision making** or **fast response** is another essential feature in proactive robotic systems. Embodiment allows for on-the-spot processing of sensory data and immediate action, which is crucial for proactivity in dynamic environments.

Large Language Models (LLMs) are performing well in generating situation-cautious responses for present robots. However, the proactivity of a robot might require physical presence or form in the real world, allowing it to interact directly with its environment. LLMs, on the other hand, exist only in digital environments. They process and generate text based on data and algorithms, but lack a physical form or the ability to physically interact with the world. Their capabilities are confined to processing and producing language-based information, hence lacking embodiment which is required for a robot's survival in certain contexts. Therefore, addressing the role of a physical body for a robot constitutes a distinct topic within the broader field of robotics.

2.3 Challenges for proactive robotic systems

Several challenges could be identified for robots to become proactive. One is the complexity of environments. Proactive behaviour requires understanding complex and often unpredictable environments. However in certain contexts, especially in outdoor environments, the capability of present sensing systems will not be adequate. Another challenge is concerns regarding safety and ethics. Ensuring that proactive actions are safe and ethically sound is a major challenge in robotics. Embodiment in robotics enhances a robot's ability to be proactive by providing the necessary sensory inputs, physical capabilities, learning opportunities, and autonomy to anticipate and respond to the needs of its environment effectively and safely. Fig. 3 illustrates the relationship between the capabilities of present robotic systems and their applications. Although robots are preferred to have humanlike capabilities in terms of body and mind, some applications restrict their body and/or mind to be limited to human scale. For instance, industrial robotic hands require much larger scale operation and higher payload while dexterity can be limited in handling a particular set of objects. There are mutual interactions between a robot's physical body, mind, the environment and the purpose-the task of the robot. Currently, a robot's task plays a major role in the physical capabilities of the robot and technical advancements act as a bottleneck for the cognitive capabilities. It can be argued that the purpose and the environment a robot is being used determine a great percentage of the make of a robot. Hence user preferences, capabilities and diversity of robots have to be grounded in terms of technology and real-world applicability.

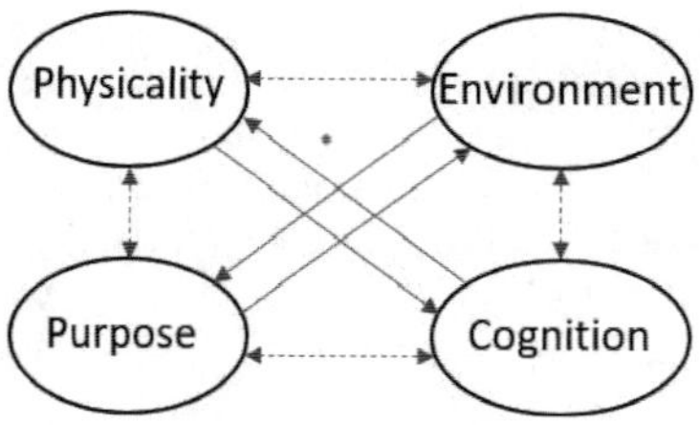

Fig. 3. Trade-off between body-mind and applications in present robotic systems. A robot's cognition and physicality represent the robot itself and purpose and environment refer to its intended task and where it's deployed respectively. These four factors influence each other when determining the capabilities and design of a robot. **physicality** represent the hardware components included in the design of the robot to achieve the **purpose**. However, one or more aspects could be prioritised over others depending on the technical constraints and other real-world problems.

A comparison of the cognitive and physical capabilities of different types of robots deployed in various environments is shown in Fig. 4. While distinguishing clear divisions between these categories can be challenging, it is important to recognise that certain applications and environments favour a robot's cognitive abilities, while others prioritise physical capabilities. Nonetheless, with some exceptions, the majority of current robotic deployments tend to benefit from the integration of both cognitive and physical capabilities, emphasising the synergy between these two aspects. However, sometimes robots with lesser physical complexity (lesser DOFs) could still provide much higher intelligence in particular tasks. The work in [31] explains how the character "Baymax" from the movie "Big Hero" could provide more insight into this aspect. This is related to physical embodiment' which is discussed above as well.

IOP Conf. Series: Materials Science and Engineering **1321** (2024) 012004 doi:10.1088/1757-899X/1321/1/012004

3 Social intelligence and social embodiment

Social intelligence and social embodiment are distinct yet interconnected concepts, particularly in the context of robotics and AI. Social intelligence includes the ability to understand and navigate social environments effectively, which involves recognising and interpreting social cues, understanding emotions, and responding appropriately in social interactions.

3.1 Key elements of social intelligence in Robotics and AI

We identify the following key elements in the process of improving the social intelligence of a robot.

Emotional Recognition The ability to identify and interpret human emotions through facial expressions, voice intonations, and body language.

Adaptive Interaction Adjusting behaviour based on the social context and the reactions of human counterparts. Humanlike continual learning during long term interactions is another aspect of adaptive interaction, for instance, personalized adaptation to their distinct teaching styles in educational robots [25].

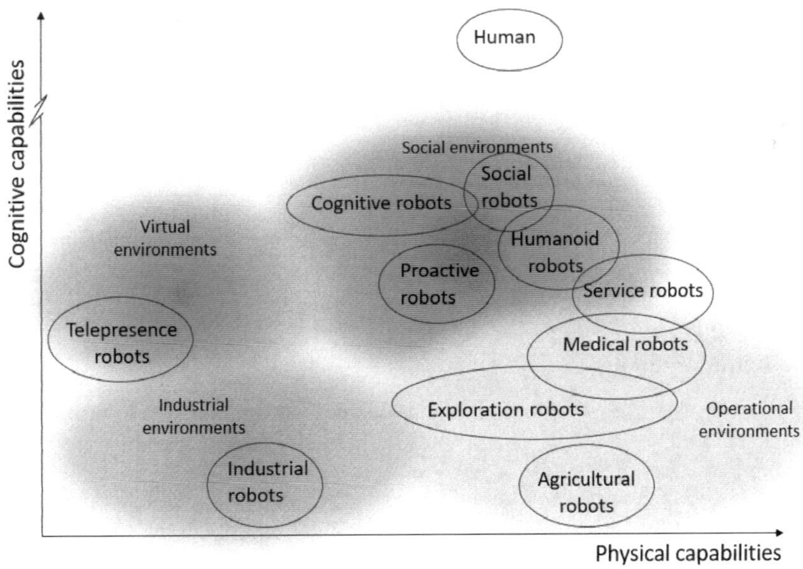

Fig. 4. An approximate comparison of cognitive and physical capabilities of different types of robots deployed in diverse environments. Physical capabilities include dexterity, mobility, sensory-motor control, speed, precision and accuracy etc. and cognitive capabilities include perception, learning, adaptation, navigation and autonomy etc.

Empathy and Responsiveness Demonstrating an understanding of human emotions and responding in a way that shows empathy and consideration.

Decision Making Making choices that take into account social norms, expectations, and the impact on human participants.

Application In robotics, social intelligence is crucial for creating machines that can interact with humans seamlessly and naturally. This includes service robots, therapeutic robots, educational robots, and more. Towards that end, robots and AI tools have been used to investigate emotional and physical distress and various other aspects of human health [26, 27]. Robot's realtime analytical skills could enhance the interaction between the robot and human, making therapeutic tasks more effective and observant.

Physical Appearance The design of the robot, which may include human-like features or other forms, is intended to be relatable and approachable. Studies have found that life-inspired, responsible co-design of robots can be used as rationale models in teaching children [28].

Non-Verbal Communication Use of gestures, facial expressions (if applicable), and body language to convey messages and emotions. Presence in Space: How the robot occupies and navigates physical space, respecting personal boundaries and using positioning to facilitate communication.

Application Social embodiment in robotics is particularly important in contexts where the physical presence of the robot is integral to the interaction, such as in caregiving, companionship, or collaborative work environments.

Social embodiment includes the physical manifestation of a robot or AI system that enables it to participate in social interactions and often involves its design, movements, and the way it uses physical space.

3.2 Relationship Between Social Intelligence and Social Embodiment

There are several facts regarding the relationship between social intelligence and social embodiment. Firstly, these two are interdependent. Social embodiment in robots can often be seen as a medium through which their social intelligence is expressed. For instance, a socially intelligent robot would need to physically manifest gestures or expressions (social embodiment) for effective communication. Secondly, these two concepts complement each other. While a robot can have a high degree of social intelligence (e.g., understanding human emotions), its effectiveness in social interactions is enhanced when it also has an appropriate level of social embodiment (e.g., being able to express empathy through a comforting gesture). Fig. 5 depicts this scenario. This portrays how the capabilities of a robot allow more details of its surroundings, providing a more approachable experience for its user.

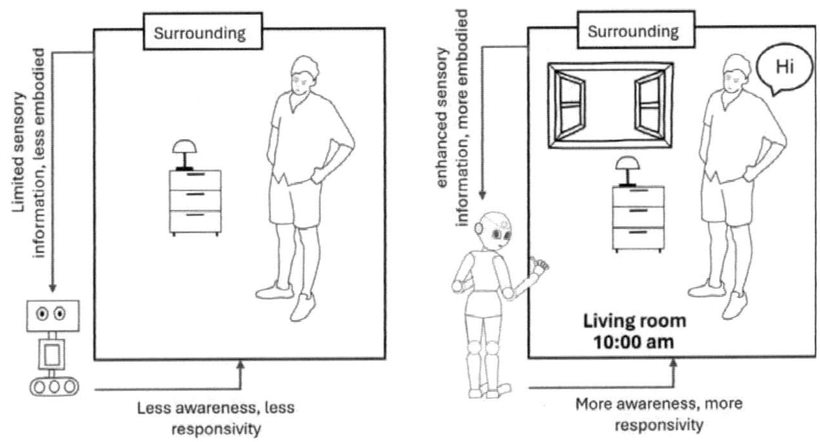

Fig. 5. This figure illustrates how the two concepts: social embodiment and social intelligence complement each other. Robots of simple implementation (the one on the left) have limited capability with their hardware and get limited information from their surroundings. Thus has less embodiment which leads to less awareness of their surroundings and phenomena. In contrast, robots with higher physical and cognitive capabilities (the one on the right) get more awareness of their surroundings. Thus making it more responsive towards its counterparts.

In essence, social intelligence generates the cognitive and emotional capabilities required for social interactions, while social embodiment realises this using physical presence and manifestation that supports these interactions. Both are crucial for creating robots that engage effectively and meaningfully in social contexts.

3.3 Effect of social embodiment and social intelligence on proactive behaviour of robots

The effect of social intelligence on proactive behaviour can be seen in various aspects. Firstly, in anticipating human needs, socially intelligent robots can interpret human emotions and social cues, allowing them to proactively address human needs. For example, a robot might notice signs of distress in a person and offer assistance or comfort. Secondly, social intelligence enhances adaptive communication. With social intelligence, a robot can tailor its communication style to fit the social context or the individual it's interacting with. This proactive adaptation can improve the effectiveness of the interaction. Furthermore, this enhances predictive interaction which plays a major role, specifically in social domains. By understanding social norms and patterns, a robot can anticipate the likely course of an interaction and prepare responses or actions in advance, thereby smoothing the flow of social exchanges. Next, social intelligence optimises conflict avoidance and resolution. Social intelligence enables a robot to recognise potential conflicts or discomfort in social settings and proactively take steps to mitigate or resolve these issues.

Similarly, the impact of social embodiment on proactive behaviour has multiple facets. Firstly, social embodiment can enhance expressive gestures and actions. Through social embodiment, a robot can use physical gestures or actions to convey understanding, empathy, or intention, proactively enhancing the quality of social interactions. In addition, with the involvement of robots in social domains, non-verbal communication has become crucial for natural interaction with its human counterparts. The ability to use body language or facial expressions (if applicable) allows the robot to communicate proactively in ways that are intuitive and reassuring to humans. A socially embodied robot can move and position itself in ways that are conducive to interaction, such as approaching to offer help or distancing to give space, based on the social context. Hence navigating physical environments becomes easier with social embodiment. In terms of demonstrating empathy and understanding, physical actions, like a comforting gesture or an attentive posture, can be used by the robot to proactively show empathy and understanding in social interactions.

3.4 Social robots

Social robots make use of proactive behaviour more than any type of robot. Hence these robots interact with the environment and respond to cues from the surroundings based on its capabilities based on sensory-motor capabilities. Social robots and the concept of social embodiment are at the forefront of human-robot interaction. Social robots are designed to engage with humans in a manner that feels natural and intuitive [29]. The work presented in [30] studies the believability of implementation through two standpoints: character animation and technical capabilities. For example, simple head motion in a character may be believable from a distance, but as individuals move closer to a robot, the illusion of life breaks, and eye gaze becomes essential. This is an example of an embodiment of sufficient features and skills to sustain the interactivity between two parties. Social embodiment in this context refers to the robot's physical form and behaviours that facilitate social interaction.

Key aspects of social robots related to social embodiment can be listed as follows.

Physical Appearance: The design of a social robot often mimics human or animal-like features to facilitate relatable interactions. This can include facial expressions, gestures, and body movements that are familiar to humans. While materials perspective and mechanical functionality are important, the emotional representation of the robot's perception of the environment decides more on the social acceptance of the robot.

Communication Skills: Social robots are equipped with sophisticated communication technologies. They use verbal cues (speech generation and recognition) and non-verbal cues (gestures, eye contact) to interact effectively.

Emotional Intelligence: Advanced AI enables these robots to recognize and respond to human emotions. They can interpret facial expressions, voice tone, and body language to tailor their interactions.

Autonomous Behaviour: Social robots can initiate actions and conversations based on their understanding of the social context and human needs. This proactivity makes interactions more engaging and natural.

Learning and Adaptation: Over time, social robots learn from interactions and adapt to individual preferences and behaviours. This personalisation enhances the quality of the social exchange.

Ethical Considerations: Designing social robots involves ethical considerations, such as privacy, emotional dependency, and the implications of human-like robots in social roles.

Applications: Social robots are used in various settings, including education (teaching assistants), healthcare (companions for the elderly or individuals with disabilities), customer service, and personal companionship.

We identify three major roles of social embodiment. *Firstly*, a socially embodied robot facilitates more trust and acceptance compared to a normal robot. A robot's physical embodiment can make interactions more trustworthy and acceptable to humans. A relatable form can lower barriers to communication and cooperation. This can be considered a main motive in designing social robots. *Secondly*, such capabilities in robots enhance the quality of interaction. The physical aspects of the robot, like its size, appearance, and movement, are tailored to suit specific interaction contexts, thus enhancing the quality and effectiveness of the social interaction. *Thirdly*, expressiveness of socially embodied robot will be much higher than ordinary robots and much closer to humans. Social embodiment allows for a range of expressions that can convey emotions, attentiveness, and empathy, which are crucial in social interactions.

Challenges for socially embodied robots Creating robots that can fluidly and accurately mimic human social cues is technologically challenging. This happens as a result of technological limitations. In addition, there are concerns about the impact of social robots on human relationships and societal structures. While the physical embodiment of these robots is crucial in making interactions feel natural, building trust, and enhancing communication effectiveness, this field also faces significant technological challenges and ethical considerations. However, according to the uncanny valley in social robotics, emotion recognition rates; hence interactivity, decreases with an increasing degree of anthropomorphism [32]. Recognising the trade-offs between the aspects shown in Fig. 3 might be a solution for this.

3.5 Synergy in proactive behaviour and social embodiment

The combination of social intelligence and embodiment allows a robot to engage in more holistic and effective interactions. The robot not only understands what action might be appropriate (social intelligence) but also how to physically execute that action in a socially acceptable and effective manner (social embodiment). Hence the synergy between social intelligence and social embodiment facilitates a **holistic interaction** between a robot and a human.

Together, these aspects enable a robot to respond more quickly and appropriately to changes in human behaviour or the environment, making its proactive actions more relevant and timely. Hence an **enhanced responsiveness** could be expected from a robot in this context, facilitating meaningful interactions with humans. When it comes to embodied proactive robots, it is worth considering the nature of the connection they are supposed to build up between themselves and their environment. What defines the essence of genuine human connection in an increasingly digital world will require timely answers with the expansion of robot's proactivity in human-robot domains. However with this, reconciling authenticity in social interactions with societal norms and expectations will still be a question of technology and time. Timely considerations will have to be given on such matters. What ethical responsibilities accompany the power of social influence, particularly in the age of social media and large language models (LLMs) will play a major role in determining this.

Furthermore, from a robot's perspective, the boundaries of self, have to be defined in social contexts. An important question in this regard is how do we establish the boundaries between an individual identity and the collective social environment robots are embedded. Benchmarking a robot's perception of social cues and embodiment which influence our reality and decision-making is also an important milestone in this regard.

When social environments are considered cultural dynamics decide some percentage of human behaviours in an encounter. Hence robots having an understanding of how the culture shapes the behaviour of their human counterparts, will be an expression of social intelligence and embodiment. Cultivating empathy will enrich the effectiveness of such encounters. Finally, better modelling of mind-body connection of a robot, although not human-inspired, will have a positive impact upon its capabilities and survival in social domains. These aspects touch on the complex interplay between robots' social selves, understanding of others, and the philosophical implications embedded in these concepts.

In conclusion, in the context of socially intelligent and embodied robots, yet there are some concerns and challenges to overcome. Following are some aspects we recognised which require benchmarking and grounding.

In terms of **ethical boundaries**, it is worth addressing what ethical considerations arise when imbuing robots with social intelligence. In other terms, this defines appropriate behaviour and interactions for AI-driven entities. When considered **human-robot relationships**, there still are questions regarding how robots can truly comprehend and reciprocate emotions, and what impact this has on forming meaningful relationships with humans.

Is it ethical or meaningful for robots to simulate emotions or exhibit social behaviours, even if they don't truly experience them? The extent to which robots have to replicate emotions or whether it is similar to humans still remains a challenge. Behaviours of robots change according to their applications. Hence defining the personality of a robot might solve this issue which has been impossible yet due to many reasons, including the ones below. We further have fewer regulations on robotic decision making, for instance, Issac Asimov's three laws of robotics. Hence who holds responsibility when a socially intelligent robot makes decisions or interacts autonomously in a social context is yet an unanswered question in this domain.

Cultural Adaptation: How should robots adapt to different cultural norms and social expectations, and what challenges arise in achieving this adaptation?

Perception of Personhood: At what point, if any, does a socially intelligent robot cross into being perceived as having personhood, deserving of certain rights and considerations?

Empathy in Machines: Can robots authentically exhibit empathy, and how might this impact their role in caregiving or support services?

Physical Embodiment: How does the physical design and appearance of a robot affect human interaction and the development of trust or rapport?

These questions delve into the ethical, emotional, and societal implications of integrating social intelligence and embodiment into artificial entities like robots, reflecting on the complex interplay between technology and human social dynamics.

4 Conclusions

Recent studies on robots suggest that social intelligence and embodiment are powerful concepts in the development of proactive robotic systems. Exploiting the dynamics provided by the surroundings as well as the interaction between physical and information processes promises to extend the capabilities of robot design methodologies to cater to multiple domains, including social robots. Although bridging the gap between robots and humans requires addressing many conceptual and technological challenges, we believe that the first important step is the abstraction of a set of design principles. Such principles will form a knowledge base on how robotics systems must be designed and distinguished in social domains rather than adopting humanlikeness. These will further make human-robot environments diverse but still acceptable, ethically and behaviourally. Thrilling adventures in robots await us.

References

1. C. Sirithunge, A.G.B.P. Jayasekara, and D.P. Chandima, 2019. Proactive robots with the perception of nonverbal human behaviour: A review. IEEE Access, 7, pp.77308-77327.

2. Niedenthal, P.M., Barsalou, L.W., Winkielman, P., Krauth-Gruber, S. and Ric, F., 2005. Embodiment in attitudes, social perception, and emotion. Personality and social psychology review, 9(3), pp.184-211.
3. Sirithunge, C. and Dahn, N., 2023, October. How Do We Trust Our Robots?. In IOP Conference Series: Materials Science and Engineering (Vol. 1292, No. 1, p. 012014). IOP Publishing.
4. Pfeifer, R., Lungarella, M. and Iida, F., 2007. Self-organization, embodiment, and biologically inspired robotics. science, 318(5853), pp.1088-1093.
5. Maroto-Gómez, M., Alonso-Martín, F., Malfaz, M., Castro-González, Á., Castillo, J.C. and Salichs, M.Á., 2023. A systematic literature review of decision-making and control systems for autonomous and social robots. International Journal of Social Robotics, 15(5), pp.745-789.
6. Buyukgoz, S., Grosinger, J., Chetouani, M. and Saffiotti, A., 2022. Two ways to make your robot proactive: Reasoning about human intentions or reasoning about possible futures. Frontiers in Robotics and AI, 9, p.929267.
7. Kraus, M., Wagner, N., Callejas, Z. and Minker, W., 2021. The role of trust in proactive conversational assistants. IEEE Access, 9, pp.112821-112836.
8. Cifuentes, C.A., Pinto, M.J., Céspedes, N. and Múnera, M., 2020. Social robots in therapy and care. Current Robotics Reports, 1, pp.59-74.
9. van Ewijk, G., Smakman, M. and Konijn, E.A., 2020, June. Teachers' perspectives on social robots in education: an exploratory case study. In Proceedings of the interaction design and children conference (pp. 273-280).
10. Avelino, J., Paulino, T., Cardoso, C., Nunes, R., Moreno, P. and Bernardino, A., 2018. Towards natural handshakes for social robots: human-aware hand grasps using tactile sensors. Paladyn, Journal of Behavioural Robotics, 9(1), pp.221-234.
11. Kyrarini, M., Lygerakis, F., Rajavenkatanarayanan, A., Sevastopoulos, C., Nambiappan, H.R., Chaitanya, K.K., Babu, A.R., Mathew, J. and Makedon, F., 2021. A survey of robots in healthcare. Technologies, 9(1), p.8.
12. Porkodi, S. and Kesavaraja, D., 2021. Healthcare robots enabled with IoT and artificial intelligence for elderly patients. AI and IoT-Based Intelligent Automation in Robotics, pp.87-108.
13. Wood, L.J., Zaraki, A., Robins, B. and Dautenhahn, K., 2021. Developing kaspar: a humanoid robot for children with autism. International Journal of Social Robotics, 13, pp.491-508.
14. Wang, Y.J., Zhang, B., Chen, J. and Sreenath, K., 2023. Prompt a robot to walk with large language models. arXiv preprint arXiv:2309.09969.
15. Qi, W., Ovur, S.E., Li, Z., Marzullo, A. and Song, R., 2021. Multi-sensor guided hand gesture recognition for a teleoperated robot using a recurrent neural network. IEEE Robotics and Automation Letters, 6(3), pp.6039-6045.
16. Yang, G.Z., Bellingham, J., Dupont, P.E., Fischer, P., Floridi, L., Full, R., Jacobstein, N., Kumar, V., McNutt, M., Merrifield, R. and Nelson, B.J., 2018. The grand challenges of science robotics. Science robotics, 3(14), p.eaar7650.
17. Sirithunge, C., Jayasekara, A.B.P. and Chandima, D.P., 2019. Proactive robots with the perception of nonverbal human behaviour: A review. IEEE Access, 7, pp.77308-77327.
18. Bremner, P., Dennis, L.A., Fisher, M. and Winfield, A.F., 2019. On proactive, transparent, and verifiable ethical reasoning for robots. Proceedings of the IEEE, 107(3), pp.541-561.
19. Grant, A.M. and Ashford, S.J., 2008. The dynamics of proactivity at work. Research in organizational behaviour, 28, pp.3-34.

20. Buyukgoz, S., Pandey, A.K., Chamoux, M. and Chetouani, M., 2021. Exploring behavioural creativity of a proactive robot. Frontiers in Robotics and AI, 8, p.694177.
21. Buyukgoz, S., Grosinger, J., Chetouani, M. and Saffiotti, A., 2022. Two ways to make your robot proactive: Reasoning about human intentions or reasoning about possible futures. Frontiers in Robotics and AI, 9, p.929267.
22. Manninen, T. and Kujanpää, T., 2002. Non-verbal communication forms in multiplayer game session. In People and Computers XVI-Memorable Yet Invisible: Proceedings of HCI 2002 (pp. 383-401). Springer London.
23. Sirithunge, C., Priyanayana, K.S., Ravindu, H.M., Bandara, T., Dahn, N., Jayasekara, B.P. and Chandima, D.P., 2020, August. Tell me more! A Robot's Struggle to Achieve Artificial Awareness. In 2020 29th IEEE International Conference on Robot and Human Interactive Communication (RO-MAN) (pp. 60-66). IEEE.
24. Bastian, M., Heymann, S. and Jacomy, M., 2009, March. Gephi: an open source software for exploring and manipulating networks. In Proceedings of the international AAAI conference on web and social media (Vol. 3, No. 1, pp. 361-362).
25. Ayub, A., Mehta, J., De Francesco, Z., Holthaus, P., Dautenhahn, K. and Nehaniv, C.L., 2023. How do Human Users Teach a Continual Learning Robot in Repeated Interactions?. In 2023 32nd IEEE International Conference on Robot and Human Interactive Communication (RO-MAN) (pp. 1975-1982). IEEE.
26. Dennler, N., Cain, A., De Guzman, E., Chiu, C., Winstein, C.J., Nikolaidis, S. and Matarić, M.J., 2023. A metric for characterizing the arm nonuse workspace in poststroke individuals using a robot arm. Science Robotics, 8(84), p.eadf7723.
27. Spitale, M., Silleresi, S., Garzotto, F. and Matarić, M.J., 2023. Using Socially Assistive Robots in Speech-Language Therapy for Children with Language Impairments. International Journal of Social Robotics, 15(9), pp.1525-1542.
28. Boulicault, M., Phillips-Brown, M., Kory-Westlund, J. M., & Nguyen, S. 2021. Authenticity and Co-design: On Responsibly Creating Relational Robots for Children. Algorithmic Rights and Protections for Children.
29. Henschel, A., Hortensius, R. and Cross, E.S., 2020. Social cognition in the age of human–robot interaction. Trends in Neurosciences, 43(6), pp.373-384.
30. Pan, M.K., Choi, S., Kennedy, J., McIntosh, K., Zamora, D.C., Niemeyer, G., Kim, J., Wieland, A. and Christensen, D., 2020, October. Realistic and interactive robot gaze. In 2020 IEEE/RSJ International Conference on Intelligent Robots and Systems (IROS) (pp. 11072-11078). IEEE.
31. Gavero, G.C. and Guerrero, A.P., 2017. My Teacher Baymax: Lessons from the Film Big Hero 6. Academic Psychiatry, 41, pp.601-605.
32. Mende, M.A., Fischer, M.H. and Kühne, K., 2019. The use of social robots and the uncanny valley phenomenon. AI love you: developments in human-robot intimate relationships, pp.41-73.

Embodied Intelligence in Additive Manufacturing

Jie Xu, Yue Xie, Fumiya Iida

Bio-Inspired Robotics Laboratory, Department of Engineering, University of Cambridge, Trumpington Street, Cambridge CB2 1PZ, United Kingdom

E-mail: {jx322, yx388, fi224}@cam.ac.uk

Abstract. Embodied intelligence reflects the ability of an agent to interact with its environment through its body to gain intelligence and autonomy. Such a concept has become popular in recent years for the design of robotic systems. This paper generalises the embodied intelligence concept to Additive Manufacturing (AM) to provide a new understanding of the technology with inspirations from biological production processes including organism growth and animal construction. Morphology-based embodied intelligence is analysed on the design and manufacturing aspects for delivering intelligent products (including robots) while embodied artificial intelligence (AI) (foundation models) is discussed for the printer itself. The paper opens a new perspective for AM research.

1. Introduction

The creation of matters is the fundamental dynamics for the physical universe: atoms are synthesised to form molecules; organisms (e.g., animals, plants, etc.) emerge and grow in nature; and numerous products are manufactured daily to serve human society. No matter what a process of matter creation is, and whether it is happening on microscale or macroscale, it always follows a fundamental principle - being driven by and adapted to certain environments. Such adaptability reflects the intelligence of a system to be qualified and/or maintain its presence in a particular environment, which lies in the physical embodiment of this system. This in a sense corroborates Hegel's 'What is real is rational, and what is rational is real'.

The so-called 'embodied intelligence' broadly observed in nature such as fishes using fins to swim or birds using wings to fly, have been mimicked and applied to the design of innovative robot morphologies by roboticists [1, 2]. The intention is to enable the robot to perform a specific task or behaviour through learning and development in an unknown environment, which creates intelligence and autonomy [3]. If we consider that every artificial product (e.g., a desk, a car, a building, etc.) in human society is a robot (being able to react to external stimulus in certain ways and/or at certain levels, etc.), their embodied intelligence, despite being possibly too low-level to be called 'intelligent', allow them to fulfil expected functionalities for specific applications/scenarios which are their operational environments.

Additive Manufacturing (AM), also called 3D printing, has been a rapidly developed manufacturing technology across industries since its emergence in the 1980s [4]. It actualises the goal of matter creation through an automated, normally layerwise material-adding process with which any geometries can be theoretically attained without the need for moulds. With the advancement of materials, AM has been acknowledged as the game-changing manufacturing technology for the future society for its incomparable (with conventional methods as subtractive or formative processes [5]) capability to flexibly and economically deliver innovative functional geometries and bespoke customisations (e.g., artificial organs, etc.) [6]. Our vision for AM with embodied intelligence includes both the printed product and the printer aspects: 1) high-level morphologically intelligent products to create a highly efficient, sustainable and human-centred society, 2) intelligent 3D printers for autonomously planning and executing printing activities tailored/adaptive to their in-situ environment.

This paper aims to establish and clarify the embodied intelligence concept in AM for future society development by drawing lessons from biological production processes. Section 2 gives an overview of mapping biological production processes including organism growth and animal construction to AM to establish the understanding of the generation of embodied intelligence in nature. Section 3 discusses the definition of morphological intelligence of products and the pathway to attain high-level intelligent products from both the design and manufacturing aspects. Section 4 describes the embodied artificial intelligence (AI) for 3D printers to achieve autonomous printing planning and coding with an emphasis on construction-scale AM in remote harsh environments.

2. Mapping biological production processes to Additive Manufacturing

2.1. Organism growth: From genotype to phenotype

The most universal biological production process is the growth of a living organism itself, which changes the organism's morphology and dimensions. Through millions of years of evolution, the morphology of a living organism gains its intelligence, so-called 'morphological intelligence' - a form of embodied intelligence, to adapt to its living environment. Such morphological intelligence is represented by certain functions (e.g., strength, motion, etc.) tailored to the environment, taking effects from micro level (i.e., cells and tissues) to macro level (i.e., bodies and limbs)[7, 8].

The growth of a living organism can be seen as an additive production process primarily through cell proliferation: numerous cells exponentially create relevant tissues and then the macroscopic morphology of the organism [9]. The objective of this process is to generate the desired morphological intelligence (functions) which is 'designed' and 'optimised' by Natural Selection. If we were to analogize the AM process to organism growth, the objective of this AM process should be achieving the morphological intelligence for a product to adapt to/be qualified for specific applications.

The organism growth process reflects a genotype-to-phenotype ('GtP') chain through which a set of genetic information is translated into a physical manifestation [10] (see Figure 1). The genotype refers to the genetic makeup of DNA that stores the information of the target morphology (phenotype). In AM, the genotype corresponds to the digital design parameters that define the characteristics of the desired output. These parameters could include information about material composition, microstructure geometry, macrostructural geometry, density distribution, and other relevant attributes. Essentially, the genotype in AM is a comprehensive dataset that encodes all potential variations and configurations of the product being designed.

The translation from genotype to phenotype is the process where the encoded design parameters are converted into cell proliferation instructions to create a tangible morphology. In AM, this translation process is carried out by professional processing software (e.g., slicers) embedded in AM machines (printers). The outcome is the specified building instructions (e.g., toolpaths) for guiding the material placement to create the object with all its physical and functional characteristics (the phenotype). Learning the lessons from organism growth to create morphologically intelligent AM products is discussed in Section 3.

2.2. Animal construction: From genotype to extended phenotype

Another interesting biological production process is the animal construction (manufacturing) behaviour. This has many forms in nature, for example, spiders weave webs, corals make reefs, and ants and birds build nests (Figure 1). Animal construction behaviour is considered as a type of extended phenotype for certain animals which is encoded in their genotype and generates changes to their living environment when it happens [11].

The genotype-to-extended phenotype ('GtEP') chain describes the extra process by which a set of genetic information is translated into a physical form outside of the organism's body. The traditional phenotype refers to the observable traits of the organism itself while the concept of the extended phenotype goes further to encompass the organism's influence on the environment. One of such influences refers to structures or changes an organism creates in its surroundings, which is why animal construction behaviours are commonly observed.

Mapping the animal to a 3D printer or printing robot out of an AM process, the genotype-to-extended phenotype chain defines how the design parameters of the printer itself dictate its physical products that can make a difference to their operational environment. Particularly, printers or printing robots can manufacture objects in the field at the construction scale, well known as 'construction 3D printing', which has been extensively explored and applied over the last decade [12, 13]. Learning from animal construction methods to inform the planning and implementation of construction-scale AM has elicited some interest from researchers [14], leading to process-oriented biomimetics research. This is further discussed in Section 4.

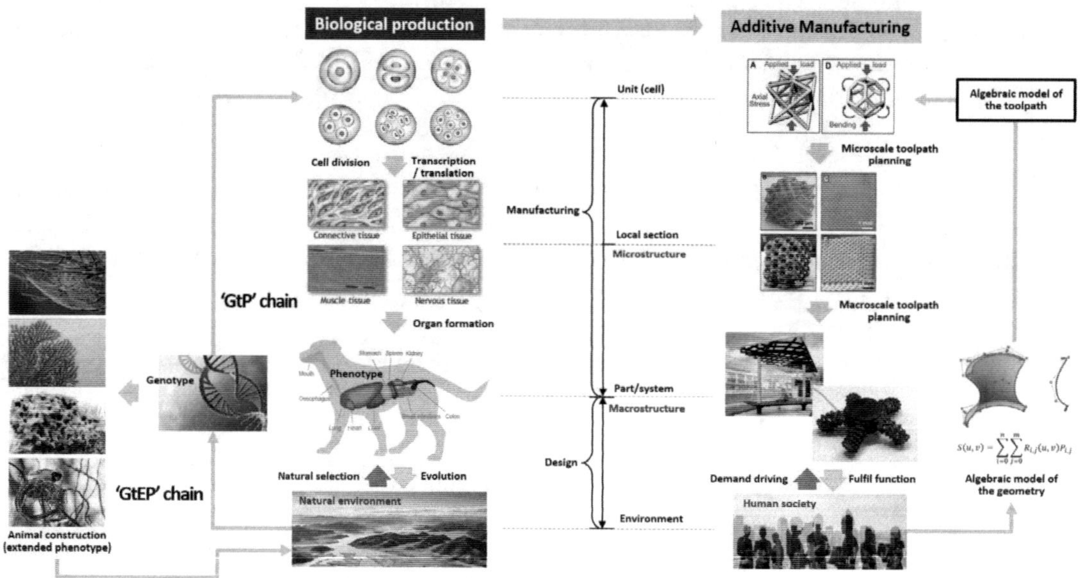

Figure 1. Mapping organism growth to Additive Manufacturing (images: [15] and google.com).

3. Towards morphologically intelligent AM products

3.1. Morphological intelligence in products

The nature of a physical product is to satisfy the desire or need of a customer [16]. We consider morphological intelligence as a measure of a product's functional performance based on its morphology for the satisfaction of human needs. In this context, every product has a certain morphological intelligence, e.g., the intelligence of a desk refers to it having several legs to stand on and bearing some loads. However, we usually don't recognise such morphological intelligence as it is too low-level to meet our common understanding of intelligence. Therefore, high-level morphological intelligence can be defined as innovative and superb functions or features that are not commonly seen or never exist in the current market. The generation of such high-level morphological intelligence can be expensive or impossible for conventional manufacturing processes while AM has the potential to deliver this goal economically and efficiently.

The desired highly intelligent morphology can refer to both macro and micro scales. Recent advances have shown the benefits of manipulating morphologies in micro [15] and macro [17] scales separately. Much microscopic intelligent morphology aims to create functional materials (e.g., graded functional materials) using bio-inspired structures. A comprehensive summary of such biological material structures is shown in Figure 2. Hereinto, one cellular/lattice structure called Triply Periodic Minimal Surface (TPMS), has been the focus [18]. Similarly, the macroscopic intelligent morphology refers to topology-optimised functional or material-saving geometries - examples of these can be seen in Figure 3.

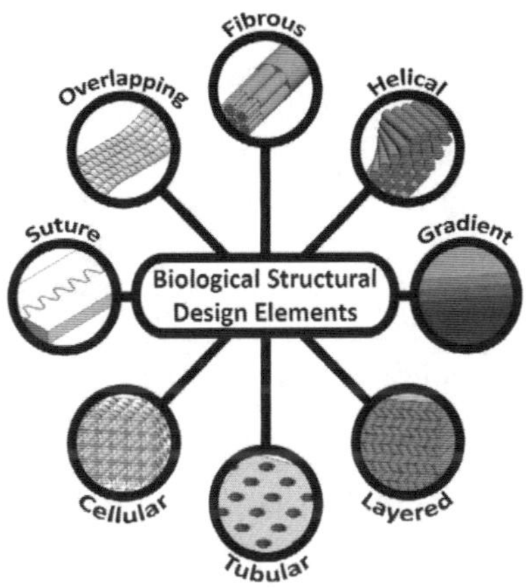

Figure 2. Common biological material structures [19].

Figure 3. Examples of the macroscopic intelligent morphology [20].

It should be noted that the macro and micro scales here are defined as per the manufacturing resolution/precision for the sake of practicality, and are relative concepts with the actual values varying according to the dimensions of the product itself. For instance, the micro scale of a protective helmet is on the order of microns up to submillimetres, while the micro scale of a building refers to millimetres up to centimetres, which represents the macro scale for the helmet. Due to their distinct performances, biological material structures are commonly used for designing both the micro and macro morphology of a product.

When the produced object can autonomously change its morphology according to the ambient environment, it reflects an even higher level of intelligence. This is how 4D printing enables a printed product to interact with the environment, often through biomimetic morphology-changing behaviours [21]. Currently, some materials have been engineered to become 4D printable and more are expected in the future [22].

3.2. Robots with morphological intelligence

Robots are a special category of products. In robotics, morphology, which includes a robot's physical form and its functional adaptation to tasks and environments, plays a critical role in determining its capabilities and efficiency. Iida et al. [23] offers an integrative perspective on perception from a biologically inspired robotics standpoint, stressing the critical role of sensor morphology in robotic systems. This insight underscores the relevance of physical form in robotics, demonstrating how structural design can impact a robot's functionality and interaction with its environment.

AM significantly releases the freedom of robotic design, particularly in morphological innovation. AM enables the creation of complex geometries and structures, previously unfeasible, leading to a new era in robotic morphology customization. Vujovic et al. [1] highlight this in their study on evolutionary developmental robotics, demonstrating how AM techniques improve robot physical structures, emphasizing the importance of AM in innovative robotic design. Further research could delve deeper into this area, perhaps exploring specific case studies or new materials.

The integration of adaptive sensory systems into a robot's morphology marks a leap in embodied intelligence. These systems, ranging from tactile sensors to advanced vision systems, are crucial for effective environmental interaction. The challenge is integrating these sensors into the robot's body to enhance functional capabilities without disrupting its overall morphology. Nurzaman et al. [24] explore this in their study on active sensing systems with adjustable sensor morphology, providing practical examples of how sensory adaptation can be integrated into robotic design, furthering the narrative of morphological adaptation enhancing functionality.

3.3. Design: Evolution-based cross-scale morphology optimisation

On the topic of morphology design and optimisation of AM products, topology optimization [25] are usually employed in product design to enhance performance and efficiency. Among the most renowned topology optimisation algorithms are the Bi-directional Evolutionary Structural Optimization (BESO) [26] and the Solid Isotropic Modeling with Penalization (SIMP) [27], both of which are integral to achieving sophisticated and functionally enhanced design outcomes. Within this context, evolutionary algorithms [28, 29] have been utilised to find forms that yield desired intelligence. Such optimisation strategies usually closely resemble the natural selection processes that govern the growth of organisms. These strategies translate genetic design information into tangible physical forms, enabling the genotype-to-phenotype chain. This translation process continuously refines the digital design parameters of an AM product through the iterative application of evolutionary algorithms, leading to the optimisation of either micro or macro-structural designs.

Our vision is to combine the evolution-based optimisation of microstructure and macrostructure (simultaneous computation) to achieve co-evolution [30, 31]. The results of this cross-scale evolutionary design are concurrently optimised morphology both in micro (local) and macro (global) levels [32, 33]. Fast and effective testing and validation of the resultant cross-scale optimised morphology through simulation (e.g., finite element analysis) will still remain a challenge for design automation.

3.4. Manufacturing: Non-layerwise volumetric forming strategies

The nature of AM lies in the material-adding mechanism. As the cross-scale intelligent morphology generally refers to discrete and non-dense material placement at the micro scale related to the ordered cavity and void generation, the printing resolution must be able to formulate the microscopic material structures with both accepted accuracy and speed. This leads to the elimination of the conventional layerwise processes utilising a fixed higher printing resolution to form the microscopic material structures (e.g., 1-micron printing resolution for a 100-micron micro structure) as it will be extremely time-consuming for forming the macro structure (the whole body) of the product. Improved layerwise slicing methods such as varying-thickness / multi-directional parallel slicing or non-parallel slicing [34] can provide better performance but the effectiveness is still limited.

A possible solution will be non-layerwise volumetric additive forming. Instead of layerwise fixed-resolution material accumulation, this forming strategy aims to create local geometries through one-time adaptive material placement with on-demand dynamically varying printing resolution at the target locations to form a specific geometry. Examples of such forming strategy can be seen in [35, 36]. The planned toolpaths (e.g., medial axes) can directly follow/capture the topology (skeleton) of the target object, enabling varying-cross-section voxels as material units to mathematically represent the volumetric forming process. Such forming strategy allows the printing toolpaths to be directly designed or together designed with the object morphology in a parametric way, leading to a much simplified and efficient Design for Manufacturing (DfM) workflow for creating intelligent morphology at both micro and macro scales.

4. Can a 3D printer have embodied intelligence?

Considering a 3D printer to gain embodied intelligence, it should be able to interact with the environment like the organism. For all the stationary 3D printers (either desktop or industrial classes), such an environment is mostly a built (artificial) environment of the human society (Figure 1). The embodied intelligence of these 3D printers would be enabled by foundational AI models such as the Large Language Model to allow the automatic generation and comprehension of toolpaths, machine codes and material application of any forms given by humans as environmental inputs [37, 38]. This will help to eliminate the barrier of demanding expert knowledge in planning and coding for executing printing activities (behaviours), significantly contributing to the efficiency and universality of AM processes.

The embodied AI can be even more essential for mobile 3D printers which usually undertake large-scale AM processes for construction purposes. Inspired by animal construction, a construction 3D printer is a mobile 3D printing robot acting like an animal to construct structures in the natural environment. Preliminary lab-based examples can be seen in the University of Stuttgart ICD/ITKE research pavilions where spider webs [39] and moth webs [40] are mimicked for robotic additive construction of pavilions. The vision is for such 3D printing robots to interact with / take in information (of limitations) from their surroundings and perform construction using in-situ material resources with printing toolpaths and manufacturing strategies adaptive to the local landscape[41]. The latest example can be seen in [42] where a robot is autonomously selecting and placing local stones to additively construct large-scale walls.

The above-mentioned mobile 3D printers with embodied AI are particularly needed for construction activities in a remote location which is hard or impossible for human beings to access (see Figure 4). Such locations can mainly include three types:

- Extreme environments which are harsh or dangerous for human beings such as polar areas [43], deep sea [44] and extraterrestrial spaces (e.g., Moon and Mars [45, 46]) for potential human habitats in the future - it should be noted that lunar construction has been a promising research and application area for large-scale AM using lunar regolith [47];
- Natural disaster sites often with risk of potential continuous damages where emergency shelters and blindages are needed for saving human lives and protecting properties [48];
- Battlefields or battlefronts where military facilities need to be rapidly constructed for the purposes of crossing gullies or shielding enlisted men and weapons [49].

Figure 4. Examples of large-scale mobile 3D printing in a remote location: a) printing at polar areas [43], b) moon base printing [50], c) emergency shelter printing [48], and d) military bridge printing [49].

5. Conclusions

Additive Manufacturing has been rapidly developing since its birth creating innovative products for modern society and will continue to grow with expanding materials and scales. This paper opens a new perspective on evaluating the performance of AM products, morphology-based embodied intelligence taking inspiration from robotics. Learning from organism growth, the paper defines high-level morphological intelligence as superb functions for AM products and discusses evolution-based cross-scale morphology optimisation (design aspect) and non-layerwise volumetric forming strategies (manufacturing aspect) for achieving such intelligence. Besides the products, 3D printers, both stationary ones and mobile ones (field printing robots) also need to be equipped with embodied intelligence (foundational AI models) to conduct autonomous environment-driven planning and coding for AM tasks. Animal construction behaviours can be learned for mobile 3D printers. The authors believe that embodied intelligence provides a promising direction for the future development of AM. One of the most important next steps is establishing metrics for measuring and benchmarking the embodied intelligence in AM to drive performance comparison and improvement.

References

[1] Vujovic V, Rosendo A, Brodbeck L and Iida F 2017 *Artificial life* **23** 169–185
[2] Howison T, Hauser S, Hughes J and Iida F 2020 *Artificial Life* **26** 484–506
[3] Pfeifer R, Lungarella M and Iida F 2007 *science* **318** 1088–1093
[4] Wong K V and Hernandez A 2012 *International scholarly research notices* **2012**
[5] Zhu Z, Dhokia V G, Nassehi A and Newman S T 2013 *International Journal of Computer Integrated Manufacturing* **26** 596–615
[6] Gibson I, Rosen D W, Stucker B, Khorasani M, Rosen D, Stucker B and Khorasani M 2021 *Additive manufacturing technologies* vol 17 (Springer)
[7] Carstairs-McCarthy A 2010 *The evolution of morphology* (OUP Oxford)
[8] Gupta A, Savarese S, Ganguli S and Fei-Fei L 2021 *Nature communications* **12** 5721
[9] Golias C, Charalabopoulos A and Charalabopoulos K 2004 *International journal of clinical practice* **58** 1134–1141
[10] Benfey P N and Mitchell-Olds T 2008 *Science* **320** 495–497
[11] Laland K N 2004 *Biology and Philosophy* **19** 313–325
[12] Buswell R A, De Silva W L, Jones S Z and Dirrenberger J 2018 *Cement and Concrete Research* **112** 37–49
[13] Ma G, Buswell R, da Silva W R L, Wang L, Xu J and Jones S Z 2022 *Cement and Concrete Research* **156** 106774
[14] Petersen K H, Napp N, Stuart-Smith R, Rus D and Kovac M 2019 *Science Robotics* **4** eaau8479
[15] Zheng X, Lee H, Weisgraber T H, Shusteff M, DeOtte J, Duoss E B, Kuntz J D, Biener M M, Ge Q, Jackson J A *et al.* 2014 *Science* **344** 1373–1377
[16] Kotler P, Armstrong A, Brown L and Adam S 2006 *Marketing* (Pearson Education Australia)
[17] Liu J, Gaynor A T, Chen S, Kang Z, Suresh K, Takezawa A, Li L, Kato J, Tang J, Wang C C *et al.* 2018 *Structural and multidisciplinary optimization* **57** 2457–2483
[18] Han L and Che S 2018 *Advanced Materials* **30** 1705708
[19] Naleway S E, Porter M M, McKittrick J and Meyers M A 2015 *Adv. Mater.* **27** 5455–5476
[20] Https://all3dp.com/ accessed: 2024-2-2
[21] Sydney Gladman A, Matsumoto E A, Nuzzo R G, Mahadevan L and Lewis J A 2016 *Nature materials* **15** 413–418
[22] Kuang X, Roach D J, Wu J, Hamel C M, Ding Z, Wang T, Dunn M L and Qi H J 2019 *Advanced Functional Materials* **29** 1805290
[23] Iida F and Nurzaman S G 2016 *Interface focus* **6** 20160016
[24] Nurzaman S G, Culha U, Brodbeck L, Wang L and Iida F 2013 *PLoS One* **8** e84090
[25] Pinskier J, Wang X, Liow L, Xie Y, Kumar P, Langelaar M and Howard D 2024 *Advanced Intelligent Systems* 2300505
[26] Xiong Y, Zhao Z L, Lu H, Shen W and Xie Y M 2023 *Advances in Engineering Software* **176** 103389
[27] Brackett D, Ashcroft I and Hague R 2011 Topology optimization for additive manufacturing *2011 international solid freeform fabrication symposium* (University of Texas at Austin)
[28] Bäck T and Schwefel H P 1993 *Evolutionary computation* **1** 1–23
[29] Giraud-Moreau L and Lafon P 2002 *Engineering Optimization* **34** 307–322
[30] Thompson J N 2014 *Interaction and coevolution* (University of Chicago Press)
[31] Ma X, Li X, Zhang Q, Tang K, Liang Z, Xie W and Zhu Z 2018 *IEEE Transactions on Evolutionary Computation* **23** 421–441
[32] Hoang V N, Tran P, Nguyen N L, Hackl K and Nguyen-Xuan H 2020 *Computer-Aided Design* **129** 102918
[33] Hoang V N, Tran P, Vu V T and Nguyen-Xuan H 2020 *Composite Structures* **252** 112718
[34] Xu J, Gu X, Ding D, Pan Z and Chen K 2018 *Rapid Prototyping Journal* **24** 1012–1025
[35] Xu J, Ding L, Cai L, Zhang L, Luo H and Qin W 2019 *Automation in Construction* **104** 95–106
[36] Thijssen Q, Toombs J, Li C C, Taylor H and Van Vlierberghe S 2023 *Progress in Polymer Science* 101755
[37] Jignasu A, Marshall K, Ganapathysubramanian B, Balu A, Hegde C and Krishnamurthy A 2023 *arXiv preprint arXiv:2309.02465*
[38] Chandrasekhar A, Chan J, Ogoke F, Ajenifujah O and Farimani A B 2024 *arXiv preprint arXiv:2406.00031*
[39] Dörstelmann M, Knippers J, Koslowski V, Menges A, Prado M, Schieber G and Vasey L 2015 *Archit. Des.* **85** 60–65
[40] Solly J, Frueh N, Saffarian S, Prado M, Vasey L, Felbrich B, Reist D, Knippers J and Menges A 2018 ICD/ITKE research pavilion 2016/2017: integrative design of a composite lattice cantilever *Proc. of the IASS Symp. 2018: Creativity in Structural Design* (International Association for Shell and Spatial Structures (IASS)) pp 1–8
[41] Oxman N, Laucks J, Kayser M, Tsai E and Firstenberg M 2013 *Green design, materials and manufacturing processes* **479** 479–483

[42] Johns R L, Wermelinger M, Mascaro R, Jud D, Hurkxkens I, Vasey L, Chli M, Gramazio F, Kohler M and Hutter M 2023 *Sci. Robot.* **8** eabp9758
[43] Keating S J, Leland J C, Cai L and Oxman N 2017 *Science robotics* **2** eaam8986
[44] Takahashi K and Kobayashi M 2023 *ce/papers* **6** 1291–1294
[45] Wilkinson S, Musil J, Dierckx J, Gallou I and de Kestelier X 2016 Autonomous additive construction on mars *15th Biennial ASCE Conference on Engineering, Science, Construction, and Operations in Challenging Environments* (American Society of Civil Engineers Reston, VA) pp 343–353
[46] Zhou C, Chen R, Xu J, Ding L, Luo H, Fan J, Chen E J, Cai L and Tang B 2019 *Automation in Construction* **104** 66–79
[47] Isachenkov M, Chugunov S, Akhatov I and Shishkovsky I 2021 *Acta Astronautica* **180** 650–678
[48] Subramanya K and Kermanshachi S 2021 Exploring utilization of the 3d printed housing as post-disaster temporary shelter for displaced people *Construction Research Congress 2022* pp 594–605
[49] Kreiger M A 2021 *Women in 3D Printing: From Bones to Bridges and Everything in Between* 71–85
[50] Https://singularityhub.com/2022/12/01/nasa-gave-icon-57-million-to-build-a-3d-printer-for-structures-on-the-moon/ accessed: 2024-2-2

Emotional Alignment for Human-Robot Cooperation in Musical Tasks

Huijiang Wang and Fumiya Iida
Bio-Inspired Robotics Lab, Department of Engineering, University of Cambridge, Trumpington Street, Cambridge CB2 1PZ, United Kingdom
E-mail: hw567@cam.ac.uk, fi224@cam.ac.uk

Abstract.
Current artificial intelligence (AI) studies have witnessed the widespread use of large language models (LLMs), significantly impacting human-machine interaction (HMI) by endowing machines with emotional subtleties. Despite the efficacy of LLMs, it remains a challenge for the learning-based agents to develop reliable computational emotion models, particularly for real-world human-robot teaming, due to the lack of embodiment. This paper investigates emotional alignment within physical human-robot interaction (pHRI) scenarios. We propose a theoretical framework that enables the robot to cooperate with a human pianist in accomplishing piano-playing tasks. By considering the human users' emotional states, the robot regulates keystroke dynamics and fingering arrangements, crucial factors affecting the resultant musical quality. Utilizing multimodal sensors and sensor fusion, the robot can capture, interpret and infer the emotional states of the human user. Consequently, the robot's decision-making process for action selection mirrors human-like artistic patterns.

1. Introduction

In recent years, there has been significant development in artificial intelligence (AI), particularly in natural language processing (NLP). Large language models (LLMs) have become prominent in discussions on machine intelligence, with models like chatGPT [1] and LaMDA [2] leading the way. These LLMs have demonstrated the ability to capture emotional context from consecutive sentences [3], advancing affective computing at the text-to-speech level. Emotions play a crucial role in decision-making processes, especially in human-robot interaction (HRI) applications, where they enhance social competence [4, 5]. There is a growing need to enable intelligent systems to interact emotionally. However, aligning emotions between robots and humans remains a grand challenge. In this paper, we refer to emotional alignment as the problem of training autonomous robotic systems to make decisions aligned with human values or preferences.

Thanks to the LLMs' capability in speech emotion recognition, AI agents are equipped with certain degrees of context-aware emotional alignment. However, this is far from enough in physical human-robot interaction (pHRI) scenarios, where interactions extend beyond the textual dimension. One such scenario is musical performances, where performers exhibit expressive features such as facial expressions, voices, body language, and other physiological cues. Audiences rely on their senses, primarily visual and auditory feedback, to perceive the performance. Despite advancements in robotics and AI, replicating the same level of interaction in musical performances remains particularly challenging for robots.

We consider a typical pHRI scenario where human-robot collaborative piano playing is achieved in a duet setting. The human engages with the upper voices on the piano, while the robot captures, interprets and comprehends the performance generated by the human. Then the robot infers the expressiveness of its human counterpart and takes actions accordingly. The feasibility of the physical implementation lies in advanced manufacturing techniques such as 3D printing [6, 7] and multilateral composition [8], enabling the creation of a dexterous robot end-effector for piano keystrokes. Various control methods like PID [9], MPC [10], fuzzy logic [11], and machine learning model-based control [12] have been used to control the system. However, emotional alignment with the human partner is crucial for the robot's sequential decision-making, especially in the context of the artistic and creative task (collaborative music performance). Challenges emerge in determining accompanying chords for human improvisation, requiring the intelligent system to gather feedback from multiple sensors, interpret emotions through sensor fusion-based affective computing, and consider the physical constraints of the robotic platform.

This paper introduces a conceptual framework for achieving emotional alignment in human-robot cooperative musical playing. The framework consists of three main components: Firstly, a comprehensive set of multimodal sensors is essential for capturing human emotional cues. Secondly, sensor fusion technology is required to integrate and process the captured signals effectively. Thirdly, temporal synchronization should be considered for seamless emotional alignment within the interaction. The proposed framework envisions the robot collaborating with a human pianist to perform piano-playing tasks. By considering the emotional states of human users, the robot adjusts keystroke dynamics and fingering arrangements, both crucial factors influencing the resulting musical quality.

2. Multi-modality Sensors

Figure 1 provides an overview of the framework, illustrating both the human and robot/agent engaged in piano playing. To comprehensively capture human emotional cues, a variety of hardware devices and apparatus must be applied to the human body, considering a range of sensory dimensions. Table 1 outlines the multiple sensors and corresponding physiological cues sensed. In this cooperative piano-playing scenario, one primary channel is human facial expressions, directly reflecting the pianist's emotional variations. Practical approaches such as facial electromyography (fEMG) [13] and camera-based methods employing facial emotion recognition (FER) algorithms are initially considered. However, camera occlusion may hinder effectiveness. In such cases, emotional body gesture recognition (EBGR) is performed to detect the upper-limb pose

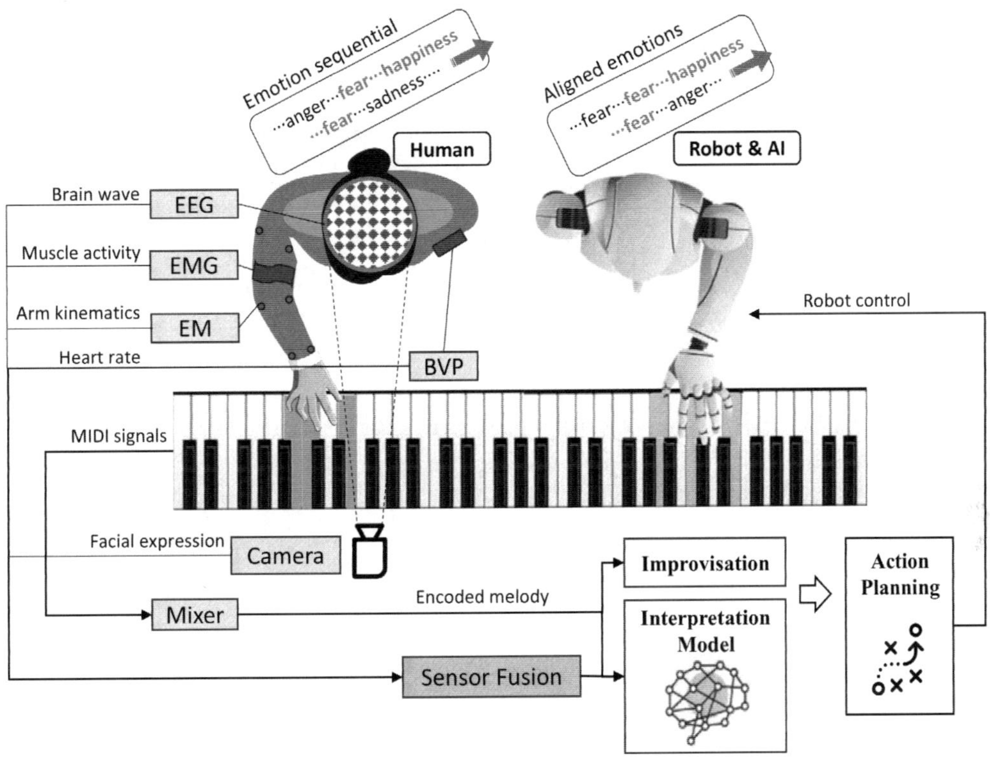

Figure 1. Overview of the conceptual framework. The emotional alignment process for human-robot cooperation involves the utilization of multiple sensors to capture various human cues, including facial expressions, body kinematics and other physiological signals. Neural networks are employed for regression and classification tasks on these sensing modalities to interpret emotions. Subsequently, the robot is planned and controlled to execute chord playing, providing harmonic support to the melody played by the human. Throughout the interaction, the robot continuously strives to achieve emotional alignment with the human partner by leveraging sensor fusion techniques and advanced control algorithms.

and kinematics [14]. Optical tracking or electromagnetic (EM) based methods [15] can be employed to track the trajectory of the pianist's arm motion and joint trajectory. In a piano-playing scenario, limb kinematics are closely linked to players' emotional variations and piano performance intonation and fluctuation. Additionally, brain activity with in piano performance can be monitored using electroencephalography (EEG) [16], although this method, while straightforward, may not always be accurate and should be complemented by other sensing modalities. Finally, other physiological cues such as blood volume pulse (BVP) [17] and galvanic skin response (GSR) [18] can be tracked to extract information such as heart rate and skin conductivity, which is also related to

Table 1. Multimodality sensors are used in human emotion recognition.

Dimension		Sensors	Modality
MIDI		Instrumental MIDI mixer	Keystroke dynamics, articulations
Facial emotion recognition (FER) [13]		Camera; Facial Electromyography (fEMG)	Facial expressions
Emotional body gesture recognition (EBGR) [14]		Electromagnetic markers, optical tracking;	Upper-limb kinematics and pose
Brain activity [16]		Electroencephalography (EEG)	Frequency analyses ; Statistical values
Peripheral physiological signals	Muscle activity	EMG	electromyograph; muscle expansion/contraction
	Blood volume pulse (BVP) [17]	photoplethysmography	Heart rate
	Galvanic skin response (GSR) [18]	biofeedback smart band	Skin conductivity; electrodermal activity (EDA)
	Other: temperature, respiration...	Thermometer; Resistive sensors	Abdominal expansion/contraction

the player's emotional arousal. While electromyography (EMG) devices can be utilized to monitor muscle activity, such as the extension and tension status of the upper limbs during piano playing, it requires a learning-based method for the classification of detected patterns.

3. Emotional Alignment in Musical Tasks

The performance of a human pianist can be parameterized through the MIDI protocol. This protocol encompasses essential music-related details such as the volume, intensity, and decay of each keystroke, as well as aspects like rhythm, tempo and articulation within the musical progression. The sensed MIDI serves as the foundational data utilized for AI-driven improvisation, wherein an agent processes the melody input from the human and determines suitable chord accompaniments to harmonize musically with its human teammate. However, before the next stage of robot execution (planning and control), it is imperative to consider the emotional cues detected by the human. This necessitates sensor fusion. In this paper, sensor fusion is the process of bringing together inputs from multiple devices including EEG, EMG and cameras to form a single model of the human's emotional state. Employing multimodal sensors coupled with corresponding processing algorithms can reduce the uncertainty associated with emotional recognition. Depending on the musical genre, different sensors are weighted differently within the fusion process. Subsequently, the robot's selection of potential actions takes into account both the improvisational context and interpreted emotions - an expressive musical robot.

3.1. Dataset and Encoding

It is important to utilize the collected data and establish a protocol for interpreting the dataset from the emotional recognition system. As previously mentioned, various physical devices such as cameras (for visual data), microphones (for auditory data), and wearables (for physiological data including heart rate and skin conductance) can be employed to capture emotional cues. A large supervised learning framework can be implemented by mapping these signals to labeled human emotions. Accurate labeling is essential for training an effective emotional recognition system. However, due to the abstract nature of emotions, encoding them into a format understandable by robots is challenging.

One potential protocol could involve a coarse-to-fine strategy. Existing emotion recognition models [19, 20] can be used to pre-label the data, which can then be refined manually. Alternatively, a comprehensive but straightforward approach involves having psychologists and trained annotators manually label emotions based on the captured physical data. Depending on the emotional model, encoding can be done using categorical or dimensional models. Advanced machine learning methods can effectively process these signals. For instance, convolutional neural networks (CNNs) are well-suited for facial expressions and body posture, while recurrent neural networks (RNNs) excel at handling sequential data like speech and physiological signals. To account for subjective variations, a weighting mechanism can be applied, adjusting the importance of each modality based on context and reliability, ensuring that user variability has minimal impact on the experimental validation accuracy.

3.2. Motion Synchronization

Another problem is the temporal synchronization between the human and the robotic partner during their collaborative performance. As the robot is providing accompaniment to the human, there is always a theoretical delay between the emotional sequences of the two partners. To address this issue, we introduce a sliding-window technique, as illustrated in Figure 2. This technique presents a fixed lag of one bar between the human's melody input and the robot's chord accompaniment. Despite this temporal delay, the music remains harmonious within a small duration, allowing the chord accompaniment to slightly adjust to the target melody flow during the music progression [21].

4. Challenges and Future Directions

One primary challenge lies in the emotional model. The widely accepted models can be discrete [19] or dimensional [20]. When considering multiple sensors, it becomes necessary to interpret some captured data as discrete signals while others are viewed as continuous. This introduces complexity in designing a hybrid emotional model suitable for such a comprehensive system. Furthermore, the integration of sensor fusion and the overall system imposes a significant computational load, leading to a considerable amount of computation.

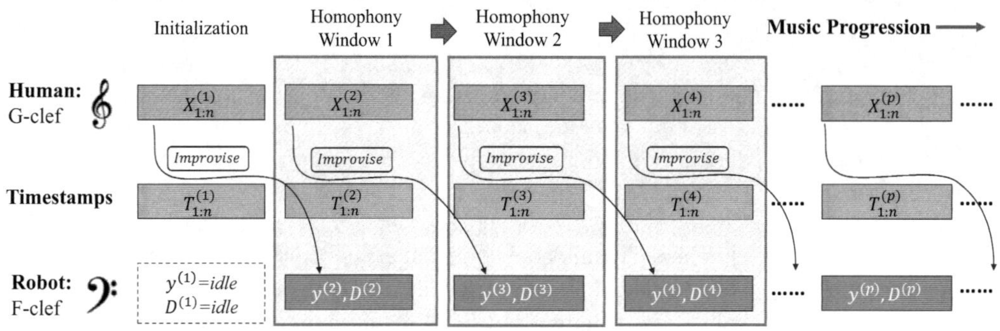

Figure 2. The sliding-window technique for the alignment between the human performance and the improvised performance.

The second significant challenge is the limitations of physical implementation. Despite the existence of advanced robots and capable end-effectors, the current state-of-the-art often falls short of adequately expressing desired emotions. For instance, in the case of anthropomorphic robotic hands, the fingers may lack the subtlety and agility required to play the piano expressively, thereby hindering their ability to differentiate between nuanced emotions such as anger and fear, which may both result from a heavy keystroke. In other scenarios, the end-effector may be incapable in executing actions in a timely or precise manner, despite receiving such instructions from the AI agent.

The third challenge is the hallucination of the agent, similar to challenges faced by large language models (LLMs). Despite the assistance of AI, the system may not always accurately discern aligned emotions. Consequently, the outcome of emotional alignment could damage the artistry of the musical piece and lead to the creation of nonsensical forms.

5. Conclusion

This article introduces perspectives on a framework facilitating synchronized emotional interaction between humans and robots. Our proposed conceptual framework aims to enable human-robot cooperation with aligned emotions, emphasizing the significance of sensor fusion techniques for comprehensive capture of human emotions. The interpretation of sensed data is processed using artificial neural networks, similar to approaches used in large language models. Multimodal sensors are encoded and tokenized in a suitable format, then input into the networks to train a proper agent based on state-of-the-art machine learning advancements. Considering improvisation and learning-based emotional interpretation, the robots need delicate planning and control of the end-effectors. The planning algorithm and controller are designed to enable the physical robot to act on the piano with a comparable level of expressiveness, achieving aligned emotional cues despite consistency challenges.

IOP Conf. Series: Materials Science and Engineering **1321** (2024) 012006 doi:10.1088/1757-899X/1321/1/012006

6. Acknowledgment
This work has received funding from the European Union's Horizon 2020 research and innovation programme under the Marie Sklodowska-Curie grant agreement No. 860108 (SMART).

7. Reference
[1] OpenAI. Chatgpt: A large-scale generative model for open-domain chat. https://github.com/openai/gpt-3, 2021.

[2] Romal Thoppilan, Daniel De Freitas, Jamie Hall, Noam Shazeer, Apoorv Kulshreshtha, Heng-Tze Cheng, Alicia Jin, Taylor Bos, Leslie Baker, Yu Du, et al. Lamda: Language models for dialog applications. *arXiv preprint arXiv:2201.08239*, 2022.

[3] Hyun-Wook Yoon, Ohsung Kwon, Hoyeon Lee, Ryuichi Yamamoto, Eunwoo Song, Jae-Min Kim, and Min-Jae Hwang. Language model-based emotion prediction methods for emotional speech synthesis systems. *arXiv preprint arXiv:2206.15067*, 2022.

[4] Thomas M Moerland, Joost Broekens, and Catholijn M Jonker. Emotion in reinforcement learning agents and robots: a survey. *Machine Learning*, 107:443–480, 2018.

[5] Oliver Damm, Karoline Malchus, Frank Hegel, Petra Jaecks, Prisca Stenneken, Britta Wrede, and Martina Hielscher-Fastabend. A computational model of emotional alignment. In *5th Workshop on Emotion and Computing*, 2011.

[6] JAE Hughes, P Maiolino, and Fumiya Iida. An anthropomorphic soft skeleton hand exploiting conditional models for piano playing. *Science Robotics*, 3(25):eaau3098, 2018.

[7] Huijiang Wang, Toby Howison, Josie Hughes, Arsen Abdulali, and Fumiya Iida. Data-driven simulation framework for expressive piano playing by anthropomorphic hand with variable passive properties. In *2022 IEEE 5th international conference on soft robotics (RoboSoft)*, pages 300–305. IEEE, 2022.

[8] Huijiang Wang, Seppe Terryn, Zhanwei Wang, Guy Van Assche, Fumiya Iida, and Bram Vanderborght. Self-regulated self-healing robotic gripper for resilient and adaptive grasping. *Advanced Intelligent Systems*, 5(12):2300223, 2023.

[9] Guy Hoffman and Gil Weinberg. Gesture-based human-robot jazz improvisation. In *2010 IEEE international conference on robotics and automation*, pages 582–587. IEEE, 2010.

[10] Huijiang Wang, Xiaoping Zhang, and Fumiya Iida. Human-robot cooperative piano playing with learning-based real-time music accompaniment. *IEEE Transactions on Robotics*, Under Review(XXX):XXX, 2023.

[11] Dajun Zhou, Minghui Shi, Fei Chao, Chih-Min Lin, Longzhi Yang, Changjing Shang, and Changle Zhou. Use of human gestures for controlling a mobile robot via adaptive cmac network and fuzzy logic controller. *Neurocomputing*, 282:218–231, 2018.

[12] Kevin Zakka, Philipp Wu, Laura Smith, Nimrod Gileadi, Taylor Howell, Xue Bin Peng, Sumeet Singh, Yuval Tassa, Pete Florence, Andy Zeng, et al. Robopianist: Dexterous piano playing with deep reinforcement learning. In *Conference on Robot Learning*, pages 2975–2994. PMLR, 2023.

[13] Myunghoon Suk and Balakrishnan Prabhakaran. Real-time mobile facial expression recognition system-a case study. In *Proceedings of the IEEE conference on computer vision and pattern recognition workshops*, pages 132–137, 2014.

[14] Fatemeh Noroozi, Ciprian Adrian Corneanu, Dorota Kamińska, Tomasz Sapiński, Sergio Escalera, and Gholamreza Anbarjafari. Survey on emotional body gesture recognition. *IEEE transactions on affective computing*, 12(2):505–523, 2018.

[15] Huijiang Wang, Tetsushi Nonaka, Arsen Abdulali, and Fumiya Iida. Coordinating upper limbs for octave playing on the piano via neuro-musculoskeletal modeling. *Bioinspiration & Biomimetics*, 18(6):066009, 2023.

[16] Prashant Lahane, Jay Jagtap, Aditya Inamdar, Nihal Karne, and Ritwik Dev. A review of recent trends in eeg based brain-computer interface. In *2019 International Conference on Computational Intelligence in Data Science (ICCIDS)*, pages 1–6. IEEE, 2019.

[17] Azadeh Kushki, Jillian Fairley, Satyam Merja, Gillian King, and Tom Chau. Comparison of

blood volume pulse and skin conductance responses to mental and affective stimuli at different anatomical sites. *Physiological measurement*, 32(10):1529, 2011.

[18] Deger Ayata, Yusuf Yaslan, and Mustafa Kamaşak. Emotion recognition via galvanic skin response: Comparison of machine learning algorithms and feature extraction methods. *IU-Journal of Electrical & Electronics Engineering*, 17(1):3147–3156, 2017.

[19] Paul Ekman. An argument for basic emotions. *Cognition & emotion*, 6(3-4):169–200, 1992.

[20] James A Russell. A circumplex model of affect. *Journal of personality and social psychology*, 39(6):1161, 1980.

[21] Ian Simon, Dan Morris, and Sumit Basu. Mysong: automatic accompaniment generation for vocal melodies. In *Proceedings of the SIGCHI conference on human factors in computing systems*, pages 725–734, 2008.

Sustaining Evolution for Shallow Embodied Intelligence

W. B. Langdon and Daniel Hulme

Department of Computer Science, University College London, Gower Street, WC1E 6BT, UK

E-mail: `w.langdon@cs.ucl.ac.uk`, `daniel@satalia.com`

Abstract. Lenski's experiments with E. Coli show Biology can sustain continual evolutionary improvement. However long term evolutionary experiments (LTEE) with evolutionary computing find that information theory's failed disruption propagation (FDP) in deeply nested genetic programming (GP) hierarchies can greatly slow adaptation. We propose that researchers aiming at embodied artificial intelligence should control software robustness by using porous high surface area geometrical architectures, perhaps composed of many shallow mangrove like tree structures intimately linked to their data rich fitness environment.

1. Introduction

Rich Lenski has shown in biology evolution can retain its ability for continued change even after 75 000 generations [1, 2]. (Homo Sapiens appears to be about 9300 generations old [3].) He started the long term evolutionary experiment (LTEE) in 1998 and it is still running. We have shown genetic programming (GP) [4, 5] can do similarly when run for a million generations and experiments can be run in days or weeks rather than years [6, 7]. However, information theory (Section 3) explains why the rate of fitness improvement falls as very deep GP trees become resistant to crossover. Mutation testing shows that human written software can also be resilient to many source code changes and that there is a tendency for deeply nested code to be more robust [8].

There are examples of software systems that exceed a billion lines of source code. Information theory's failed disruption propagation (FDP) [9] helps to explain why maintaining, testing and debugging deeply nested codebases is hard and why software companies prefer unit testing of modules (each of which is typically only shallowly nested) rather than system testing of complete functional hierarchies. There is already SBSE [10] work on automatically optimising test oracles [11]. FDP suggests software systems aught to be built with many densely packed test agents so that disruption caused by bugs has little distance to travel before being discovered by an oracle.

For evolutionary computing and artificial life experiments aiming for sustained innovation, we propose the use of porous high surface area architectures composed of many small trees which are intimate with their environment (see Figure 10, page 11). For continuous innovative evolution the fitness function needs to be able to measure on average if genetic changes are good or not. That is, they must have made a difference. This means we must overcome robustness, without introducing chaos. We suggest this might be met by systems where the bulk of the code remains close to the fitness environment and the disruption caused by most mutations and crossovers has only a short distance to propagate in order to have a measurable fitness impact.

The interested reader can find Koza's early experiments and introduction to genetic programming in his book [4]. Whilst the Poli's "field guide" [5] is specifically aimed at students and includes some more recent work.

In evolutionary computing "complexity analysis" has several meanings. For example it can mean traditional computer science (Big O) complexity analysis, which aims at deriving formal mathematics saying important things about solving particular problems such as how long it will take on average and how runtime scales with the dimensions of the problem [12, 13]. Complexity can also be in terms of information theory. With complex solutions containing more information. Although strictly not computable, Kolmogorov defines the complexity of a string as the size of shortest program able to reproduce the string. Vitanyi has shown in practise it can be estimated using every day compression tools [14, 15, 16, 17]. Finally the size of solutions is often taken as related to their complexity. However, without counter measures there is a tendency for solution size to increase [18, 19]. Sometimes it is assumed that smaller solutions will be more human comprehensible and so preferred and nowadays there are a variety of sophisticated anti-bloat tools [20, 21]. Although size can be measured and Kolmogorov complexity can be estimated for GP solutions [22], the information needed to reproduce the evolved program can fail to usefully capture our notion of the complexity of the solution. For example, programs in a GP population can all be different (i.e. their genotypes are not the same) and yet their behaviour (phenotypes) are the same [7].

2. Sustaining "Thin" Evolutionary Machine Learning

One lesson we can take from the recent successes of artificial intelligence (AI) is that huge AI systems can be created automatically when they are based on learning. That is, when we automatically extract knowledge from training data, rather than manually encoding knowledge into a computer based repository. We suggest in the hunt for the next level artificial intelligence or machine consciousness [23], we should consider systems which, instead of assembling all the training data and learning everything the system can do in one huge bout of training, we should investigate systems which continually learn.

The next section gives a quick introduction to information theory and applying it to software, showing that entropy loss leads to failed disruption propagation (FDP) and so to robust software that is error tolerant. Section 4 shows an example of recent measurements of FDP in hand written C, where deeper code tends to be more robust. This is followed in Section 5 by studies of FDP in artificially evolved (genetic programming) code which shows, as expected, deep nested hierarchies are also robust. The discussion (Sections 6 and 7) suggests FDP is not only widespread in existing systems but that future AI systems must take it into account and suggests the adoption of "thin" shallow data rich architectures. The second part of the discussion (Section 7) considers the creativity of evolving systems and the widespread adoption of genetic programming, before we conclude in Section 8 that to demonstrate automatic Embodied Intelligence we need systems that continue to evolve and so recommend architectures that are not too robust but instead permit continuous change. Therefore we should investigate architectures composed of shallow code (i.e. the opposite of deeply nested code) which have a large exposed surface facing a data rich environment.

3. Information Theory and Software

All computer operations are irreversible. Meaning in general given an operation's output we cannot infer its inputs. Figure 1 show this graphically. The left hand side of Figure 1 depicts adding together two 32 bit floating point values to give a single 32 bit output. Because addition is not reversible, given the output, we cannot infer the inputs. For example, $1.0 + 1.0 = 2.0$ and $1.5 + 0.5 = 2.0$, so given the output is 2.0 we cannot say what the two inputs were. That is, we cannot reverse the computation and so information about the inputs has been lost.

We can say how much information there is by calculating the entropy of a distribution of values. For example, the entropy of the input distribution and the entropy of the output distribution. The difference between the two says how much information has been destroyed by the operation. In fact, this can be done not just for a single operation but also for whole programs. The right hand side of Figure 1 contains two plots. The solid line (red) shows the distribution of integer values, which are added together to give the output (dashed blue). In the example, both inputs are digits, i.e., the numbers 0 to 9. Here digits 0 and 1 are more common than the other eight. (The example comes from C source code used in Section 4.) The graph assumes each input is independent of the other and so the output is one of the nineteen integers 0 to 18. The dashed blue plot shows their distribution. The information content (entropy) of either input is 2.88 bits. As they are independent, the information of them together is $2 \times 2.88 = 5.76$ bits. Note the information content (entropy) of the output is 3.75 bits. So this (irreversible) addition has lost $5.76 - 3.75 = 2.01$ bits of information. Entropy loss is inherent in irreversible operations. I.e. all computation loses entropy.

In [9] we consider the informational impact of all forms of disruptions to the smooth execution of software. We treat together all types of disruption, be they coding errors (software bugs), source code changes introduced by mutation testing [24], subtree swaps made by genetic programming crossover, transients due to cosmic rays, EM radiation inserted by a hostile opponent, etc., as long as they change the state of the program during execution. For such a disruption to be effective, that state change has to propagate from its origin to the program's output. If it does not, the program is robust to the error. From an information theory point of view, information about the disruption has to propagate to the output. In nested hierarchies, information loss is progressive. So that even partial information lost cannot be recovered. Hence when information has to pass through many levels of nesting, all information about the disruption may be lost, giving rise to failed disruption propagation (FDP). But as we have seen, because digital computation is not-reversible, entropy (i.e. information) loss is inherent in calculation. So we expect FDP in any deeply nested calculation. In fact, FDP is common even in languages with side-effect (e.g. global variables in C/C++) that might allow information to by-pass function call nesting. Thus in non-trivial software small transients tend not to have any knock on consequences (Figure 2). Leading ordinary software to be robust to many errors [25].

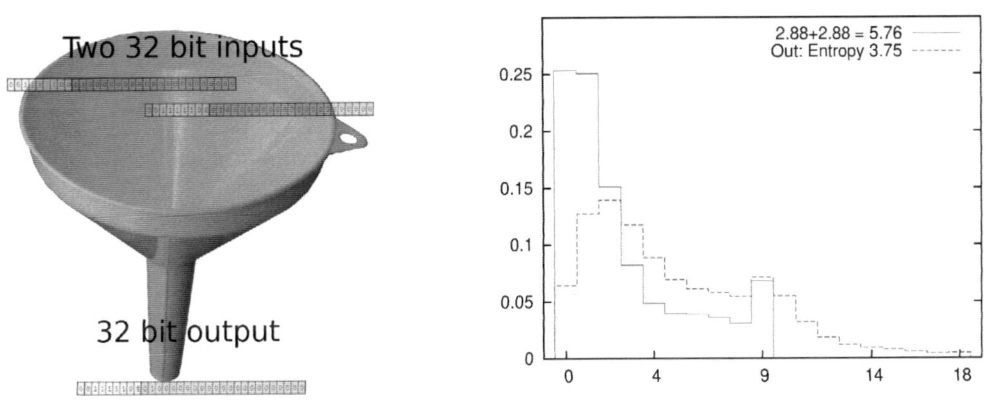

Figure 1. Left: Information Funnel. Computer operators are irreversible. Meaning input state cannot be inferred from outputs. Information is lost. Right: red 0–9 actual distribution of 0–9 digits in C code (Section 4). Dashed blue 0–18 distribution when they are added. Although the output of + is wider and has higher entropy (information content, 3.75 bits), it is smoother and has less entropy than the combined entropy (5.76 bits) of the two inputs to + [8].

IOP Conf. Series: Materials Science and Engineering **1321** (2024) 012007 doi:10.1088/1757-899X/1321/1/012007

4. Failed Disruption Propagation (FDP) in Nested Software

In order to judge the variation in Failed Disruption Propagation (FDP) with depth, we took some highly nested benchmark C code (i.e. VIPS) and inserted mutations into it at random [8]. We measured how deep the mutation was, and followed Voas' PIE framework [27] to measure which mutations were indeed **E**xecuted, which did change (**I**nfect) state and then did that state change **P**ropagate from the mutation's location to impact the outputs. Figure 4 shows in this example, random C code changes deeper than 30 levels of function nesting tend to show FDP, i.e. deeper human written code can be more robust.

VIPS is well established C code which in addition to function's arguments and return values makes heavy used of pointers and shared data. This means information flows both through the nested hierarchy of function calls (like in genetic programming, next section) but also short circuits the hierarchy by passing information via global data. Nevertheless we do see a weak relationship between run time nesting depth and degree of failed disruption propagation. If we exclude mutations which stopped the program immediately, e.g. with a segmentation error, Figure 4 shows there is indeed some variation between mutations with FDP and those without and their execution depth.

The PARSEC suite of benchmarks has been used to test parallel super computers for NASA. It mostly consists of numeric algorithms but also includes some image processing tools, including VIPS [28, 29]. VIPS is an image processing library of 90 000 lines of C source code. We chose the vipsthumbnail application as it was known to contain some highly nested code. vipsthumbnail runs in parallel threads (hence it is non-deterministic). It takes a large image and create a small image (a thumbnail). By default the output is 128 pixels wide. We used a combination of Linux perf and GDB [8] to profile vipsthumbnail and select just the parts of the VIPS library which are heavily used in thumbnail generation (Figure 3). Almost all the unused C code was removed so it could not be mutated This reduced the total library from 90 000 to 7 328 lines of code spread over 37 files. We used the genetic improvement tool Magpie[1] [30] to select uniformly at random (independent of their depth) 1000 locations in the active C code to mutate and record their runtime execution depth and their impact, if any, on the program's output. The results are summarised in Table 1.

Table 1 shows the code is robust to most mutations. Only 16.4% of mutations caused a runtime error. Most of these stopped the program immediately, and there were only 37 where

[1] `https://github.com/bloa/magpie`

Figure 2. Lorenz Butterfly considered harmful. Lorenz [26] viewed the world's weather as being chaotic and posed the question "Does the Flap of a Butterfly's Wings in Brazil Set Off a Tornado in Texas?" but entropy loss makes software robust not chaotic.

Figure 3. Left: 3264×2448 input image. Right: 128×96 thumbnail image generated by VIPS

Table 1. 1000 random Magpie VIPS mutants

Compiled, ran correct output	526	Correct output	438
		Mutation is identical to original code	88
Failed to compile	302		
Failed to run correctly or gave incorrect output	164	exception	127
		output error	37
Magpie TypeError	8		

the program terminated as usual but it generated output that was not identical to the unmutated code. We know these 37 satisfy the PIE frame work. We selected at random 25 of them and looked for 25 mutants amongst the 438 which changed the code and gave the right output. In both cases we select mutants which are executed and which do change the program's internal state and we measure their execution depth. In most cases the mutants are executed more than once, typically thousands of times, and at more than one depth. The 25 mutants which encountered FDP and so gave the right answer (red) and the 25 (of 37) where the mutants disruption reached the output (blue dashed) are plotted in Figure 4. Notice deep mutations ($y > 30$) tend not to impact VIPS thumbnail output.

5. Convergence/Stasis in Genetic Programming (GP)

In a series of experiments we have run genetic programming (GP) on benchmark problems which show GP populations continue to evolve and find better solutions even when run for up to one million generations and evolving trees with up to two billion nodes. Due to a series of innovations [31, 6, 32, 33, 34, 35, 19, 36, 37] in Andy Singleton's GPquick [38] these experiments can be run in days or weeks (rather than years or decades needed by Lenski).

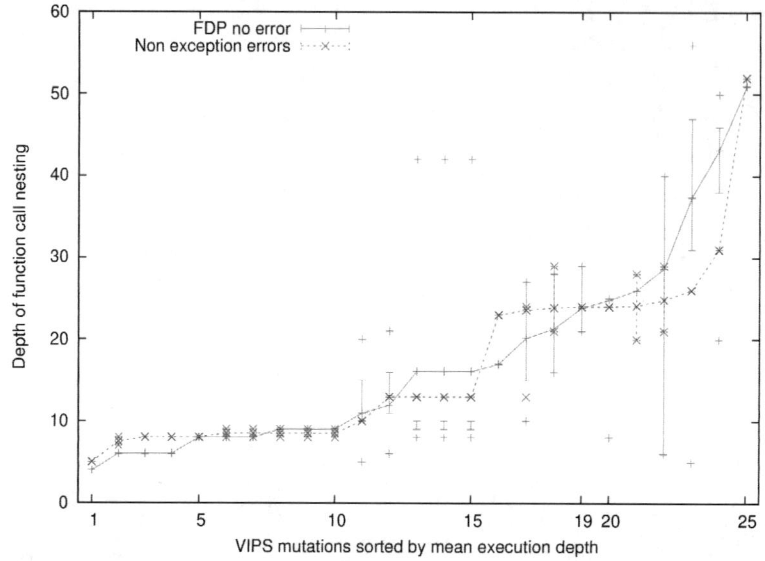

Figure 4. 25 VIPS mutations with no impact (mean depth +) and 25 which change output (mean depth ×). Error bars show interquartile range. +× also show min and max depth. Notice mutations with greater average depth tend not to impact VIPS thumbnail output.

These experiments show that GP runs may find thousands of innovations during an extended run and although fitness improvement continues, it slows (Figure 5). These runs were allowed to produce enormous trees and the rate of improvement is inversely proportional to the size of the trees because failed disruption propagation in these strictly hierarchical systems means only genetic events near the tree's output make any difference (worse or better).

In tree GP programs are usually drawn as inverted trees (e.g. Figures 6 and 7), with the output being taken from the root node, at the top of the diagram and the program's inputs are fed into it via the tree's leafs. The leafs are shown towards the bottom of the diagram.

Figure 6 shows failed disruption propagation where the disruption is introduced at runtime in an integer problem. The shaded nodes indicate the parts of the tree where inserting runtime disruptions has some impact on the program's output. FDP means in the bulk of the tree the disruption has no impact (not shaded). As trees get bigger the sensitive area near the output shrinks as a fraction of the whole tree.

Figure 7 shows an evolved floating point tree where the disruption is to change the program and then re-evaluate it on all the test cases. The coloured oval nodes show the difference between the evaluation in the original tree and the modified tree. The large black oval immediately above the code disruption shows initially the internal evaluation is disrupted on all test cases. As with Figure 1, the system is hierarchical and once information about the disruption is lost it is not recovered higher up the tree. Thus if on a given test case the evaluation between the original (mother) program and her offspring becomes the same, the evaluation above that point (i.e. closer to the output) will remain the same. In particular if FDP occurs for a particular test case then the evaluation of the two programs at the root node will be the same on that test case. So they have the same output. If FDP occurs somewhere on all test cases, then it will succeed in hiding the disruption and both programs will have the same output. Typically this means they have the same fitness scores.

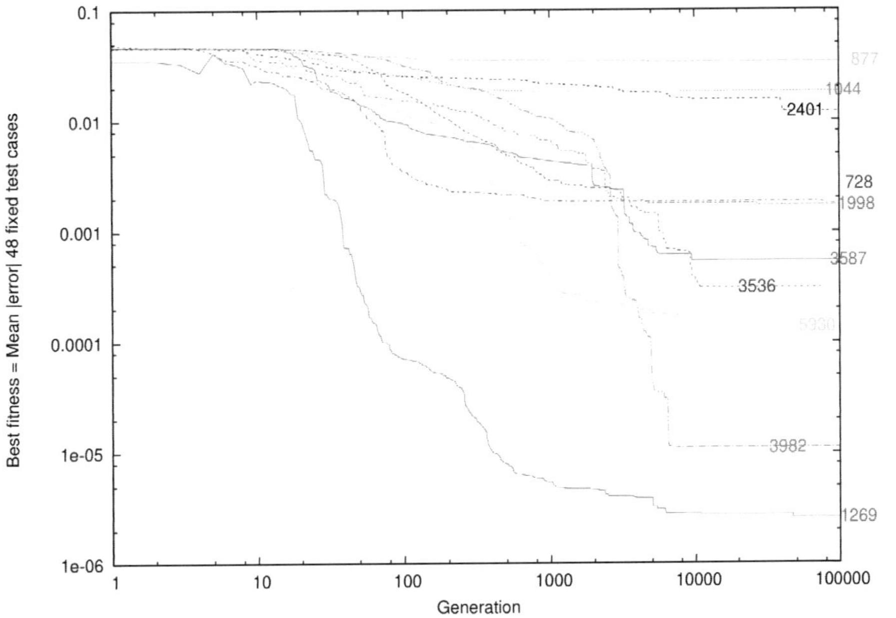

Figure 5. Evolution of mean absolute error in ten GP runs of Sextic polynomial [4] with population of 500. Runs to 100 000 generations (2 stopped early). Labels give number of generations when fitness got better [6].

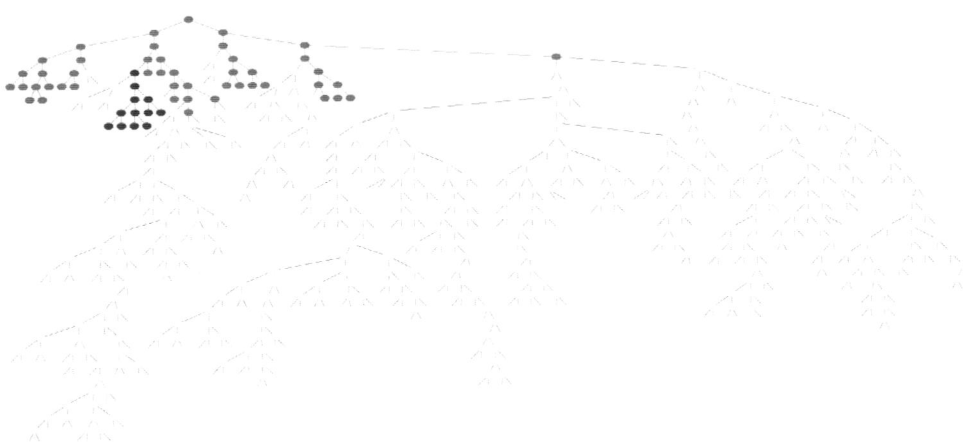

Figure 6. Example evolved solution (size 771) to Fibonacci problem [39]. Coloured shading shows external impact of disrupting internal runtime evaluation by increasing it by 1 (smallest possible change). Disruption at red nodes cause 16-20 test cases to fail. Blue 1 test case fails. As predicted by information theory, failed disruption propagation (FDP) means inserting run time errors far from the output (top) has little effect. Thus almost the whole program is not affected on any test case (white).

Figure 7. Trace of impact of crossover (red) in evolved tree. Replaced and new code are both evaluated on the test set (48 tests) [33, 40]. The size and number in each node gives the number of test cases where the evaluation of the parent and child are not identical. Their average evaluation difference is indicated on a log scale by the node's colour. Average differences greater than 0.01 are shown with dark colours, less than 0.01 by brighter colours. Brightest yellow shows smallest non-zero difference (RMS 3.1 10^{-10}). If, as here, parent and child evaluations are identical before reaching the root node, the remainder of the evaluation is not needed (gray nodes) [35].

In Figure 7 the number of test cases where evaluation is different in the original and child program is given as a number in the coloured nodes. As expected, this falls monotonically as we move away from the disrupted code towards the root node. Although information theory does not guarantee this (and counter examples do occur), in our floating point trees the size of the difference in evaluation tends to fall. Figure 7 shows this using the colour of the shading, with dark (black or blue) colours indicating large differences and fainter (yellow) showing evaluation in the two programs are almost identical. If, as in Figure 7, at any point evaluation of the parent and child programs become identical, evaluation at their root nodes must be identical, and so their outputs must be identical. Hence evaluation can stop early and we can simply copy the results from the first program to be the outputs of the new program.

Typically there is considerable variation between different trees in different GP runs, nevertheless by taking averages across a large number of disruption locations Figure 8 is able to plot a regression line showing the rapid exponential impact of FDP even when internal evaluation is totally randomised. With floating point, information loss is slower but still rapid and so we again see exponential fall off so that deep disruptions are unlikely to have any impact at all.

6. Implications for Future AI

In nested systems information once lost cannot be recovered. In particular as learning adapts existing programs, information about the adaptation once lost during evaluation of the new program variant can have no impact of the program's outputs. That is, there is no learning signal; externally we cannot tell if the adaptation was good or not. Without feedback, learning becomes random and so continual improvement requires good information flow from the adaptation sites to the program's outputs.

IOP Conf. Series: Materials Science and Engineering **1321** (2024) 012007 doi:10.1088/1757-899X/1321/1/012007

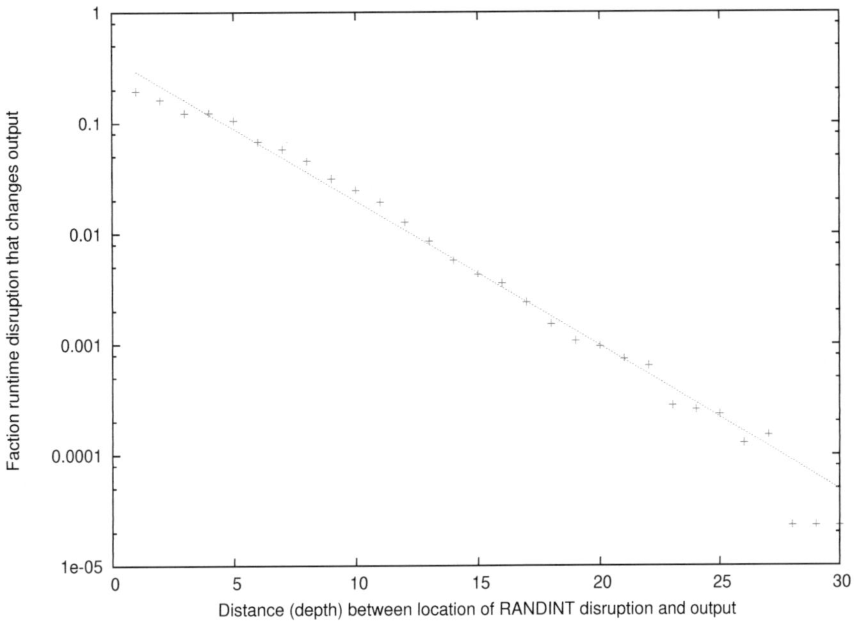

Figure 8. Complete information loss (FDP) becomes almost certain as depth increases when disrupting a large solution to Fibonacci problem with a random value [39]. Dotted line $\approx 0.4\, e^{-0.3}$ shows exponential fall in fraction of run time disruptions reaching output. Meaning the chance of disruption falls by $\frac{1}{2}$ as depth increases by 2.3 levels.

In simple functional systems (previous section) we saw rapid and complete information loss. Similar to AC electrical power flow there is a skin effect: we can make the electrical conductor as thick as we like but once it is wider than the skin depth, increasing the dimensions of the conductor makes very little difference to its ability to carry power. So we can make code as deep as we like, but any change we make will not have impact unless it is near the program's outputs. Unless it has external impact, it will be hard to measure if the change was a learning step in the right direction or not.

In practice in digital computing the "skin depth" can be surprisingly small. In the previous section, for a simple 32 bit integer experiment it was 2.3 function nesting levels. Of course in systems we construct we can bias the location of changes to be near the surface, but then we risk our system growing in complexity and size and so containing large volumes of code which is out of the reach of future changes. Such systems might eventually carry the burden of huge fossilised legacy code. Perhaps it was useful in the past but cannot now be maintained.

We suggest instead of having populations of huge trees, which resemble redwood sequoia trees and are composed almost entirely of inert heartwood, perhaps we should investigate populations which resemble mangroves (left Figure 10). Such systems might be composed of many evolving trees, each being small and shallow but they would be embedded in a data rich sea. Each new tide brings new inputs and moves outputs from one part of the forest to other data processing trees. Figure 10 suggests other naturally occurring porous systems with a high surface area, which might inspire development of thin learning systems, where adaptable code is closely linked to a data rich environment.

Information theory is universal and so unless counter measures are taken we anticipate in future scaling AI systems will have to take into account how to ensure most changes have some measurable impact by ensuring there are short or at least low entropy loss routes from the site of the change to some measurable point. In evolutionary computing terms, these are the mutation or crossover points and the survival of the individual. If we are to evolve embodied intelligence [41], as with Biology, we may distinguish between mutations etc. to the design (analogous to DNA) and how they lead to changes of form (analogous to the organisms body or phenotype). Again if feedback on the new design is to be available the information chain from design changes to body/mind changes to measurable impact in the environment cannot lose all the entropy. In digital devices such as software, we suggest, even in large systems, the route from genotype (design) change to fitness measurement (phenotype survival) needs to be short.

Large complex systems are not inherently robust but if they are not they cannot be built or survive. If a large system relies on all of its components working but they are fragile. It will never be assembled. As it gets bigger the construction team will spend more and more of its time fixing the components that have just broken, rather than completing construction.

In terms of exactly how we will scale AI the hope remains it can be assembled (or self assemble) from small shallow structures. We can easily foresee a danger that such composites themselves lose entropy and so become robust and hence do not easily adapt. Perhaps the fitness environment (e.g. the sea or atmosphere in Figure 10) will not be sufficient and our evolving architecture will have to include low entropy loss high fidelity channels (analogous to nerve cells or optical fibres) embedded in the digital matrix. However, as always, there is a danger that mandating this means we remain wedded to human designed systems where evolution makes only short term adaptations rather than facilitating long term progress.

7. Artificial Creativity and Uptake of Evolutionary Artificial Intelligence

The creative power of AI in the form of evolutionary computing has been known for several decades [42]. For example, *Creative Evolutionary Systems* [43] contains chapters from computer generated music to novel fighter jet combat [44, p272–276], whilst Koza *et al.*'s 2003 book [45] stresses genetic programming as a routine invention machine. Indeed the annual EvoMusArt conference has been demonstrating computer creativity in the arts since its inception in 2012. Similarly game playing has long been an application of AI, e.g. Checkers (draughts) [46, 47]. The creation of learning based machines which surpass the performance of all humans, has not diminish human interest in games or sports. For example, people still run, even though they know they will never be the best human runner in the world nor outpace a motorcycle. Similarly people play chess and other games despite knowing that superhuman AI performance is readily available. As with art, there are long established conferences and journals dedicated to evolutionary and other learning based approaches to software based game playing, e.g. the IEEE Conference on Games, which started (as CIG) in 2005.

A question raised recently [48, 49] was how much is GP used? Of course most industrialists are not interested in papers. Indeed they may have sound commercial reasons for not publicising their results or even saying what they are working on. This means numbers based on published work will always be an underestimate [50]. Nevertheless data sampled from the genetic programming bibliography for last year (2023) suggests $38{\pm}5\%$ of published papers are primarily on applications which just happen to use GP (Figure 9). Many applications relate to health (e.g. [51, 52, 53, 54, 55, 56, 57, 58, 59]), civil engineering (e.g. [60, 61, 62, 63, 64, 65, 66, 67, 68]) or solid state materials, e.g. batteries [69, 70, 71, 72, 73].

IOP Conf. Series: Materials Science and Engineering **1321** (2024) 012007 doi:10.1088/1757-899X/1321/1/012007

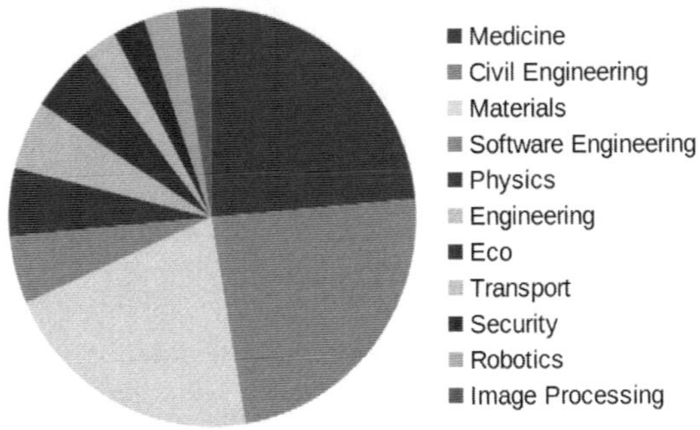

Figure 9. An estimated 38±5% of papers from 2023 in the GP bibliography are on applications

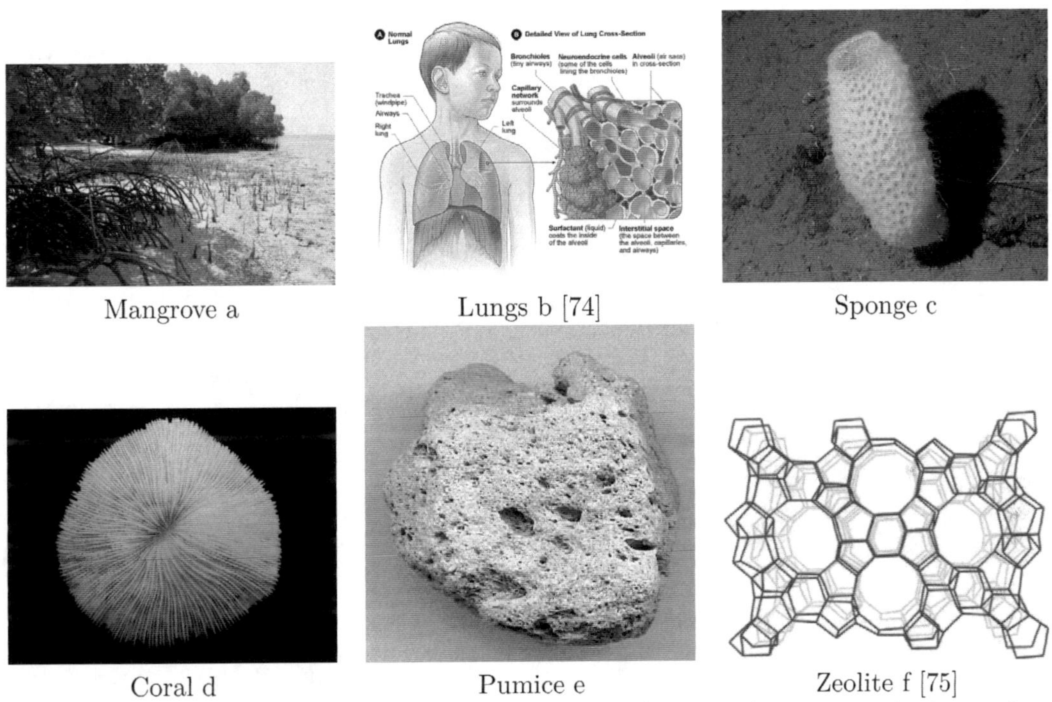

Figure 10. Possible natural sources of inspiration for software architectures suited to prolonged evolution. They have a large surface area so that their volume lies close to a surface. Such structures could provide a large space for evolving software whilst keeping most of the code near to its environment (Wikimedia a b c d e, [75]f)

8. Conclusions

Information theory predicts that code changes will suffer failed disruption propagation FDP, which makes both deeply nested automatically generated code evolved by genetic programming (GP) and human written software resistant to change. In GP with an integer representation the impact of even very large run time disruptions on average halved for each increase in depth of 2.3 levels.

Therefore we advocate research into porous high surface area shallow code "mangrove" architectures, where the adapting code is closely connected to a data rich environment. To ensure that updates continue to have impact, the system must have limited robustness and so limited code depth. Figure 10 shows a few natural structures with high surface areas, which might inspire continual adaptation EI architectures.

References

[1] Lenski R E *et al.* 2015 *Proceedings of the Royal Society B* **282** ISSN 0962-8452 URL http://dx.doi.org/10.1098/rspb.2015.2292
[2] Good B H *et al.* 2017 *Nature* **551** 45–50 URL http://dx.doi.org/10.1038/nature24287
[3] Wang R J *et al.* 2023 *Science Advances* **9** eabm7047 URL http://dx.doi.org/10.1126/sciadv.abm7047
[4] Koza J R 1992 *Genetic Programming: On the Programming of Computers by Means of Natural Selection* (Cambridge, MA, USA: MIT Press) ISBN 0-262-11170-5 URL http://mitpress.mit.edu/books/genetic-programming
[5] Poli R *et al.* 2008 *A field guide to genetic programming* (Published via http://lulu.com and freely available at http://www.gp-field-guide.org.uk) (With contributions by J. R. Koza) URL http://www.gp-field-guide.org.uk
[6] Langdon W B and Banzhaf W 2022 *Artificial Life* **28** 173–204 ISSN 1064-5462 invited submission to Artificial Life Journal special issue of the ALIFE'19 conference URL http://dx.doi.org/10.1162/artl_a_00360
[7] Langdon W B 2022 *Genetic Programming and Evolvable Machines* **23** 71–104 ISSN 1389-2576 URL http://dx.doi.org/10.1007/s10710-021-09405-9
[8] Langdon W B and Clark D 2024 *13th International Workshop on Genetic Improvement @ICSE 2024* ed An G *et al.* (Lisbon: ACM) pp 1–8 best paper URL http://dx.doi.org/10.1145/3643692.3648259
[9] Petke J *et al.* 2021 *ESEC/FSE 2021, Ideas, Visions and Reflections* ed Avgeriou P and Zhang D (Athens, Greece: ACM) pp 1475–1478 URL http://dx.doi.org/10.1145/3468264.3473133
[10] Harman M and Jones B F 2001 *Information and Software Technology* **43** 833–839 ISSN 0950-5849 URL http://dx.doi.org/10.1016/S0950-5849(01)00189-6
[11] Jahangirova G *et al.* 2016 *Proceedings of the 25th International Symposium on Software Testing and Analysis (ISSTA'16)* (Saarbruecken, Germany: ACM) pp 247–258 URL http://dx.doi.org/10.1145/2931037.2931062
[12] Beyer H G 2001 *The Theory of Evolution Strategies* Natural Computing Series (Springer) ISBN 3-540-67297-4
[13] Reeves C R and Rowe J E 2003 *Genetic Algorithms–Principles and Perspectives: A Guide to GA Theory* (Kluwer Academic Publishers)
[14] Li M and Vitanyi P M B 1992 *Journal of Computer and System Sciences* **44** 343–384
[15] Cilibrasi R *et al.* 2004 *Computer Music Journal* **28** 49–67 URL http://homepages.cwi.nl/~paulv/papers/music.pdf
[16] Cilibrasi R and Vitanyi P M B 2005 Automatic meaning discovery using Google v2 URL http://www.arxiv.org/abs/cs.CL/0412098
[17] Cilibrasi R L and Vitanyi P M B 2007 *IEEE Transactions on Knowledge and Data Engineering* **19** 370–383 ISSN 1041-4347 URL http://dx.doi.org/10.1109/TKDE.2007.48
[18] Langdon W B and Poli R 1997 *Soft Computing in Engineering Design and Manufacturing* ed Chawdhry P K *et al.* (Springer-Verlag London) pp 13–22 ISBN 3-540-76214-0 URL http://dx.doi.org/10.1007/978-1-4471-0427-8_2
[19] Langdon W B 2022 A trillion genetic programming instructions per second ArXiv URL https://arxiv.org/abs/2205.03251
[20] Silva S *et al.* 2012 *Genetic Programming and Evolvable Machines* **13** 197–238 ISSN 1389-2576 URL http://dx.doi.org/10.1007/s10710-011-9150-5
[21] Poli R and McPhee N F 2013 *Theory and Principled Methods for the Design of Metaheuristics* Natural Computing Series ed Borenstein Y and Moraglio A (Springer) pp 181–204 URL http://dx.doi.org/10.1007/978-3-642-33206-7_9
[22] Card S W 2010 *GECCO 2010* ed Card S W and Borenstein Y (Portland, Oregon, USA: ACM) pp 1851–1854 URL http://dx.doi.org/10.1145/1830761.1830815
[23] Hulme D *et al.* 2024 Solving machine consciousness: Theory and approach Conscium
[24] DeMillo R A *et al.* 1978 *IEEE Computer* **11** 31–41 URL http://dx.doi.org/10.1109/C-M.1978.218136
[25] Langdon W B and Petke J 2015 *Complex Systems Digital Campus E-conference, CS-DC'15* Proceedings in Complexity ed Parrend P *et al.* (Springer) pp 203–211 invited talk URL http://dx.doi.org/10.1007/978-3-319-45901-1_24
[26] Lorenz E N 1993 *The Essence of Chaos* Jessie and John Danz lectures (Seattle, USA: University of Washington Press) chap Appendix 1, pp 181–184 URL https://uwapress.uw.edu/book/9780295975146/the-essence-of-chaos/
[27] Voas J M and Miller K W 1995 *IEEE Software* **12** 17–28 ISSN 0740-7459 URL http://dx.doi.org/10.1109/52.382180
[28] Martinez K and Cupitt J 2005 *Proceedings of the 2005 International Conference on Image Processing, ICIP* (Genoa, Italy: IEEE) pp 574–577 URL http://dx.doi.org/10.1109/ICIP.2005.1530120
[29] Langdon W B and Clark D 2024 *EuroGP 2024: Proceedings of the 27th European Conference on Genetic*

Programming (*LNCS* vol 14631) ed Giacobini M *et al.* (Aberystwyth: Springer Verlag) pp 209–226 URL http://dx.doi.org/10.1007/978-3-031-56957-9_13

[30] Blot A and Petke J 2022 MAGPIE: Machine automated general performance improvement via evolution of software arXiv 2208.02811 URL http://dx.doi.org/10.48550/arxiv.2208.02811

[31] Langdon W B 2019 *GECCO '19 Companion* ed Doerr C (Prague, Czech Republic: ACM) pp 63–64 URL http://dx.doi.org/10.1145/3319619.3326770

[32] Langdon W B 2020 Multi-threaded memory efficient crossover in C++ for generational genetic programming ArXiv 2009.10460 URL http://arxiv.org/abs/2009.10460

[33] Langdon W B 2021 *EuroGP 2021: Proceedings of the 24th European Conference on Genetic Programming* (*LNCS* vol 12691) ed Hu T *et al.* (Virtual Event: Springer Verlag) pp 229–246 URL http://dx.doi.org/10.1007/978-3-030-72812-0_15

[34] Langdon W B 2021 *Proceedings of the Genetic and Evolutionary Computation Conference Companion* GECCO '21 ed Chicano F *et al.* (Internet: Association for Computing Machinery) pp 253–254 URL http://dx.doi.org/10.1145/3449726.3459437

[35] Langdon W B 2021 *Genetic Programming Theory and Practice XVIII* Genetic and Evolutionary Computation ed Banzhaf W *et al.* (East Lansing, MI, USA: Springer) pp 143–164 URL http://dx.doi.org/10.1007/978-981-16-8113-4_8

[36] Langdon W B 2022 *ACM Transactions on Evolutionary Learning and Optimization* **2** ISSN 2688-299X URL http://dx.doi.org/10.1145/3539738

[37] Langdon W B 2022 *Complex Systems* **31** 287–309 ISSN 0891-2513 URL http://dx.doi.org/10.25088/ComplexSystems.31.3.287

[38] Singleton A 1994 *BYTE* 171–176 ISSN 0360-5280 URL http://www.assembla.com/wiki/show/andysgp/GPQuick_Article

[39] Langdon W B 2022 *Proceedings of the Genetic and Evolutionary Computation Conference Companion* GECCO '22 ed Trautmann H *et al.* (Boston, USA: Association for Computing Machinery) pp 574–577 URL http://dx.doi.org/10.1145/3520304.3528878

[40] Langdon W B *et al.* 2021 *5th Workshop on Landscape-Aware Heuristic Search* GECCO 2021 Companion ed Veerapen N *et al.* (Internet: ACM) pp 1683–1691 URL http://dx.doi.org/10.1145/3449726.3463147

[41] Hughes J *et al.* 2022 *IOP Conference Series: Materials Science and Engineering* **1261** 012001 URL http://dx.doi.org/10.1088/1757-899X/1261/1/012001

[42] Langdon W B 2024 *Communications of the ACM* **67** 8 ISSN 0001-0782 letter to the editor URL http://dx.doi.org/10.1145/3654698

[43] Bentley P J and Corne D W 2001 *Creative Evolutionary Systems* ed Bentley P J and Corne D W (Morgan Kaufmann) pp 1–75 ISBN 1-55860-673-4 URL http://dx.doi.org/10.1016/B978-155860673-9/50035-5

[44] Smith R E 2019 *Rage Inside the Machine–the prejudice of algorithms, and how to stop the internet making bigots of us all* (Bloomsbury business) ISBN 9781472963888

[45] Koza J R *et al.* 2003 *Genetic Programming IV: Routine Human-Competitive Machine Intelligence* (Kluwer Academic Publishers) ISBN 1-4020-7446-8 URL http://dx.doi.org/10.1007/0-387-26417-5_1

[46] Samuel A L 1959 *IBM Journal* **3** 210–229

[47] Fogel D B 2001 *Blondie24: Playing at the Edge of AI* (Morgan Kaufmann) ISBN 1-55860-783-8 URL https://en.wikipedia.org/wiki/Blondie24

[48] Langdon W B 2023 *Genetic Programming and Evolvable Machines* **24** Article number: 19 ISSN 1389-2576 special Issue: Thirtieth Anniversary of Genetic Programming: On the Programming of Computers by Means of Natural Selection URL http://dx.doi.org/10.1007/s10710-023-09467-x

[49] Bartoli A *et al.* 2023 *Genetic Programming and Evolvable Machines* **24** Article number: 23 ISSN 1389-2576 special Issue: Thirtieth Anniversary of Genetic Programming: On the Programming of Computers by Means of Natural Selection URL http://dx.doi.org/10.1007/s10710-023-09471-1

[50] Langdon W B 2023 *Genetic Programming and Evolvable Machines* **24** Article number: 26 ISSN 1389-2576 special Issue: Thirtieth Anniversary of Genetic Programming: On the Programming of Computers by Means of Natural Selection URL http://dx.doi.org/10.1007/s10710-023-09474-y

[51] Nguyen S *et al.* 2023 *International Journal of Disaster Risk Reduction* **97** 104004 ISSN 2212-4209 URL http://dx.doi.org/10.1016/j.ijdrr.2023.104004

[52] Andelic N and Baressi Segota S 2023 *Cancers* **15** article no. 3411 ISSN 2072-6694 URL http://dx.doi.org/10.3390/cancers15133411

[53] Guidetti V *et al.* 2023 *2023 IEEE 10th International Conference on Data Science and Advanced Analytics (DSAA)* URL http://dx.doi.org/10.1109/DSAA60987.2023.10302622

[54] Hurta M *et al.* 2023 *2023 IEEE International Conference on Bioinformatics and Biomedicine (BIBM)* pp 3782–3787 ISSN 2156-1133 URL http://dx.doi.org/10.1109/BIBM58861.2023.10385615

[55] Bartlett L K *et al.* 2023 *Proceedings of the 45th Annual Meeting of the Cognitive Science Society* ed Goldwater

M *et al.* (Sydney, Australia) pp 2833–2839 URL http://hdl.handle.net/2299/27181

[56] Romano J D *et al.* 2023 *Computational Toxicology* **25** 100261 ISSN 2468-1113 URL http://dx.doi.org/10.1016/j.comtox.2023.100261

[57] Javed N *et al.* 2023 *AISB 2023 convention proceedings. The Society for the Study of Artificial Intelligence and Simulation Behaviour* ed Mueller B (Swansea, UK) pp 43–50 URL http://eprints.lse.ac.uk/id/eprint/118805

[58] MacLachlan J *et al.* 2023 *Proceedings of the 2023 Genetic and Evolutionary Computation Conference* GECCO '23 ed Silva S *et al.* (Lisbon, Portugal: Association for Computing Machinery) pp 1409–1417 silver 2023 HUMIES URL http://dx.doi.org/10.1145/3583131.3590434

[59] Hurta M *et al.* 2023 *2023 26th International Symposium on Design and Diagnostics of Electronic Circuits and Systems (DDECS)* pp 155–160 ISSN 2473-2117 URL http://dx.doi.org/10.1109/DDECS57882.2023.10139399

[60] Degefa A B *et al.* 2023 *Sustainability* **15** Article No. 15471 ISSN 2071-1050 URL http://dx.doi.org/10.3390/su152115471

[61] Ozbayrak A *et al.* 2023 *Arabian Journal for Science and Engineering* **48** 5347–5370 ISSN 2193-567X URL http://dx.doi.org/10.1007/s13369-022-07445-6

[62] Martin-Alcantara A *et al.* 2023 *Sustainable Energy Technologies and Assessments* **56** 103053 ISSN 2213-1388 URL http://dx.doi.org/10.1016/j.seta.2023.103053

[63] Yin Z *et al.* 2023 *Ocean Engineering* **285** 115372 ISSN 0029-8018 URL http://dx.doi.org/10.1016/j.oceaneng.2023.115372

[64] Ismail M K *et al.* 2023 *Engineering Structures* **295** 116806 ISSN 0141-0296 URL http://dx.doi.org/10.1016/j.engstruct.2023.116806

[65] Althoey F *et al.* 2023 *Case Studies in Construction Materials* **18** e01774 ISSN 2214-5095 URL http://dx.doi.org/10.1016/j.cscm.2022.e01774

[66] Sadat Hosseini A *et al.* 2023 *Ocean Engineering* **279** 114465 ISSN 0029-8018 URL http://dx.doi.org/10.1016/j.oceaneng.2023.114465

[67] Al-Aghbari M and M Gujarathi A 2023 *Geoenergy Science and Engineering* **228** 211967 ISSN 2949-8910 URL http://dx.doi.org/10.1016/j.geoen.2023.211967

[68] Xue X *et al.* 2023 *Alexandria Engineering Journal* **81** 599–619 ISSN 1110-0168 URL http://dx.doi.org/10.1016/j.aej.2023.09.053

[69] Di Capua G *et al.* 2023 *26th International Conference, EvoApplications 2023* (*LNCS* vol 13989) ed Correia J *et al.* (Brno, Czech Republic: Springer Verlag) pp 461–474 URL http://dx.doi.org/10.1007/978-3-031-30229-9_30

[70] Di Capua G *et al.* 2023 *2023 IEEE International Symposium on Circuits and Systems (ISCAS)* ISSN 2158-1525 URL http://dx.doi.org/10.1109/ISCAS46773.2023.10181456

[71] Vandana *et al.* 2023 *2023 IEEE 3rd International Conference on Sustainable Energy and Future Electric Transportation (SEFET)* URL http://dx.doi.org/10.1109/SeFeT57834.2023.10244776

[72] Milano F *et al.* 2023 *2023 IEEE International Workshop on Metrology for Automotive (MetroAutomotive)* pp 35–40 URL http://dx.doi.org/10.1109/MetroAutomotive57488.2023.10219104

[73] Hosseinhashemi S *et al.* 2023 *Journal of Energy Storage* **73** 109046 ISSN 2352-152X URL http://dx.doi.org/10.1016/j.est.2023.109046

[74] Langdon W B 2022 *Proceedings of 2022 International Conference on Embodied Intelligence, EI-2022* (*IOP Conference Series: Materials Science and Engineering* vol 1292) ed Iida F *et al.* (Internet, Cambridge: IOP Publishing) p 012021 URL http://dx.doi.org/10.1088/1757-899X/1292/1/012021

[75] Castro M *et al.* 2016 *Chem. Eur. J.* **22** 15307–15319 URL http://dx.doi.org/10.1002/chem.201600511

An Information-Theoretic Perspective on Acting Agents

Carlotta Langer[1] **and Nihat Ay**[1,2,3]

[1] Hamburg University of Technology, Hamburg, Germany
[2] Santa Fe Institute, Santa Fe, USA
[3] Leipzig University, Leipzig, Germany

E-mail: `carlotta.langer@tuhh.de`

Abstract. Every embodied intelligent agent constantly interacts in its own way with its environment. It perceives information about the world, processes this information in its controller, e.g. its brain, and sends signals to the body, which then interacts with its surroundings. Here we give a thorough introduction to an information-theoretic approach that aims at quantifying these information flows among an agent's body, brain and environment. Thereby we measure how much an information flow impacts the behavior of an agent. For instance, we might ask the question whether the brain is essential for a certain behavior or whether it is a reflexive stimuli response. In order to measure these interactions we model the system using the sensorimotor loop and assess the different information flows by information-theoretic measures. This includes measures related to Morphological Computation and the complexity of the controller. Morphological Computation describes the reduction of computational cost for the controller that results from the interaction with the environment. We include results from an implementation of this framework with simplistic simulated robots. Experiments with these robots show, for instance, that an agent with a well-adapted body does not rely as much on its controller. Furthermore, learning a new task requires an increased information integration in the controller.

Keywords: Information Theory, Morphological Computation, Integrated Information

1. Introduction

Embodied agents, whether they are biological or artificial, constantly interact with their environments. In Figure 1 a sketch of such an agent is depicted. There the agent senses information about its environment through its eyes and processes this information in its controller, which is the brain in case of biological agents and could be an artificial neural network in case of a robot. The information from the sensors can then lead to a direct stimuli response, a reflexive action such as a movement of a hand, or the controller can send commands to the hands, which in turn influence the environment.

When an intelligent embodied agent, like an animal, is faced with a changed environment it has to adapt and learn new behaviors. It has to figure out how to best use the properties of its own body to cope with the new surroundings. There are various theories that aim to explain how exactly animals learn. Here we take a step back from this problem and do not ask "how" an agent learns, but instead which parts

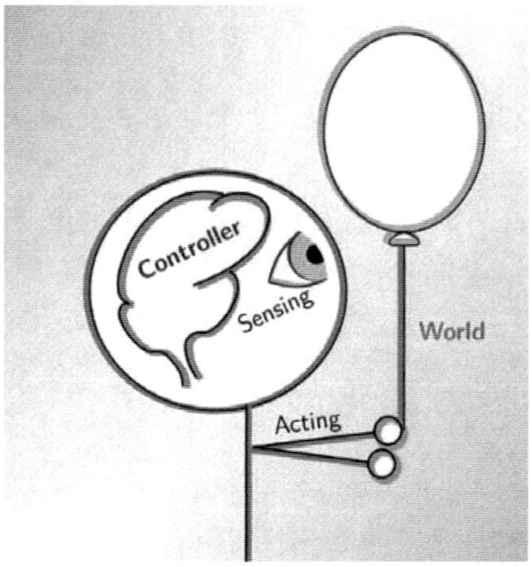

Figure 1. Sketch of an embodied agent that interacts with its environment.

of an agent are most important for its behavior. If the agent is catching an object, for example, is this behavior mostly governed by the commands from the brain of the agent or is this a reactive stimuli response? Does this change during the process of learning this behavior?

In our daily experiences we have an intuition of how important the influences from the brain or our body are. Consider a skilled musician playing an instrument. The musician does not need to actively think about where to place each finger on the instrument, since the movements that translate to the desired melody have already been learned and can be applied almost automatically. On the other hand, someone who is new to an instrument needs to think carefully about every movement, hand placement and learn through repetitions. So, our hypothesis would be that an agent that learns a new task needs first a high involvement of the brain, which then gradually decreases.

Now assume we have a robot learning to play the same instrument. There we are able to directly observe the signals processed in the controller, e.g. a neural network, and measure them using information theory. In order to learn we would expect it to go through similar stages of development. This could be an indication of how natural the learning process of an artificial agent is or in which way it differs from our expectations. In [34] we discuss a simplistic setting with learning simulated robots and there we observe the dynamics hypothesized above. There we conclude that an agent first needs a high involvement from a complex controller to understand its environment and the more it can exploit the interactions between body and environment, the less controller complexity is needed.

In this article we give a thorough introduction to the information-theoretic methodology for tracing information flows within sensorimotor interactions, thereby following the Ph.D. thesis of one of the authors [32]. In the following sections we therefore first give a quick introduction to information theory and then we define the sensorimotor loop in the Section 3. The sensorimotor loop models the different parts of an acting agent in its environment. Afterwards, in Section 4, we introduce a way to measure the importance of different information flows. Hereby, we consider pair-wise as well as synergistic influences. Two interesting concepts, that can be measured in this way, are Morphological Computation and Integrated Information, introduced in Section 5 and 6, respectively. The former describes the reduction of computational cost for the controller that results from the interaction of the body with its environment and the latter aims to quantify consciousness by a system's ability to integrate information. In Section 7 we review previous results that we gained by applying this framework to simple simulated agents. We conclude with a discussion in Section 8.

2. Information Theory

In 1948 C. E. Shannon published the article "A Mathematical Theory of Communication" [51], developed further into a book with W. Warren as "The Mathematical Theory of Communication" [52], which provides the foundation of information theory. We give a short introduction in this section, a more thorough treatment of information theory can be found in, for example, [15, 35].

The formalization of a communication system is sketched in Figure 2. A message gets passed from an information source through a transmitter, then the signal might be disturbed by noise before it reaches the receiver and then its destination.

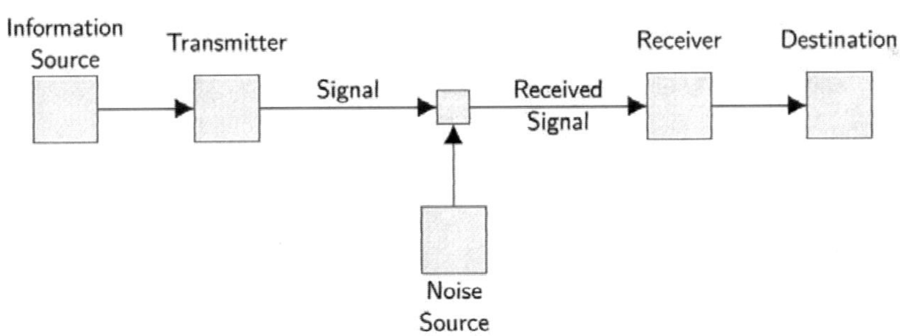

Figure 2. Reproduction of Shannon's general communication system, Figure 1 in [51].

Within his theory Shannon defines the optimal rate of data transportation and the maximal compression without loss of information in order to improve the transmission of information. One important concept in this context is the entropy.

Definition 2.1 (Entropy)**.** The entropy H of a discrete random variable X with the state space $\mathcal{X}$ and the probability mass function $p(X)$ is defined as follows:

$$H(X) = -\sum_{x\in\mathcal{X}} p(x)\log p(x)$$

with the convention $0\log 0 = 0$.

The entropy is non-negative and is upper bounded by the entropy of the uniform distribution over $\mathcal{X}$. The base of the logarithm merely determines in what units we measure information, bits for the base 2 or nats for the base e, and this has no further implication on the properties of H.

The entropy is a lower bound for the compression and it can be thought of as the average reduction of uncertainty from observing an outcome x, when we initially know the probabilities $p(x)$. If we have a random variable with a uniform distribution, for example given by a fair die, then observing each side of the die is equally likely. Hence, the reduction of uncertainty would be large, roughly 2.585 bits. If, however, we have a weighted die that falls on one side 95% of all throws, then we would expect to observe this side and our average reduction of uncertainty is low, in this case roughly 0.402 bits.

In order to quantify the difference between two probability distributions we now define the Kullback-Leibler-divergence, also called KL-divergence or relative entropy. It measures how much the uncertainty of the random variable increases if we use q as probability distribution instead of the true distribution p.

Definition 2.2 (Relative Entropy)**.** The relative entropy, or KL-divergence D of two probability mass functions p and q with the state space $\mathcal{X}$ is defined as follows:

$$D(p \parallel q) = \sum_{x \in \mathcal{X}} p(x) \log \frac{p(x)}{q(x)}$$

with the conventions that $0 \log \frac{0}{q(x)} = 0$ for $q(x) \geq 0$ and $p(x) \log(\frac{p(x)}{0}) = \infty$ for $p(x) > 0$.

The KL-divergence is non-negative and equals zero if and only if the two arguments are equal, meaning that $p(x) = q(x)$, for all $x \in \mathcal{X}$. Note that the KL-divergence is not symmetric, hence this is not a metric.

Shannon defines the channel capacity by using the mutual information, which is the relative entropy between a distribution $p(x, y)$ and the product distribution $p(x)p(y)$ on $\mathcal{X} \times \mathcal{Y}$. The maximum of this quantity with respect to $p(x)$ leads to the maximal rate at which information can be sent through a communication channel, characterized by $p(y|x)$, with a vanishingly low probability of error.

In the following sections we will also use the conditional mutual information.

Definition 2.3 (Conditional Mutual Information)**.** Let (X_1, X_2, X_3) be a random vector on $\mathcal{X} = \mathcal{X}_1 \times \mathcal{X}_2 \times \mathcal{X}_3$ and let $p(x_1, x_3) > 0$, for all $(x_1, x_3) \in \mathcal{X}_1 \times \mathcal{X}_3$. The conditional mutual information of the random variables X_1 and X_2 given X_3 is defined as

$$I(X_1; X_2|X_3) = \sum_{(x_1,x_2,x_3) \in \mathcal{X}} p(x_1, x_2, x_3) \log \frac{p(x_1, x_2|x_3)}{p(x_1|x_3)}.$$

The conditional mutual information is 0 if and only if X_1 and X_2 are independent given X_3, this can be written as $X_1 \perp\!\!\!\perp X_2|X_3$.

The main theorems of classical information theory, like the noisy channel coding theorem, are asymptotic results, but the information-theoretic quantities above give also rise to many advanced theories. Here, we will only mention a few of them to give an intuition to the variety of applications of information theory.

In machine learning the evidence lower bound of the variational auto-encoder (VAE) is given by relative-entropy and cross-entropy terms [25]. In cybernetics, for instance, Ashby defines his law of requisite variety in terms of the Shannon entropy in [5]. He postulates that a system needs to have a greater than or equal number of states in its control mechanism compared to the system being controlled in order to be stable.

Furthermore, the well-known "active inference" framework aims to explain the behavior of sentient agents as them minimizing the free energy, which is defined using relative entropy terms. The free energy function is a variational bound on the surprisal, given by $-\log p(x)$, also called Shannon information, [42].

In Section 6 we additionally introduce the Integrated Information Theory, which aims at measuring the consciousness of a system.

3. Sensorimotor Loop

In order to analyze an acting agent in its environment we use the sensorimotor loop to model the interactions among an agent's sensors, actuators, its controller and the environment. This model is also called perception-action loop or action perception cycle [26, 27, 49, 53].

The agent receives information about the world through its sensors and can then process this information in its controller. It can also send signals from the sensors to the actuators, which corresponds to a direct stimuli response. The actuator can additionally receive commands from the controller and the signals from the internal actuators lead to actions and influence the environment of the agent. A sketch of the sensorimotor loop is given in Figure 3 on the left.

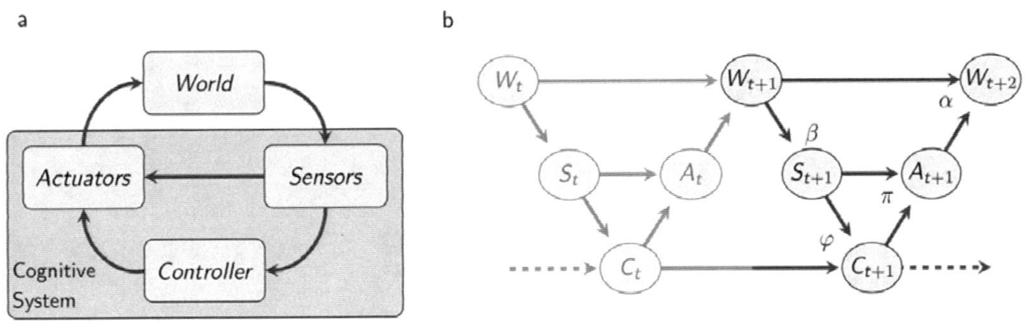

Figure 3. (a) The sensorimotor loop and (b) the graphical representation of the sensorimotor loop in discrete timesteps.

Here, we further model the system as a Bayesian network, similar to [12, 10, 28]. We assume discrete timesteps, which leads to the graphical representation on the right hand side of Figure 3. The different parts of the sensorimotor loop are modeled as discrete random variables and the interactions among them are given by the Markov kernels α, β, π and φ, as noted on the right of Figure 3. These Markov kernels are stochastic maps that determine the mechanisms of the variables. The kernel α, for instance, is called the world dynamics kernel in [18] and describes the probability of a state w_{t+2} given w_{t+1} and a_{t+1} by $\alpha(w_{t+2}|w_{t+1}, a_{t+1})$.

The whole system is then described by a discrete Markov Process $(X_t)_{t\in\mathbb{N}}$ with

$$X_t = (W_t, S_t, A_t, C_t)$$

and the state space $\mathcal{X} = \mathcal{W} \times \mathcal{S} \times \mathcal{A} \times \mathcal{C}$. The transition probability of the whole system from time step t to $t+1$ can be factorized as follows

$$p(x_{t+1}|x_t) = \alpha(w_{t+1}|w_t, a_t)\beta(s_{t+1}|w_{t+1})\varphi(c_{t+1}|c_t, s_{t+1})\pi(a_{t+1}|s_{t+1}, c_{t+1})$$

for all $(x_t, x_{t+1}) \in \mathcal{X} \times \mathcal{X}$. Using this description of the sensorimotor loop we can now define the measures of the different information flows.

4. Measuring an Information Flow

In the previous section we discuss the sensorimotor loop, a formalization of the information flows among the different parts of an agent and its environment. Next, we introduce a method to measure the strength of these information flows. The main idea our measures follow is that the strength of a connection is given as the minimal difference between a full system and the set of systems without this connection. We formalize this as follows.

Definition 4.1 (Measure $\Psi_{\mathcal{M}}$). Let p be a full distribution factoring according to the sensorimotor loop and let $\mathcal{M}$ be a set of positive probability distributions that factor according to the sensorimotor loop but are lacking the information flow that we want to measure. Then we define the measure $\Psi_{\mathcal{M}}$ by minimizing the KL-divergence between $\mathcal{M}$ and the full distribution

$$\Psi_{\mathcal{M}} = \inf_{q\in\mathcal{M}} D(p \parallel q).$$

This equips us with a technique to measure all the information flows among the different parts of the sensorimotor loop. Measures that use this approach to quantify an information flow are discussed in, for example, [22, 19, 20, 32]. In a similar setting, in [28], the amount of control an agent has over its environment is measured as "empowerment". The authors of [50] use such a measure to improve the efficiency of a reinforcement learning algorithm.

An alternative to this observational technique is to actively perturb the system, which would model an intervention and can be formalized by Pearl's *do*-calculus, described in [44, 43, 45]. This method is used in [11] to define causal information flows. Different approaches to quantify the causal strength of an information flow are discussed in [23].

As an example of our approach we now discuss the situation sketched in Figure 4. There we have a simplified agent without a controller and without sensors. In this case we have an agent that generates actions and influences its environment through them.

Consider a toy car with a broken remote that continuously moves forward and steers left and right with a fixed probability. Now, depending on the environment we let this car run in, the actions might have a large impact on the state of the environment, which includes the position of the car. In an open room the position of the car in the next point is determined by its last position and the selected actions.

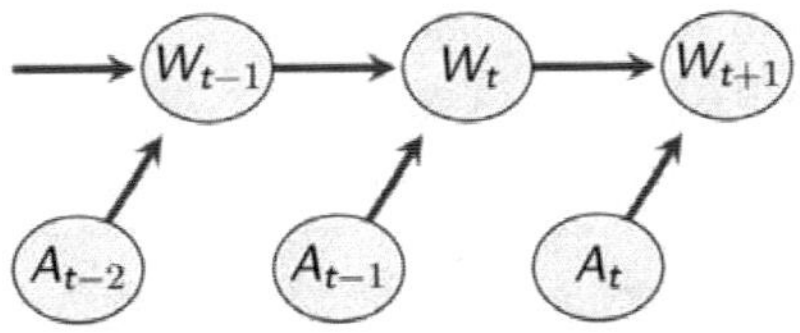

Figure 4. Acting agent influencing its environment.

If, however, this car is in a labyrinth with narrow passages, then the next position of the car is mostly determined by the environment. Now, we want to ask the question how to measure the strength of the influence of the world, the actuators or their synergistic influence. We start with quantifying a pair-wise connection.

First, we want to measure the influence of A_t on W_{t+1} given W_t and we define a set of distributions $\mathcal{M}$ that describe agents in which this connection does not exist, sketched as dashed arrow in Figure 5. This means that the kernel α only depends on W_{t+1} and W_t, hence $\alpha(w_{t+1}|a_t, w_t) = \alpha(w_{t+1}|w_t)$ for all $(w_{t+1}, w_t, a_t) \in \mathcal{W} \times \mathcal{W} \times \mathcal{A}$. The minimization in Definition 4.1 has the following closed form solution in this case

$$\Psi_{\mathcal{M}} = I(W_{t+1}; A_t|W_t).$$

The equality above can be easily proven by using that the cross-entropy is greater or equal to the entropy, as applied in a similar setting in Proposition 2 of [31].

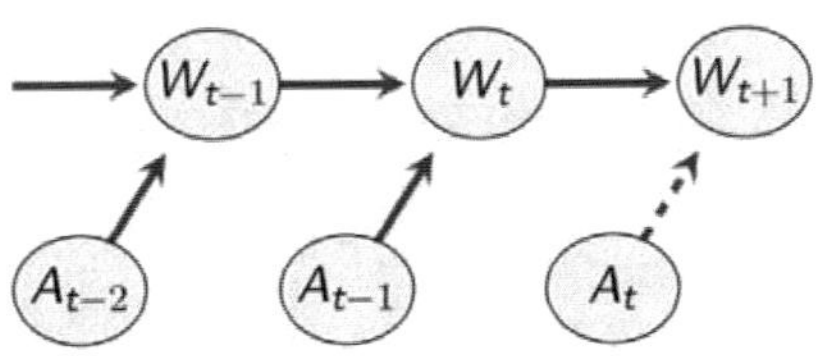

Figure 5. Acting agent for which the connection between A_t and W_{t+1} should be quantified.

Since $\Psi_{\mathcal{M}}$ can be written as a conditional mutual information we can conclude that this measure is 0 if and only if W_{t+1} is independent of A_t given W_t. In that case the action of the agent would have no influence on the environment at all.

In addition to the pair-wise interactions we are also able to define a measure for the synergistic influences based on triple-wise interactions, as sketched on the left of Figure 6. If the toy car steers into an obstacle and pushes it away, then this outcome is not determined by the action or the environment alone, but can only be explained by considering the information from both, W_t and A_t at the same time.

Now, in order to measure the synergistic influence of W_t and A_t on W_{t+1} we need to define $\mathcal{M}$ as a set of all those distributions in which there exist no three-way interactions among the variables. These are sketched on the right of Figure 6. We are able to define these distributions as those for which there exist two non-negative functions f_1, f_2 such that

$$p(w_{t+1}|w_t, a_t) = f_1(w_{t+1}, a_t) f_2(w_{t+1}, w_t),$$

for all $(w_{t+1}, w_t, a_t) \in \mathcal{W} \times \mathcal{W} \times \mathcal{A}$.

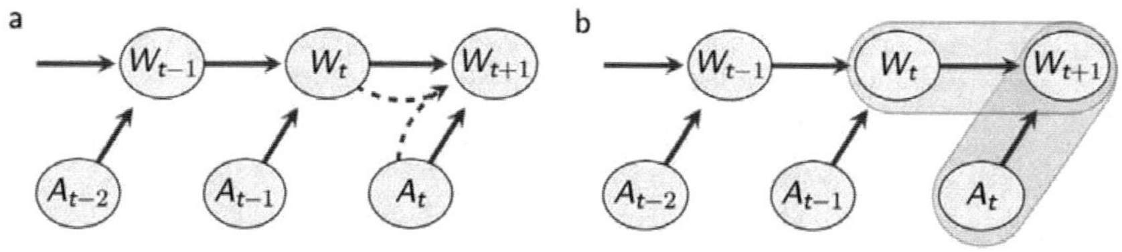

Figure 6. Sketch of an acting agent with a synergistic influence in (a) and one without such an influence in (b).

Minimizing the KL-divergence in Definition 4.1 does not lead to a closed form solution here. Instead, this can be calculated using, for instance, the iterative scaling algorithm, defined in [16]. This is discussed in more detail in the context of a synergistic measure for an internal prediction in [34] and a synergistic measure for Morphological Computation in [20]. Next we introduce the notion of Morphological Computation in more detail.

5. Morphological Computation

An embodied agent can interact with its environment in different ways depending on the properties of its body. When we grasp a fragile object, for instance, we can vary the amount of pressure that we are able to apply without damaging the object or loosing grip, because of the soft tissue of our skin. However, a robot with a metal hand needs to carefully calculate the amount of applied pressure. This intuitive concept, that the morphology of the body can alleviate the computational burden of the brain, is called Morphological Computation. There exist various definitions of Morphological Computation, especially the use of the term "computation" is debated, as discussed in more detail in, for example, [18, 38].

The importance of considering an embodied agent in a real environment in order to understand its behavior is stressed in [13, 14] and the concept of Morphological Computation was introduced in [48]. Remarkable cases of Morphological Computation can be seen in the field of soft robotics, for instance in [39] in which die authors present an octopus inspired robotic arm.

Additionally, Pfeifer and Bongard formulate in [46] the "cheap design principle". It states that the body of an agent should be constructed with the goal to exploit the properties of their environment as best as possible. The authors of [37] use the information-theoretic setting discussed here in order to formalize the notion of cheap control architectures in the context of universal approximation.

In Section 7 we discuss results of Morphological Computation measures in experiments with simple simulated agents. Hence, we now introduce one class of measures for Morphological Computation that stem from the concepts introduced in the previous section. Figure 7 is a sketch of the three different information flows that we can measure and associate with Morphological Computation. The measure used in Section 7 quantifies the information flow through the world, given the action, marked with (1) in Figure 7. This is also discussed in [21, 18].

On the other hand, we can also measure the impact that the agent has on its environment, sketched as (2) in Figure 7. Subtracting the normalized strength of this connection from one leads in [21, 18] to the definition of another Morphological Computation measure.

The connection (3) in Figure 7, the interplay of the influences of W_t and A_t on W_{t+1}, is defined in [20] as a synergistic measure for Morphological Computation.

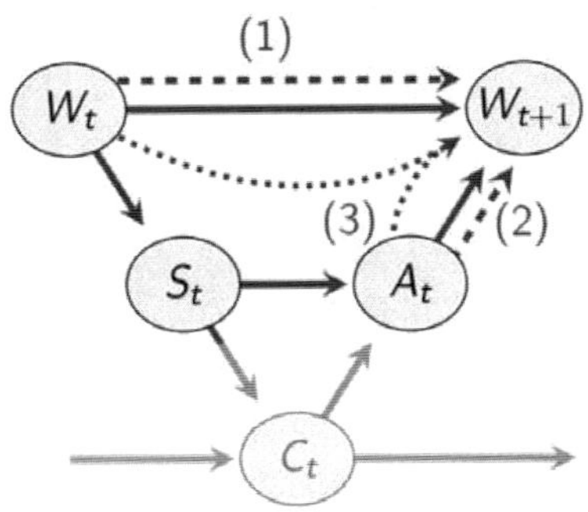

Figure 7. Different information flows that lead to a measure for Morphological Computation.

Since it is often not possible to access the world states directly, we can instead use only the information about the environment that is accessible to agents, given by the sensory states. The justification for using S instead of W is discussed in [33].

6. Controller Complexity as Information Integration

Additionally, we are able to use this information-theoretic approach to assess the complexity of the controller. In this setting the controller is given by a multivariate random variable $C_t = (C_t^1, \dots C_t^n)$ in each point in time t, as depicted for $n = 2$ on the left hand side of Figure 8. In the previous sketches we display only one node per random vector regardless of the dimension to increase readability.

Here we deem a controller to be complex if it is more than the sum of its parts. Hence, we quantify the difference between the whole system and one in which the different parts of the controller do not interact with each other. In the latter case there is no information flow from C_t^i to C_{t+1}^j if $i \neq j$, marked here by the dashed connections in Figure 8. This means that the different controller nodes do not integrate information among each other.

On the left hand side of Figure 8 we only depict the controller without explicitly including external influences, similar to the setting in [41, 4]. Instead, we now embed this measure in the sensorimodor loop, which leads to the more complicated sketch on the right hand side of Figure 8 that depicts the following measure

$$\Phi = \sum_i I(C^i_{t+1}; C^{I\setminus i}_t | C^i_t, S_{t+1}),$$

which is discussed in the context of other information-theoretic measures in [31, 32]. Further similar complexity measures can be found in [7, 8, 24, 41, 4, 36].

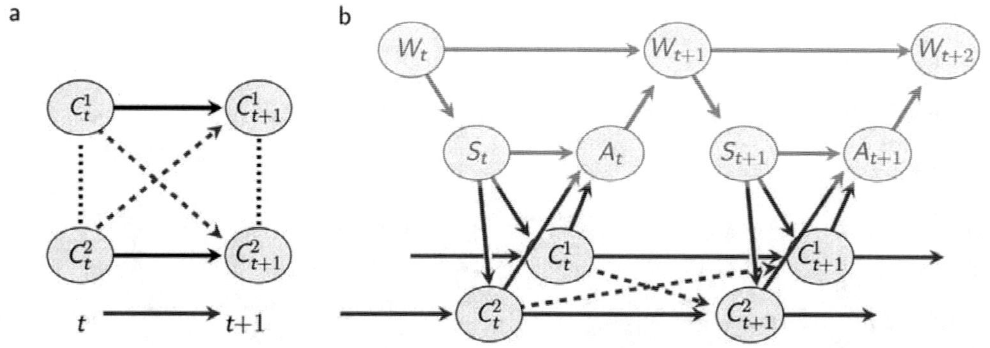

Figure 8. Sketch of the Integrated Information in the controller in (a) and the Integrated Information embedded in the sensorimotor loop in (b).

This measure of complexity is related to the Integrated Information Theory of consciousness in which the capability of a system to integrate information is equated with its amount of consciousness. This theory started as a measure for brain complexity in [55] and developed through different iteration, mainly by Tononi and his coworkers, towards a broad theory of consciousness, in, for instance, [54, 40, 2].

The different branches and iterations of the theory define various approaches to measuring the Integrated Information in a system. The information-theoretic measure discussed above can be seen in the context of [54] in which Tononi summarizes the theory as follows:

> "In short, integrated information captures the information generated by causal interactions in the whole, over and above the information generated by the parts."

The information-theoretic framework discussed here allows us to calculate the Integrated Information of a system as well as other measures such as the Morphological Computation and relate them to each other. We discuss our previous results in the next section.

Although the Integrated Information Theory is largely restricted to only considering the controller, the following selection of studies relates the Integrated Information to an agent's behavior and environment. In [17] the authors simulate evolving agents in a maze and observe that the Integrated Information increases with the fitness of the agents. The authors of [3] conclude that an agent needs an increased Integrated Information value in order to navigate in a complex environment and in [1] a high Integrated Information increases the likelihood of a rich, dynamical agent behavior.

7. Applications and Previous Results

Here we discuss two situations for which we have intuitive hypotheses that we were able to confirm using the information-theoretic approach. There we measure the information flows in simplistic settings with simulated agents.

When an embodied agent performs a task then it can make use of the properties of its body in a more or less optimal way, using higher or lower amounts of Morphological Computation. Hence, we would expect that the necessary controller complexity is inversely correlated with the amount of Morphological Computation, meaning the more an agent exploits its morphology, the less information integration in the brain is required.

This notion was summarized in [47], where the authors characterize the relationship between the morphology and the controller in the following way:

> "Second, there is a kind of trade-off or balance: the better the exploitation of the dynamics, the simpler the control, the less neural processing will be required."

We are able to test this hypothesis by calculating the information-theoretic measures for Morphological Computation and the controller complexity, discussed above. In the article [33] we use simple simulated agents that move inside a racetrack and have the goal to not touch the walls. These simulated agents consist of a round body and two forward facing binary sensors that can detect a wall. Additionally, they can be thought of as having two wheels that can spin either fast or slow, resulting in four different movements, as depicted on the left of Figure 9. Inside the agent are the different components, namely sensors, actuators and the controller, as described in Section 3. The controller consists in this case of two binary nodes such that the measure for the complexity of the controller corresponds to the one depicted in Figure 8. The environment of these agents, with five agents moving in it, can be seen in Figure 9 (b). In (c) there are seven different agents displayed with sensor length from 0.5 on the top to 2 on the bottom. For each sensor length we select an optimal behavior using the concept of planning as inference, discussed in, for instance, [6, 57, 56]. More details can be found in [33, 32] and the corresponding code can be accessed at [30].

Figure 10 depicts the result of the Morphological Computation and the controller complexity. The used Morphological Computation measure quantifies the information flowing through the environment, given the action, which corresponds to the connection marked by (1) in Figure 7. The x-axis of Figure 10 displays the quality of the sensors, in this case the reach of the binary sensors. Varying this has a direct influence on the interaction among an agent's body and its environment. In Figure 10 we can observe

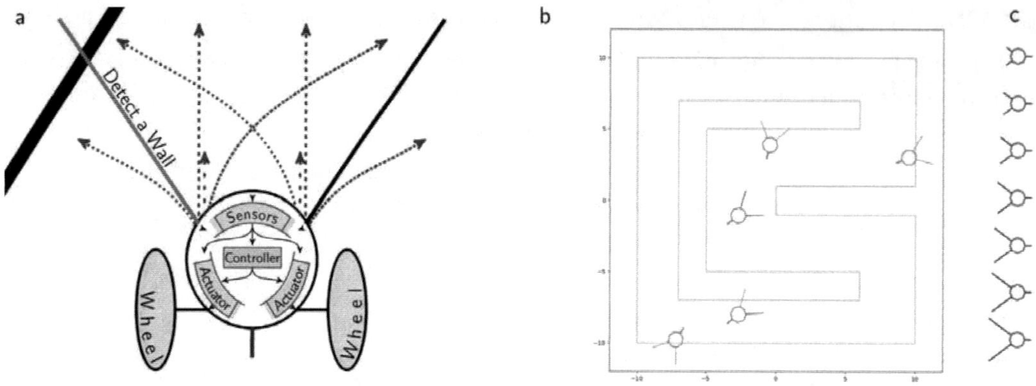

Figure 9. (a) A sketch of the two-wheeled agent with its two sensors and four different movements, (b) the environment with five different agents acting in it and (c) agents with sensor length ranging from 0.5 on the top to 2 on the bottom.

an inversely correlated relationship between the Morphological Computation and the complexity of the controller, given by the Integrated Information. The more an agent interacts with the environment, the less complexity in the controller is required, as hypothesized. So, the longer the sensors of the agents are, the more they can perceive of their environment and the better they can interact with it. This better interaction with the environment makes it possible for the agent to select appropriate actions without the need for a complex control architecture.

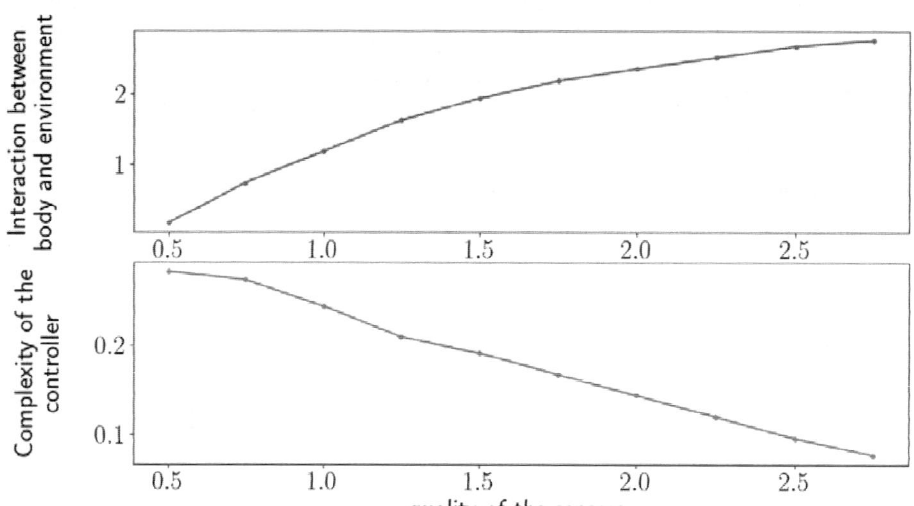

Figure 10. The arithmetic means over 100 agents of the Morphological Computation on the top and the complexity of the controller in the bottom, discussed in detail in [33].

Now the question remains why we would need a complex controller in agents that perform simple tasks at all. We could design the morphology of the agents in a way that minimizes the need for a complex controller. Naturally, we would need a higher complexity in the case of agents that are faced with more advanced settings or that have to perform multiple tasks.

Here, we want to instead highlight the challenge that learning a new task poses. As discussed in the introduction, via the example of a musician, performing a learned task is often much easier compared to acquiring a novel behavior in the first place. The agent might not know initially how to utilize the properties of the body best in order to use Morphological Computation. Hence, our hypothesis would be that a learning agent first needs a higher controller complexity, a higher integration of information, which then decreases as the model learns to exploit the properties of the body.

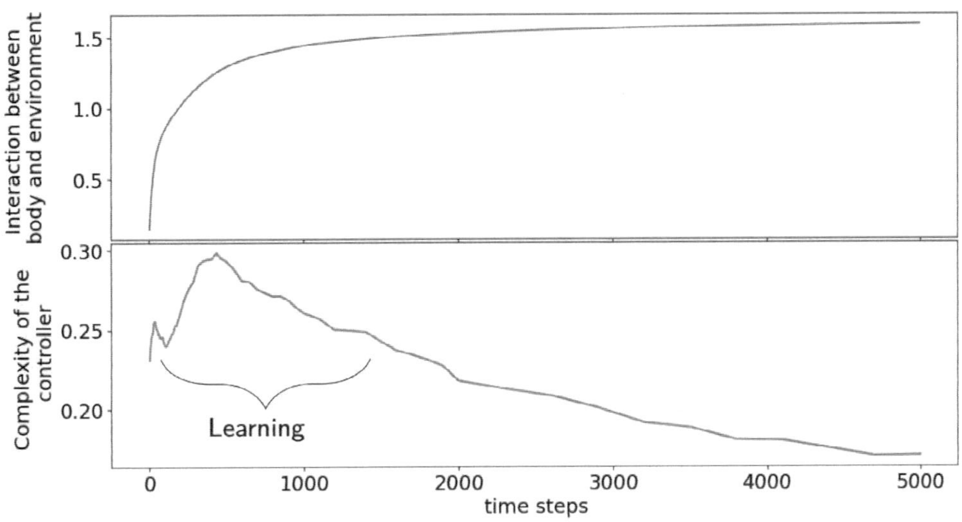

Figure 11. The results of the measures for the Morphological Computation on the top and the complexity of the controller on the bottom of learning agents, discussed in more detail in [34].

In [34] we observe simple simulated agents that learn how to move inside the racetrack depicted in Figure 9 (b) and to avoid the walls using a modified *em*-algorithm. The agents are very similar to the ones depicted in Figure 9 (a), but they additionally have an internal prediction of their next sensory state in order to enable the learning. A video of an agent in its environment and the code can be found at [29]. The learning algorithm is described in more detail in the appendix of [34] and in Section 5.4 of [32]. The results of the agents that do learn to avoid the walls are given in Figure 11. On the top we see the increasing Morphological Computation and the bottom depicts the complexity of the controller. Here the controller complexity includes an Integrated Information measure as well as a synergistic measure for the internal prediction. We observe that there is an increased controller complexity in the first approximately 1500 time steps, after which the controller complexity decreases. Note that agents which do not learn to perform the task have a lower Morphological Computation value and a constantly high controller complexity.

We conclude that an increased controller complexity is necessary for the agent to understand and exploit the environment. The agent forms an internal model of the world to predict the next sensory state. In order to learn an accurate model of the dynamics of the environment the agent needs to combine different information sources, which leads to a higher complexity of the controller. If the agent is able to form an accurate world model, then the agent can make use of its body's interaction with the environment. Hence, a higher controller complexity can facilitate a better interaction of the agent's body with its environment, leading to a higher Morphological Computation. This higher Morphological Computation then in turn reduces the necessity for a complex controller. In conclusion, Morphological Computation and the controller complexity influence each other.

8. Discussion

The information-theoretic framework discussed here enables us to measure all the information flows among the different parts of the sensorimotor loop, namely the environment and the sensor, actuator and controller nodes. Here it is possible to not only quantify the pair-wise influences, but also the synergistic impact that two or more variables have together on an additional variable.

Alternative information-theoretic approaches to measuring an information flow make use of interventional methods, as discussed in, for instance, [11, 23]. Additionally, the minimization defined here in Definition 4.1 is proposed to be changed in [9]. There the author proposes to restrict the set $\mathcal{M}$ is further to more accurately describe the underlying channel.

One caveat of the presented method is that it is only applicable for small, discrete networks at the moment. Calculating the introduced information-theoretic measures involves multiple iterations over the state spaces of the corresponding probability distributions, which is very costly for large systems. Hence, for larger networks the involved conditional probability distributions become intractable and would need to be approximated. Similarly, calculating the measures in a continuous setting is a challenge, because there the relative entropy can rarely be calculated directly but needs to be estimated. Nonetheless, extending this framework to a continuous setting would make it possible to analyze a wide range of more natural situations and is therefore a promising direction for future research.

Furthermore, the measures discussed in this setting are defined using Shannon's information theory, as introduced in Section 2. In this theory the importance of an information is solely judged by the frequency of its occurrence and not by its content, as Shannon describes in [51] as follows:

> "The fundamental problem of communication is that of reproducing at one point either exactly or approximately a message selected at another point. Frequently the messages have meaning; that is they refer to or are correlated

> according to some system with certain physical or conceptual entities. These semantic aspects of communication are irrelevant to the engineering problem."

Hence, the syntax or semantics of a message are not explicitly included in the theory, but one could argue that they would implicitly emerge over time.

Nonetheless, the introduced information-theoretic measures provide us with a framework in which we can analyze every information flow among the different parts of an acting agent. As demonstrated in Section 7 this can be used to confirm our intuitions about the influences of the controller and the environment. In the future it might lead to the discovery of interesting, counter intuitive relationships among body, brain and environment.

In conclusion, information theory provides useful tools that could serve as an additional viewpoint to analyze artificial learning and acting agents.

Funding

The authors acknowledge funding by the Deutsche Forschungsgemeinschaft Priority Programme "The Active Self" (SPP 2134).

References

[1] L. Albantakis and G. Tononi. "The Intrinsic Cause-Effect Power of Discrete Dynamical Systems—From Elementary Cellular Automata to Adapting Animats". In: *Entropy* (2015), pp. 5472–5502. DOI: 10.3390/e17085472.

[2] L. Albantakis, L. Barbosa, G. Findlay, M. Grasso, A. M. Haun, W. Marshall, W. G. P. Mayner, A. Zaeemzadeh, M. Boly, B. E. Juel, S. Sasai, K. Fujii, I. David, J. Hendren, J. P. Lang, and G. Tononi. "Integrated information theory (IIT) 4.0: Formulating the properties of phenomenal existence in physical terms". In: *PLOS Computational Biology* 19.10 (Oct. 2023), pp. 1–45. DOI: 10.1371/journal.pcbi.1011465. URL: https://doi.org/10.1371/journal.pcbi.1011465.

[3] L. Albantakis, A. Hintze, C. Koch, C. Adami, and G. Tononi. "Evolution of Integrated Causal Structures in Animats Exposed to Environments of Increasing Complexity". In: *PLOS Computational Biology* 10.12 (2014), pp. 1–19. DOI: 10.1371/journal.pcbi.1003966.

[4] S. Amari, N. Tsuchiya, and M. Oizumi. "Geometry of Information Integration". In: *Information Geometry and Its Applications*. Ed. by N. Ay, P. Gibilisco, and F. Matúš. Cham: Springer, 2018, pp. 3–17. ISBN: 978-3-030-07405-0.

[5] W. R. Ashby. *An introduction to cybernetics*. New York: J. Wiley, 1956.

[6] H. Attias. "Planning by Probabilistic Inference". In: *Proceedings of the Ninth International Workshop on Artificial Intelligence and Statistics*. 2003, pp. 9–16.

[7] N. Ay. "Information Geometry on Complexity and Stochastic Interaction". In: *MPI MIS PREPRINT 95* (2001). URL: https://www.mis.mpg.de/preprints/2001/preprint2001_95.pdf (visited on 04/28/2022).

[8] N. Ay. "Information Geometry on Complexity and Stochastic Interaction". In: *Entropy* 17 (2015), pp. 2432–2458. DOI: 10.3390/e17042432.

[9] N. Ay. "Confounding Ghost Channels and Causality: A New Approach to Causal Information Flows". In: *Vietnam Journal of Mathematics* (2021), pp. 547—576. DOI: 10.1007/s10013-021-00511-w.

[10] N. Ay and W. Löhr. "The Umwelt of an embodied agent-a measure-theoretic definition". In: *Theory in Biosciences* 134 (Dec. 2015), pp. 105—116. DOI: 10.1007/s12064-015-0217-3.

[11] N. Ay and D. Polani. "Information flows in causal networks". In: *Advances in complex systems* 11.1 (2008), pp. 17–41. ISSN: 0219-5259. DOI: 10.1142/S0219525908001465.

[12] N. Ay and K. Zahedi. "On the Causal Structure of the Sensorimotor Loop". In: *Guided Self-Organization: Inception.* Ed. by M. Prokopenko. Berlin, Heidelberg: Springer, 2014, pp. 261–294. DOI: 10.1007/978-3-642-53734-9_9.

[13] R. A. Brooks. "Intelligence without reason". In: *Proceedings of the 12th international joint conference on artificial intelligence (IJCAI-91).* Ed. by R. R. Myopoulos J. Morgan Kaufmann publishers Inc., 1991, pp. 569–595.

[14] R. A. Brooks. "Intelligence without representation". In: *Artificial Intelligence* 47.1 (1991), pp. 139–159. ISSN: 0004-3702. DOI: 10.1016/0004-3702(91)90053-M.

[15] T. M. Cover and J. A. Thomas. *Elements of Information Theory.* New York: John Wiley & Sons, Inc., 2006. ISBN: 978-0-471-06259-2.

[16] J. Darroch and D. Ratcliff. "Generalized iterative scaling for log-linear models". In: *The Annals of Mathematical Statistics* 43 (1972), pp. 1470–1480. DOI: 10.1214/aoms/1177692379.

[17] J. A. Edlund, N. Chaumont, A. Hintze, C. Koch, G. Tononi, and C. Adami. "Integrated Information Increases with Fitness in the Evolution of Animats". In: *PLoS Computational Biology* 7.10 (2011). DOI: 10.1371/journal.pcbi.1002236.

[18] K. Ghazi-Zahedi. *Morphological Intelligence.* Cham: Springer, 2019. ISBN: 978-3-030-20620-8.

[19] K. Ghazi-Zahedi, D. Häufle, G. Montúfar, S. Schmitt, and N. Ay. "Evaluating Morphological Computation in Muscle and DC-Motor Driven Models of Hopping Movements". In: *Frontiers in Robotics and AI* 2 (2016). DOI: 10.3389/frobt.2016.00042.

[20] K. Ghazi-Zahedi, C. Langer, and N. Ay. "Morphological Computation: Synergy of Body and Brain". In: *Entropy* 19.456 (2017). DOI: 10.3390/e19090456.

[21] K. Ghazi-Zahedi and N. Ay. "Quantifying Morphological Computation". In: *Entropy* 15.5 (2013), pp. 1887–1915. DOI: 10.3390/e15051887.

[22] K. Ghazi-Zahedi, R. Deimel, G. Montúfar, V. Wall, and O. Brock. "Morphological Computation: The Good, the Bad, and the Ugly". In: *2017 IEEE/RSJ International Conference on Intelligent Robots and Systems.* 2017, pp. 464–469. DOI: 10.1109/IROS.2017.8202194.

[23] D. Janzing, D. Balduzzi, M. Grosse-Wentrup, and B. Schölkopf. "Quantifying causal influences". In: *The Annals of Statistics* 41.5 (2013), pp. 2324 –2358. DOI: 10.1214/13-AOS1145. URL: https://doi.org/10.1214/13-AOS1145.

[24] M. S. Kanwal, J. A. Grochow, and N. Ay. "Comparing Information-Theoretic Measures of Complexity in Boltzmann Machines". In: *Entropy* 19.310 (2017). DOI: 10.3390/e19070310.

[25] D. P. Kingma and M. Welling. *Auto-Encoding Variational Bayes.* 2022. arXiv: 1312.6114 [stat.ML].

[26] A. S. Klyubin, D. Polani, and C. L. Nehaniv. "Representations of Space and Time in the Maximization of Information Flow in the Perception-Action Loop". In: *Neural computation* 19 (Oct. 2007), pp. 2387–432. DOI: 10.1162/neco.2007.19.9.2387.

[27] A. S. Klyubin, D. Polani, and C. L. Nehaniv. "Tracking Information Flow through the Environment: Simple Cases of Stigmergy". In: *Artificial Life IX: Proceedings of the Ninth International Conference on the Simulation and Synthesis of Living Systems.* 2004, pp. 563–568. ISBN: 978-0-262-66183-6. URL: https://uhra.herts.ac.uk/bitstream/handle/2299/3945/101985.pdf?sequence=4&isAllowed=y (visited on 08/10/2023).

[28] A. Klyubin, D. Polani, and C. Nehaniv. "Empowerment: a universal agent-centric measure of control". In: *2005 IEEE Congress on Evolutionary Computation.* Vol. 1. 2005, pp. 128–135. DOI: 10.1109/CEC.2005.1554676.

[29] C. Langer. *LearningRequiresIntInf GitHub Repository.* 2022. URL: https://github.com/CarlottaLanger/LearningRequiresIntInf (visited on 09/01/2022).

[30] C. Langer. *Morphology Shapes Integrated Information GitHub Repository.* 2021. URL: https://github.com/CarlottaLanger/MorphologyShapesIntegratedInformation (visited on 07/20/2023).

[31] C. Langer and N. Ay. "Complexity as Causal Information Integration". In: *Entropy* 22.10 (2020). DOI: 10.3390/e22101107.

[32] C. Langer. "Information Integration in Embodied Systems". PhD thesis. Hamburg University of Technology, 2024.

[33] C. Langer and N. Ay. "How Morphological Computation Shapes Integrated Information in Embodied Agents". In: *Frontiers in Psychology* 12 (2021). ISSN: 1664-1078. DOI: 10.3389/fpsyg.2021.716433.

[34] C. Langer and N. Ay. "Outsourcing Control Requires Control Complexity". In: *Artificial Life* (June 2024), pp. 1–22. ISSN: 1064-5462. DOI: 10.1162/artl_a_00443.

[35] D. J. C. MacKay. *Information Theory, Inference and Learning Algorithms.* Cambridge: Cambridge University Press, 2003. ISBN: 978-0-521-64298-9.

[36] P. A. Mediano, A. K. Seth, and A. B. Barrett. "Measuring Integrated Information: Comparison of Candidate Measures in Theory and Simulation". In: *Entropy* 21.1 (2018). ISSN: 1099-4300. DOI: 10.3390/e21010017.

[37] G. Montúfar, K. Ghazi-Zahedi, and N. Ay. "A Theory of Cheap Control in Embodied Systems". In: *PLOS Computational Biology* 11.9 (2015). DOI: 10.1371/journal.pcbi.1004427.

[38] V. Müller and M. Hoffmann. "What Is Morphological Computation? On How the Body Contributes to Cognition and Control". In: *Artificial Life* 23 (2017). DOI: 10.1162/ARTL_a_00219.

[39] K. Nakajima, H. Hauser, R. Kang, E. Guglielmino, D. Caldwell, and R. Pfeifer. "A soft body as a reservoir: case studies in a dynamic model of octopus-inspired soft robotic arm". In: *Frontiers in Computational Neuroscience* 7 (2013). DOI: 10.3389/fncom.2013.00091.

[40] M. Oizumi, L. Albantakis, and G. Tononi. "From the Phenomenology to the Mechanisms of Consciousness: Integrated Information Theory 3.0". In: *PLOS Computational Biology* (2014). DOI: 10.1371/journal.pcbi.1003588.

[41] M. Oizumi, N. Tsuchiya, and S. Amari. "Unified framework for information integration based on information geometry". In: *JournalProceedings of the National Academy of Sciences of the United States of America.* 2016, pp. 14817–14822. DOI: 10.1073/pnas.1603583113.

[42] T. Parr, G. Pezzulo, and K. J. Friston. *Active inference: the free energy principle in mind, brain, and behavior.* MIT Press, •.

[43] J. Pearl. *Causality.* Cambridge: Cambridge University Press, 2009. ISBN: 978-0-511-80316-1.

[44] J. Pearl. *Probabilistic Reasoning in Intelligent Systems.* San Francisco: Morgan Kaufmann Publishers, 1988. ISBN: 978-1-55860-479-7.

[45] J. Pearl. "On the Interpretation of do(x)". In: *Journal of Causal Inference* 7 (2019). DOI: 10.1515/jci-2019-2002.

[46] R. Pfeifer and J. Bongard. *How the body shapes the way we think: a new view of intelligence.* Cambridge: The MIT Press (Bradford Books), 2006. ISBN: 978-0-262-53742-1.

[47] R. Pfeifer and G. Gómez. "Morphological Computation – Connecting Brain, Body, and Environment". In: *Creating Brain-Like Intelligence: From Basic Principles to Complex Intelligent Systems.* Ed. by B. Sendhoff, E. Körner, O. Sporns, H. Ritter, and K. Doya. Berlin, Heidelberg: Springer Berlin Heidelberg, 2009, pp. 66–83. DOI: 10.1007/978-3-642-00616-6_5.

[48] R. Pfeifer, F. Iida, and G. Gomez. "Morphological computation for adaptive behavior and cognition". In: *International Congress Series* 1291 (June 2006), pp. 22–29. DOI: 10.1016/j.ics.2005.12.080.

[49] C. Salge, C. Glackin, and D. Polani. "Empowerment–An Introduction". In: *Guided Self-Organization: Inception.* Ed. by M. Prokopenko. Berlin, Heidelberg: Springer, 2014, pp. 67–114. DOI: 10.1007/978-3-642-53734-9_4.

[50] M. Seitzer, B. Schölkopf, and G. Martius. "Causal Influence Detection for Improving Efficiency in Reinforcement Learning". In: *Advances in Neural Information Processing Systems.* Ed. by A. Beygelzimer, Y. Dauphin, P. Liang, and J. W. Vaughan. 2021. URL: https://openreview.net/forum?id=DXJl9826dm (visited on 07/10/2023).

[51] C. E. Shannon. "A Mathematical Theory of Communication". In: *The Bell System Technical Journal* 27 (1948), pp. 379 –423. ISSN: 0005-8580.

[52] C. Shannon and W. Weaver. *The Mathematical Theory of Communication.* The University of Illinois Press, 1949. ISBN: 978-0-252-72548-7.

[53] N. Tishby and D. Polani. "Information Theory of Decisions and Actions". In: *Perception-Action Cycle: Models, Architectures, and Hardware.* Ed. by V. Cutsuridis, A. Hussain, and J. G. Taylor. New York: Springer, 2011, pp. 601–636. DOI: 10.1007/978-1-4419-1452-1_19.

[54] G. Tononi. "Consciousness as Integrated Information: a Provisional Manifesto". In: *The Biological Bulletin* 215 (2008), pp. 216–242. DOI: 10.2307/25470707.

[55] G. Tononi, O. Sporns, and G. M. Edelman. "A measure for brain complexity: Relating functional segregation and integration in the nervous system". In: *Proceedings of the National Academy of Sciences of the United States of America* (1994), pp. 5033–5037. DOI: 10.1073/pnas.91.11.503.

[56] M. Toussaint. "Probabilistic inference as a model of planned behavior". In: *Künstliche Intelligenz* 23.3 (2009), pp. 23–29.

[57] M. Toussaint, L. Charlin, and P. Poupart. "Hierarchical POMDP Controller Optimization by Likelihood Maximization". In: *Proceedings of the Twenty-Fourth Conference Annual Conference on Uncertainty in Artificial Intelligence.* 2008, pp. 562–570. DOI: 10.5555/3023476.3023543.

IOP Conf. Series: Materials Science and Engineering **1321** (2024) 012009 doi:10.1088/1757-899X/1321/1/012009

Inherent Behavioral Stochasticity in Soft Robots: Analysis and Control Strategies

Muhammad Sunny Nazeer,[1, 2] Syeda Shadab Zehra Zaidi,[2, 3] Matteo Cianchetti[2, 3] and Egidio Falotico[2, 3]

[1] College of Design and Engineering, National University of Singapore, Singapore
[2] The BioRobotics Institute, Scuola Superiore Sant'Anna, Pisa, Italy
[3] Department of Excellence in Robotics and AI, Scuola Superiore Sant'Anna, Pisa, Italy

E-mail: `m.nazeer@nus.edu.sg ; (s.zaidi, m.cianchetti, e.falotico)@santannapisa.it`

Abstract. The unique composition of soft robots, predominantly constructed from elastomers and polymers, amplifies the likelihood of unpredictability in their performance, setting the stage for significant behavioral stochasticity as compared to their rigid counterparts. In this paper, we present a control-centric perspective on the intrinsic behavioral stochasticity observed in soft robots, exploring the underlying reasons and presenting control methodologies tailored to address the associated challenges. In addition to these challenges, we also provide insights on the potential benefits that the intrinsic behavior of the soft structure can have when they come in contact with unstructural environments. Finally, we discuss the generic control schemes traditionally used with these robots and highlighted potential strategies to alleviate the performance gaps introduced by the inherent unpredictability.

1. Introduction

Literature sees the role of stochasticity central to the adaptation [1] whether it is about responding to the unstructured environment [2, 3, 4], learning by exploring [5, 6], modelling a sophisticated behavior in terms of distributions of different parameters of a complex robot [7], or automating the novel design and manufacturing process of robots with functional materials [8, 9]. In all of these studies, intentionally introducing stochasticity in the process improved the capability of the underlying application. Stochasticity may result in a favorable outcome where adaptation means reliability and compliance by exploring. However, it may lead to an unfavorable outcome when precision is required [10, 11].

The field of soft robotics lies at the intersection of mechanical engineering, material science and a combination of control theory and computer science. Soft robots are compliant, adaptable, and safe to interact with. With compliance and softness also comes the stochasticity in their behavior. The stochasticity here refers to the inherent variability and unpredictability in their movements and responses to external environment. Since soft robots are made up of flexible material, their stochasticity is traditionally considered a consequence of the plastic properties of the materials such as damping, hysteresis, nonlinear deformations, etc. However, there are additional factors that may contribute to the behavioral stochasticity, such as varying environmental conditions like temperature, humidity, pressure, manufacturing inaccuracies, varying moment of inertia due to imbalanced morphology, etc.

When it comes to the precision requirements for soft robot applications, such as delicate surgeries [12] or mimicking a biological phenomena learned from a biological system [13], whether in modeling [14] or control [6, 10, 15], stochasticity may lead to unfavorable outcomes primarily due to the factors that make this class of robotics advantageous i.e., variability, compliance, and unpredictability. Additionally, the control solutions that are inherently robust in nature like, Reinforcement Learning (RL)-based algorithms, tend to underperform with systems that exhibit stochastic behavior [6, 10], introducing additional challenges to the soft robots' control field and therefore requires more sophisticated and resource intensive solutions [16]. The unpredictability also increases the safety risks associated with the robot's own morphology and the intended application environment.

The stochasticity introduced in the behaviors of soft robots, as perceived by control algorithms, can also stem from the sensing mechanisms employed to estimate their state feedback. Factors such as sensor noise, delays, inaccuracies, and the dynamic nature of environmental conditions contribute significantly to the stochastic behaviors observed [17]. For example, sensors may provide inconsistent feedback due to changes in temperature, humidity, or pressure, leading to variability in robot response [18, 19]. Moreover, the quality and precision of sensory mechanisms play a crucial role in how effectively control algorithms can adapt and manage robot behavior [20]. Thus, while the intrinsic stochasticity of soft robots can enhance adaptability and compliance, the additional layer of variability introduced by sensing mechanisms poses challenges in achieving precision and reliability in their control.

Another factor responsible for nonrepeatable behavior in soft robots, beyond the material and control solution, is the actuation mechanism. Traditionally, soft robotic platforms are actuated using fluidic channels, tendons/wires, Shape Memory Alloys (SMAs), or Magnetorheological Materials (MRs) [21, 22], etc. The mechanisms introduce variability of their own; For example, pneumatically driven soft robotic platforms include electronic pressure regulators, and they are known for their characteristic variable response [15]. The same pneumatic regulators usually do not include negative pressure regulators, which further introduce slow or incomplete depletion of the pneumatic channels. Similarly, other associated actuation-related stochastic responses come from the variable tensions in the tendons, slow and temperature-dependent response of SMAs and magnetic hysteresis of MRs, etc.

Based on the aforementioned points, it is evident that understanding the stochastic nature of the system is essential for identifying appropriate applications, control strategies, countermeasures, and compensation techniques for them to ensure consistent, reliable, and extended performance.

In this paper, we present our perspective on the link of the stochasticity-introducing factors contributing to the recorded change in the behavior of a soft continuum manipulator and, thus, the validity of a control solution over its useful lifetime. Finally, potential remedies for addressing stochasticity in terms of control solutions are discussed, along with the impending challenges and additional factors to consider to achieve desired response.

2. Stochasticity Analysis

It is important to understand the contributing factors of stochasticity in soft robots in order to optimize the phases of their design and operation. The main reason behind the inconsistency in the behavior of soft robots is the material that is used for their fabrication. These materials are mainly composed of very flexible elastomers that are sensitive to a number of elements, as discussed below.

2.1. Material Induced Stochasticity in Soft Robots

2.1.1. Temporal behavioral variations As elastomers age, they undergo changes. Over time, their physical attributes, such as their resilience to stress and flexibility, may deteriorate. A

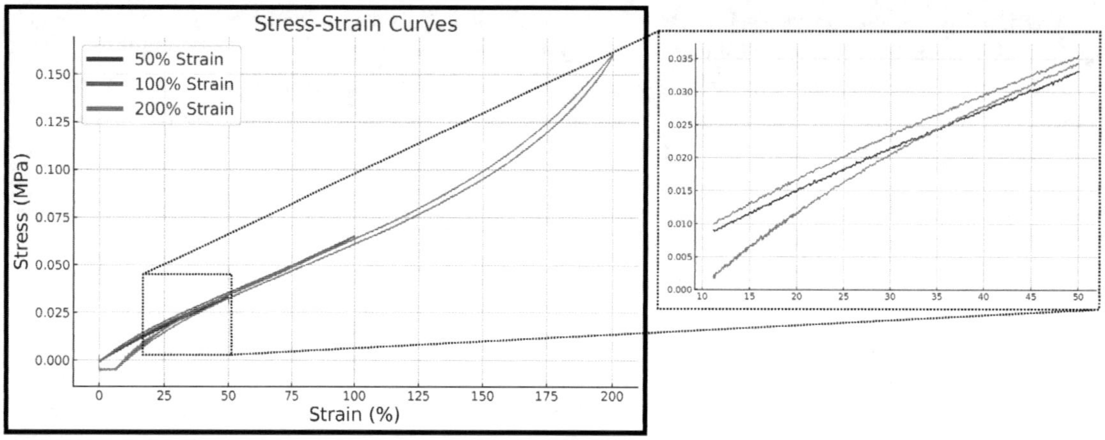

Figure 1. Stress-strain curves for Ecoflex 50 Tensile Sample; If actuators in a robot have been characterised against different levels of strain (for e.g at 50%, 100%, 200%) but then during operation have to operate at same levels of strain (for e.g 30%), there behavior would be different as can be seen from the magnified image.

robot could be impacted by it by becoming less predictable during operation.

2.1.2. Strain-induced variability This refers to the variations that occur as a result of mechanical deformation or stress. When subjected to different strains, the elastomers change their behaviour which makes it challenging to precisely forecast their reaction to changing loads particularly in dynamic scenarios, due to their nonlinear characteristics.

The strain induced variability can be observed through an example. Figure 1 shows the results of a tensile test that was conducted using an Instron (Universal Testing Machine) on a standard Ecoflex 50 tensile sample (ASTM D412). Three different samples were used in the experiment, and each of them was subjected to three different levels of strain: 50%, 100%, and 200%. Four tests were performed on each sample for each strain at a constant speed of 50 mm/min. The first cycle was ignored to account for Mullin's effect. The average of the second, third, and fourth cycles has been plotted for each curve.

From the figure, it is clear that the sample's material characteristics cause it to show hysteresis. Moreover, there are behavioural changes in response to varying strain levels. It is clear from a close inspection of every curve that the stress response differs for every one of them at the same strain value. Understanding this behavior is important to understand the stochastic nature of soft robots.

2.1.3. Temperature sensitivity Elastomers react strongly to temperature variations, in contrast to other materials. Environments with fluctuating temperatures can affect the overall mechanical properties of the elastomers. An example is that an elastomeric sample that is allowed to cure at room temperature displays different characteristics than the one cured at higher temperature.

2.1.4. Interaction with the surroundings The elastomers have a responsive behavior to their environment. The material properties of elastomers can undergo alterations as a result of exposure to various chemicals, moisture, and several other environmental factors. These changes have a direct impact on the overall performance of a soft robot.

2.1.5. Variations in mixing ratios Due to the fact that elastomers are commonly composed of a mixture of different chemicals, their synthesis depends on accurate mixing ratios. Fluctuations in the proportion of ingredients can influence the characteristics of the final result, leading to inconsistent results.

2.1.6. Shelf life and storage conditions Elastomeric materials can be influenced by factors such as light, air, temperature, and storage conditions, which can potentially impact their longevity. For instance, an elastomeric sample that is newly synthesized will have distinct characteristics compared to the one that has been stored for a period of time. It is important to maintain proper storage conditions to avoid compromising the material properties.

2.2. Stochasticity due to Wear and Degradation of Soft Robots

The unpredictability in soft robots' behavior can also arise due to their wear and degradation patterns. Its important to understand these stochastic aspects as it directly relates to the maintenance processes and robots' lifespan. Soft robots during their operation undergo complex motions and patterns unlike the rigid robots which can give rise to wear and tear over a period of time.Predictive maintenance should be performed in order to take into account this aspect of stochasticity.

2.3. Stochasticity due to Manufacturing Discrepancies

Three soft linear actuators internally reinforced with fibers made up of Dragon skin 10 shown in Figure 2a have been fabricated under exactly same conditions. When characterized against pressure, all three exhibited different behaviors (see Figure 2b). The main reason behind this is the human factor involved in the fabrication process. Although the same fabrication protocols were followed throughout the process, the fact that humans cannot replicate any two repeated steps with 100 percent accuracy leads to persistent challenges. In addition to the human error involved, the factors discussed in sections 2.1 and 2.2 contributed to the overall stochastic behavior of these actuators.

After the soft continuum manipulator was fully fabricated, testing its performance under the same actuation-space trajectories multiple times resulted in different task-space trajectories, as shown in Figure 3a. A population statistics was computed from the results where the very first trial conducted during the experiments (shown in red color in Figure 3a) was treated as a base trial. All the subsequent trials were compared with the base trial. It is important to note here that any trial out of six can be considered as a base trial. An error was computed in the individual axes (between the base trial and subsequent trials). The error (in mm) along y-axis is reported, for every trial along x-axis, as shown in Figure 3b, in the form of boxplots.

The factors responsible for the stochasticity in the soft manipulator may involve the incomplete depletion of the pneumatic chambers of the soft manipulator, variable moment of inertia along the length of the soft manipulator due to imbalanced morphology of the soft manipulator resulting from the actuators' fabrication discrepancies discussed in sections 2.1 and 2.2, and variable initial conditions (e.g., task-space tip position of the soft manipulator, etc.). We used motion capture systems (with error $\leq 0.5\ mm$) to record the soft robot movements which reduced the uncertainties due to the sensor.

2.4. Application Driven Stochasticity

Depending on the application where soft robots are deployed, the stochasticity can arise. An example can be the biomedical applications where soft robots interact with the tissues that are themselves stochastic in nature owing to their softness. When working in the areas of prosthetics, tissue engineering and other related fields, these stochastic interactions should be taken into account to better predict the behavior of soft robots during their operation.

IOP Conf. Series: Materials Science and Engineering **1321** (2024) 012009 doi:10.1088/1757-899X/1321/1/012009

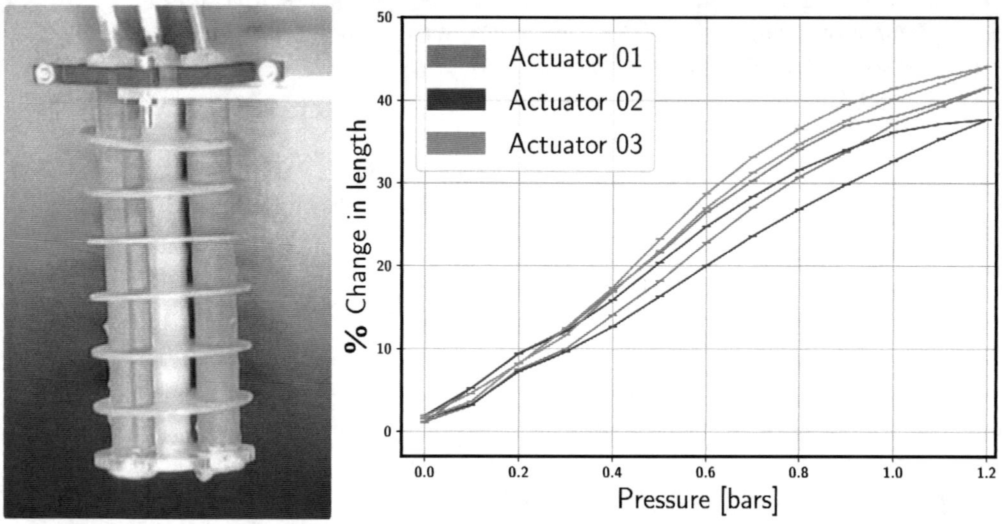

(a) (b)

Figure 2. Three internally fiber reinforced soft linear actuators were designed with Dragon-Skin 10. A soft continuum manipulator with three-independently actuated linear actuators were combined to a single module, as shown in (a). The three linear actuators were individually characterized and the pressure vs percentage elongation of the three actuators is shown in (b).

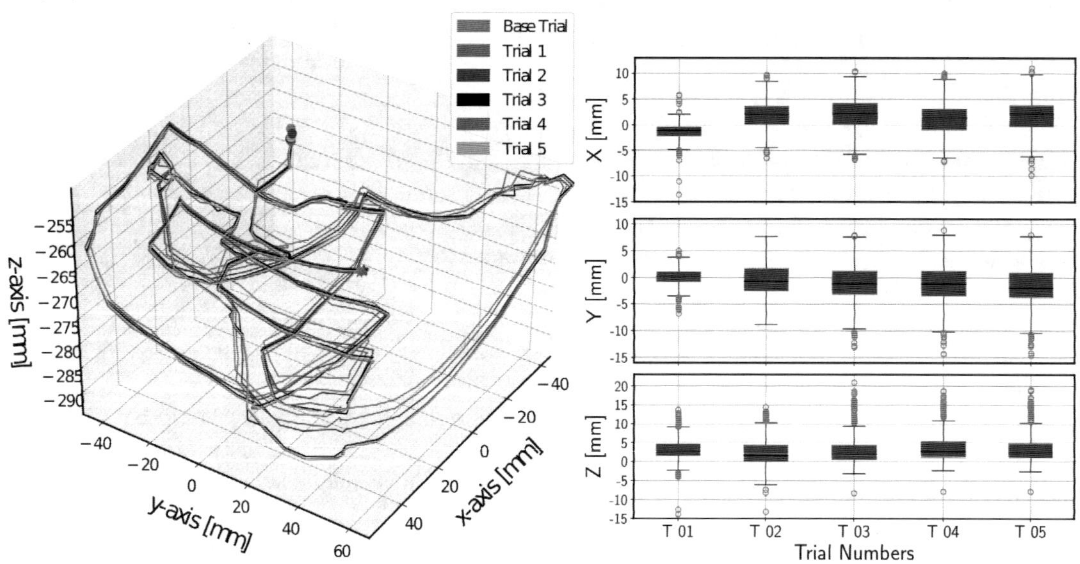

(a) (b)

Figure 3. A single-module pneumatically actuated soft continuum manipulator, with three independently actuated pneumatic chambers, was actuated six times under the same actuation-space trajectory. The 3D task-space trajectories acquired from consecutive trials are shown in (a). The population statistics, for the error between the base trials (the first trial conducted during the experiments) and the following five trials, is shown in (b). The robot was actuated in the dynamics domain.

3. Effect on the Control of Soft Robots

In the literature, the control of soft robots is proposed in one of the three ways: type 1 — modeling the behavior of the underlying platform and deriving a control solution for it [23, 24, 25], type 2 — deriving a control solution directly on the physical platform [13, 26, 27, 28, 29], and type 3 — a combination of the two [10, 30, 31, 32]. However, it is acknowledged that deriving control solutions directly on a soft robot is a relatively underexplored area due to the complex and stochastic mechanics of soft robots.

When it comes to the modeling of soft robots, there have been numerous solutions [33], including Constant Curvature, Piecewise Constant Curvature, Piecewise Smooth Curvature, Koopman operators, Kirchhoff love shell theory, Euler Bernoulli beam theory, Cosserat rod theory, Finite Element Methods, Data-driven deep learning methods, etc. Most of the methods are capable of describing the soft robots' behavior and geometry in static, kinematic and dynamic domain. However, the validity of these approaches decreases when we move from the static to the dynamic domain. Many factors that remained dormant in the static (equilibrium) domain, or had time to settle when displaced from one point to another (such as hysteresis, forces, moment of inertia), do not enjoy the same luxury in the dynamic domain. Additionally, the factors, experimented upon in Section 2, are not necessarily addressed by the modeling approaches. Consequently, the solutions presented in type 1 and 2 suffer greatly and face a substantial sim2real performance gap [6, 10].

Type 2 solutions are usually learned using sample-efficient learning-based approaches [27, 34, 35]. It is imperative that these approaches learn from a limited dataset due to the fragile morphology of soft robots and that they are learned directly on the platform. Consequently, the control solutions do not exhibit sim2real performance gap [27, 13]. However, their robustness and adaptability to variability in the application environment or material changes within the soft platform itself can be questionable owing to the limited dataset used for learning. Additionally, if the morphology of the subject platform is not fragile, one may execute the online learning scheme for extended period. Doing so, as demonstrated in Section 2, may result in the soft robot undergoing change in its behavior due to the change in the material properties and by that extension, causing the solution to underperform. A similar trend was also observed for a tactile sensor undergoing change due to a change in its material properties as a result of overnight real-world data-driven model training [36].

This aspect also highlights a gap in the field of soft robot control. While many control solutions in the literature demonstrate good performance, there is a noticeable absence of discussions regarding the adaptability and repeatability of these solutions over an extended period. In essence, if a control solution proves effective initially, how likely it is to remain applicable after prolonged use of the soft platform or as the material ages? Interestingly, this aspect is extensively studied in sensor models, as sensors tend to introduce drift in their readings over time, models typically are made to reflect this drift in sensor readings [17].

4. Potential Solutions

Based on the control solutions listed in Section 3, stochasticity can be overcome through one of the three strategies: (1) With model improvements: a control solution is derived from a model that can account for either the global properties of the underlying soft platform and forecast any changes that may potentially occur during the lifetime of the soft platform or it is sub-optimal and undergoes improvements periodically. (2) From scratch again: whenever the soft platform undergoes a significant change in performance, a control solution is derived from scratch. (3) Compensatory agent: a compliant adaptor responsible for overcoming the performance gap on the top of a control solution.

4.1. With Model Improvements
Deriving a global model of a soft robot is not a practically feasible solution due to the attributes inherent to the soft structures and the variabilities in the material properties over time. However, a sub-optimal model of the behavior of a soft robot is possible. As expected the performance of this model may degrade over time which can be mitigated by periodically improving it as presented in our previous works [13, 27, 32]. In these works, a control solution is trained using Imitation learning (IL) with a behavioral map of the underlying soft manipulator in the dynamics domain. Since, the solution was trained using IL so its robustness against stochasticity and other variations needed improvements. Combining the IL-trained solution with a periodically improving dynamic behavioral map of the soft manipulator via Transfer Learning (TL) showed not only improved performance against stochasticity but also task repeatability with minimum standard deviation. While this approach proved to be sample efficient, it can only perform qualitatively. For quantitative response, a dynamic gait controller was presented in [34] that executes, periodically, a bootstrapping algorithm on an ensemble of dynamic gaits generated by a neural oscillator.

4.2. From Scratch Again
The strategy explores the possibility of deriving or training a control solution everytime the underlying platform undergoes significant changes. Such a technique is particularly useful when employed with a learning-based control strategy where the performance of the trained solution decreases drastically [37, 38] in response to the behavioral modification of the soft robot. However, such a solution can be sample inefficient and may also become impractical if the behavior of the soft platform varies more often owing to external conditions.

4.3. Compensatory Agent
This strategy is about deriving/training an additional agent whose job is to compensate for the performance gap. The examples of such schemes are presented in our previous works in [6, 10]. RL-trained policies tend to be robust; however, if the training environment undergoes variation, the performance degrades. We presented a compliant adaptor, learned completely in online mode with the soft robot, which can compensate for the training-to-reality gap, stochasticity, and performance degradation due to external stress affecting the soft robot's performance. Additionally, the performance of these schemes was also evaluated when one or more of the actuation channels of the soft robot were damaged. The compensator was able to adapt by using the redundancy present in the robot. However, such a scheme can only work as long as the task constraints are met. If there is a change in the soft robot behavior where the current task constraints become invalid, the compensator may consider it as a new task and won't be able to adapt to it.

Other similar works presented in the literature include combining a model-plant mismatch compensator with a Model Predictive Control (MPC), by Koryakovsky et al. [39], for a set-point-reaching task. However, the validity of the approach has been tested with a seven-DoF rigid robot. A similar approach in the kinematics domain was also presented by Kalidindi et al. [40] tested with the kinematic simulation environment of a soft manipulator. Johnson et al. [41] used a first-principles model combined with a data-driven model based on machine learning (ML) for the MPC-based controller. Their study showed a 52% increase in the controller's performance compared to four different models employed for the same task with the MPC. In addition to control strategies, the other aspects that can help reduce stochasticity in soft robots are discussed below.

4.4. Manufacturing and Quality Control
The stochasticity induced due to manufacturing processes can be reduced by taking appropriate measures for quality control. Some of the suggestions that can be taken into account in this regard are:

- Use fabrication techniques that involve minimum inaccuracies like 3D printing where applicable.
- Make sure to use the same composition of material everytime in order to avoid variability.
- Perform the fabrication in environment with controlled humidity and temperature.
- Regularly calibrate the equipment and standardize the procedures involved in fabrication.

4.5. Simulate Behavior of Soft Robots
Significant development has been made on simulating the behavior of soft robots in the last decade. Some of the notable simulators in this regard are SOFA, Deformable Gym, SoMoGym, Elastica and Chrono. Simulating the behavior of soft robots in these advanced simulation environments can significantly enhance the understanding of the system by providing deeper insights on their interactions with the objects and environment.

5. Challenges and Future Work
The stochasticity in soft robots is an unavoidable phenomenon which can be mitigated through standardized processes of material synthesis, fabrication methods and controlled environmental conditions but cannot be fully eradicated. This is owing to the fact that the factors involved in inducing the stochasticity cannot be controlled to the fullest. The human error involved in the fabrication processes, the softness induced material non-linearity and the fluctuating environmental conditions will always be there to play there role in introducing the inconsistency in soft robot's behavior. This poses a challenge to predict the exact behavior of soft robot during operation.

In order to acquire reliability in the operation of this class of robotics, it is imperative that the proposed control solutions actively account for the stochasticity and compensate for the variability. Presenting adaptable control solutions has a high impact in this domain because while the stochasticity answers the question about the variability that is available right now, how it will change over time is a factor we may not be able to predict. Continual learning paradigms are best known for their ability to adapt for extensive changes by using previous knowledge. These approaches are inspired by the learning capabilities of the human brain. In future, we would like to use these learning paradigms as compensatory factors or as substitutes for models with adaptive capabilities.

6. Conclusion
This paper discusses several factors that potentially contribute to the stochastic behavior of soft structures. This randomness in behavior cannot be eradicated to the fullest, but can be mitigated through careful considerations related to materials, manufacturing, and controls. Where, on one hand, stochasticity poses challenges in determining the exact behavior of soft robots, on the other hand, it can be seen as an opportunity in a way that the compliance and softness of the material helps the robot function in unstructured environments. It contributes to additional challenges in terms of their reliable and repeatable control. Therefore, we listed the factors that should be carefully considered when proposing their control solutions. Finally, we discussed the generic control schemes traditionally used with these robots and highlighted potential strategies to alleviate the performance gaps introduced by the inherent unpredictability.

Acknowledgements
We would like to acknowledge the collaborative efforts between the BRain-Inspired Robotics Lab (BRAIR Lab) and the Soft Mechatronics for Biorobotics Lab (SMB Lab) of The BioRobotics Institute, Scuola Superiore Sant'Anna . This work was funded by the European Union's Horizon 2020 Research and Innovation Programme in the framework of the project SMART (Soft, Self-responsive Smart Materials for Robots) under Marie Skłodowska-Curie Actions (MSCA), Innovative Training Network (ITN) with grant agreement no. 860108.

References
[1] T. Lenormand, D. Roze, and F. Rousset, "Stochasticity in evolution," *Trends in ecology & evolution*, vol. 24, no. 3, pp. 157–165, 2009.
[2] T. George Thuruthel, E. Falotico, M. Manti, A. Pratesi, M. Cianchetti, and C. Laschi, "Learning closed loop kinematic controllers for continuum manipulators in unstructured environments," *Soft robotics*, vol. 4, no. 3, pp. 285–296, 2017.
[3] S. Mintchev, D. Zappetti, J. Willemin, and D. Floreano, "A soft robot for random exploration of terrestrial environments," in *2018 IEEE International Conference on Robotics and Automation (ICRA)*. IEEE, 2018, pp. 7492–7497.
[4] J. Talamini, E. Medvet, and S. Nichele, "Criticality-driven evolution of adaptable morphologies of voxel-based soft-robots," *Frontiers in Robotics and AI*, vol. 8, p. 673156, 2021.
[5] T. Haarnoja, A. Zhou, K. Hartikainen, G. Tucker, S. Ha, J. Tan, V. Kumar, H. Zhu, A. Gupta, P. Abbeel *et al.*, "Soft actor-critic algorithms and applications," *arXiv preprint arXiv:1812.05905*, 2018.
[6] M. S. Nazeer, D. Bianchi, G. Campinoti, C. Laschi, and E. Falotico, "Policy adaptation using an online regressing network in a soft robotic arm," in *2023 IEEE International Conference on Soft Robotics (RoboSoft)*. IEEE, 2023, pp. 1–7.
[7] S. Honji, H. Arita, and K. Tahara, "Stochastic approach for modeling soft fingers with creep behavior," *Advanced Robotics*, pp. 1–14, 2023.
[8] J. Pinskier and D. Howard, "From bioinspiration to computer generation: Developments in autonomous soft robot design," *Advanced Intelligent Systems*, vol. 4, no. 1, p. 2100086, 2022.
[9] J. Hiller and H. Lipson, "Automatic design and manufacture of soft robots," *IEEE Transactions on Robotics*, vol. 28, no. 2, pp. 457–466, 2011.
[10] M. S. Nazeer, C. Laschi, and E. Falotico, "Rl-based adaptive controller for high precision reaching in a soft robot arm," *IEEE Transactions on Robotics*, 2024.
[11] C. Laschi, T. G. Thuruthel, F. Lida, R. Merzouki, and E. Falotico, "Learning-based control strategies for soft robots: Theory, achievements, and future challenges," *IEEE Control Systems Magazine*, vol. 43, no. 3, pp. 100–113, 2023.
[12] M. Pourafzal, A. Talebi, and K. Rabenorosoa, "Stochastic model-based contact force estimation for concentric tube robots," *Mechanism and Machine Theory*, vol. 180, p. 105135, 2023.
[13] M. S. Nazeer, Y. T. Ansari, M. Rivière, Y. Meroz, C. Laschi, and E. Falotico, "An imitation learning-based approach for enabling plant-like tropic abilities in a redundant soft continuum arm," in *2024 10th International Conference on Control, Automation and Robotics (ICCAR)*. IEEE, 2024, pp. 101–108.
[14] E. N. Evans, A. P. Kendall, and E. A. Theodorou, "Stochastic spatio-temporal optimization for control and co-design of systems in robotics and applied physics," *Autonomous Robots*, pp. 1–24, 2022.
[15] D. Bruder, B. Gillespie, C. D. Remy, and R. Vasudevan, "Modeling and control of soft robots using the koopman operator and model predictive control," *arXiv preprint arXiv:1902.02827*, 2019.
[16] Z. Tang, P. Wang, W. Xin, Z. Xie, L. Kan, M. Mohanakrishnan, and C. Laschi, "Meta-learning-based optimal control for soft robotic manipulators to interact with unknown environments," in *2023 IEEE International Conference on Robotics and Automation (ICRA)*. IEEE, 2023, pp. 982–988.
[17] D. Kim, M. Park, and Y.-L. Park, "Probabilistic modeling and bayesian filtering for improved state estimation for soft robots," *IEEE Transactions on Robotics*, vol. 37, no. 5, pp. 1728–1741, 2021.
[18] C. Mattmann, F. Clemens, and G. Tröster, "Sensor for measuring strain in textile," *Sensors*, vol. 8, no. 6, pp. 3719–3732, 2008.
[19] A. Georgopoulou, P. Diethelm, M. Wagner, R. Spolenak, and F. Clemens, "Soft self-regulating heating elements for thermoplastic elastomer-based electronic skin applications," *3D Printing and Additive Manufacturing*, vol. 11, no. 2, pp. e828–e838, 2024.
[20] T. G. Thuruthel, B. Shih, C. Laschi, and M. T. Tolley, "Soft robot perception using embedded soft sensors and recurrent neural networks," *Science Robotics*, vol. 4, no. 26, p. eaav1488, 2019.
[21] S. Kim, C. Laschi, and B. Trimmer, "Soft robotics: a bioinspired evolution in robotics," *Trends in biotechnology*, vol. 31, no. 5, pp. 287–294, 2013.

[22] N. El-Atab, R. B. Mishra, F. Al-Modaf, L. Joharji, A. A. Alsharif, H. Alamoudi, M. Diaz, N. Qaiser, and M. M. Hussain, "Soft actuators for soft robotic applications: A review," *Advanced Intelligent Systems*, vol. 2, no. 10, p. 2000128, 2020.

[23] D. Rus and M. T. Tolley, "Design, fabrication and control of soft robots," *Nature*, vol. 521, no. 7553, pp. 467–475, 2015.

[24] C. Della Santina, C. Duriez, and D. Rus, "Model-based control of soft robots: A survey of the state of the art and open challenges," *IEEE Control Systems Magazine*, vol. 43, no. 3, pp. 30–65, 2023.

[25] J. Wang and A. Chortos, "Control strategies for soft robot systems," *Advanced Intelligent Systems*, vol. 4, no. 5, p. 2100165, 2022.

[26] D. Kim, S.-H. Kim, T. Kim, B. B. Kang, M. Lee, W. Park, S. Ku, D. Kim, J. Kwon, H. Lee *et al.*, "Review of machine learning methods in soft robotics," *Plos one*, vol. 16, no. 2, p. e0246102, 2021.

[27] M. S. Nazeer, C. Laschi, and E. Falotico, "Soft dagger: Sample-efficient imitation learning for control of soft robots," *Sensors*, vol. 23, no. 19, p. 8278, 2023.

[28] D. S. D. Cortes, G. Hwang, and K.-U. Kyung, "Imitation learning based soft robotic grasping control without precise estimation of target posture," in *2021 IEEE 4th International Conference on Soft Robotics (RoboSoft)*. IEEE, 2021, pp. 149–154.

[29] P. Oikonomou, A. Dometios, M. Khamassi, and C. S. Tzafestas, "Task driven skill learning in a soft-robotic arm," in *2021 IEEE/RSJ International Conference on Intelligent Robots and Systems (IROS)*. IEEE, 2021, pp. 1716–1723.

[30] S. Bhagat, H. Banerjee, Z. T. Ho Tse, and H. Ren, "Deep reinforcement learning for soft, flexible robots: Brief review with impending challenges," *Robotics*, vol. 8, no. 1, p. 4, 2019.

[31] K. Chin, T. Hellebrekers, and C. Majidi, "Machine learning for soft robotic sensing and control," *Advanced Intelligent Systems*, vol. 2, no. 6, p. 1900171, 2020.

[32] M. S. Nazeer, C. Laschi, and E. Falotico, "Imitation and reinforcement learning to control soft robots: a perspective," in *IOP Conference Series: Materials Science and Engineering*, vol. 1292, no. 1. IOP Publishing, 2023, p. 012010.

[33] C. Armanini, F. Boyer, A. T. Mathew, C. Duriez, and F. Renda, "Soft robots modeling: A structured overview," *IEEE Transactions on Robotics*, 2023.

[34] R. Szadkowski, M. S. Nazeer, M. Cianchetti, E. Falotico, and J. Faigl, "Bootstrapping the dynamic gait controller of the soft robot arm," in *2023 IEEE International Conference on Robotics and Automation (ICRA)*. IEEE, 2023, pp. 2669–2675.

[35] R. Morimoto, S. Nishikawa, R. Niiyama, and Y. Kuniyoshi, "Model-free reinforcement learning with ensemble for a soft continuum robot arm," in *2021 IEEE 4th International Conference on Soft Robotics (RoboSoft)*. IEEE, 2021, pp. 141–148.

[36] D. Hardman, T. G. Thuruthel, and F. Iida, "Tactile perception in hydrogel-based robotic skins using data-driven electrical impedance tomography," *Materials Today Electronics*, vol. 4, p. 100032, 2023.

[37] F. Piqué, H. T. Kalidindi, L. Fruzzetti, C. Laschi, A. Menciassi, and E. Falotico, "Controlling soft robotic arms using continual learning," *IEEE Robotics and Automation Letters*, vol. 7, no. 2, pp. 5469–5476, 2022.

[38] Y. Li, X. Wang, and K.-W. Kwok, "Towards adaptive continuous control of soft robotic manipulator using reinforcement learning," in *2022 IEEE/RSJ International Conference on Intelligent Robots and Systems (IROS)*. IEEE, 2022, pp. 7074–7081.

[39] I. Koryakovskiy, M. Kudruss, H. Vallery, R. Babuška, and W. Caarls, "Model-plant mismatch compensation using reinforcement learning," *IEEE Robotics and Automation Letters*, vol. 3, no. 3, pp. 2471–2477, 2018.

[40] H. T. Kalidindi, T. G. Thuruthel, C. Laschi, and E. Falotico, "Cerebellum-inspired approach for adaptive kinematic control of soft robots," in *2019 2nd IEEE International Conference on Soft Robotics (RoboSoft)*. IEEE, 2019, pp. 684–689.

[41] C. C. Johnson, T. Quackenbush, T. Sorensen, D. Wingate, and M. D. Killpack, "Using first principles for deep learning and model-based control of soft robots," *Frontiers in Robotics and AI*, vol. 8, p. 654398, 2021.

Embodied Intelligence in Transportation Systems: Road Network Computing

Kai-Fung Chu*, Fan Ye*†, Fumiya Iida

Department of Engineering, University of Cambridge, Cambridge, CB2 1PZ, UK
* Co-first authors
† Corresponding Author

E-mail: kfc35@cam.ac.uk, fy264@cam.ac.uk, fi224@cam.ac.uk

Abstract. This paper studied the concept of embodied intelligence within transportation systems, focusing on the interactions between purposefully designed physical structures and their dynamic environmental contexts. By drawing inspiration from the intricate complexities inherent in human-designed transportation systems, including the intricate web of road networks and vehicular components, the integration of intelligent elements is examined. Beyond a mere examination, the paper proposes a novel perspective—conceptualizing a transportation system, particularly road networks, as a physical reservoir. This perspective is substantiated through a thorough exploration of the reading mechanism of the traffic states and the echo-state property of road networks. We further dive into the traffic property to explore the inherent computational capabilities, such as fundamental logic operations, within these networks. The illustrative examples of OR, XOR, and AND logic gates within road networks serve as tangible demonstrations of their inherent computational capabilities. These examples not only showcase the computational abilities embedded in road networks but also lay the groundwork for a paradigm shift—a consideration of road networks as a manifestation of embodied intelligence within the broader framework of transportation systems. This paper contributes to the evolving discourse on the transformative potential of road networks as dynamic, intelligent entities, redefining our understanding of transportation systems.

1. Introduction

Embodied intelligence frequently manifests through the dynamic interplay between ingeniously designed physical systems and their surrounding environments. An example of such intelligence is the complexity of human-designed transportation systems, including complex road networks, infrastructural elements, and large number of vehicles dispersed across urban landscapes. The continuous enhancement of transportation systems is driven by the urgent need to facilitate the safe and efficient transportation of both individuals and goods in support of diverse human activities. This pursuit has given rise to the integration of intelligent components within transportation systems.

IOP Conf. Series: Materials Science and Engineering **1321** (2024) 012010 doi:10.1088/1757-899X/1321/1/012010

An example of intelligent intervention within transportation is the deployment of intelligent traffic lights. Beyond their conventional role in unidirectional traffic control, these systems demonstrate an adaptive capacity to engage with real-time conditions, thereby facilitating enhanced coordination (Chu, Lam & Li 2022b). Another instance is the sensing capabilities embedded in intelligent vehicles, designed for the purpose of gathering comprehensive information from their surroundings, a crucial aspect of enabling autonomous driving (Van Brummelen, O'Brien, Gruyer & Najjaran 2018). The autonomous vehicle, tasked with the responsibility of safely and efficiently navigating a complex and dynamically evolving physical environment, reflects the complex interaction inherent in embodied intelligence. As these intelligent components become integral to the transportation system, the traditional paradigm is poised to evolve into a truly intelligent transportation system. The synergistic fusion of ingeniously designed physical systems and adaptive, context-aware functionalities is promoting the transportation domain towards a future characterized by heightened intelligence and efficiency.

The human brain has always been acknowledged as the most intelligent processing unit. From the neuroscience point of view, the human brain is comprehended as a composition of nervous tissue forming a biological neural network that facilitates the transmission and reception of chemical and electrical signals throughout the entirety of the body. Drawing inspiration from this intricate biological neural network, computer scientists have devised artificial neural networks (ANNs) that emulate intelligent processes. It is contended that ANNs constitute a pivotal component catalyzing the proliferation of numerous state-of-the-art intelligent applications, including robots learning from demonstrations (Argall, Chernova, Veloso & Browning 2009). Remarkably, certain applications have transcended human capabilities. For instance, the application of ANNs in solving complex challenges such as the Rubik's Cube (Akkaya, Andrychowicz, Chociej, Litwin, McGrew, Petron, Paino, Plappert, Powell, Ribas et al. 2019). Both biological and artificial neural networks hinge upon the complexity inherent in their network structures. If network complexity is posited as a fundamental determinant of intelligence, then it is plausible to assert that a sophisticated road network interwoven with vehicular entities within the transportation system may inherently embody a certain degree of intelligence.

To observe, evaluate, and enhance the intelligence of a physical system, the frameworks of reservoir computing (Jaeger 2001b) (Maass, Natschläger & Markram 2002) and fluidic circuits (Preston, Rothemund, Jiang, Nemitz, Rawson, Suo & Whitesides 2019) are introduced, utilizing the inherent physical characteristics for computational purposes. Reservoir computing represents a machine learning paradigm that seeks to replicate chaotic systems by leveraging the unique attributes of a reservoir. Initially conceptualized as a set of randomly connected nodes, the reservoir demonstrates proven nonlinearity and memory (Jaeger 2001a), essential for its computational efficacy in emulating chaotic systems such as the Lorenz system (Khovanov 2021) and the Mackey-Glass systems (Jaeger & Haas 2004). Furthermore, diverse physical systems, including a silicon arm in a water tank (Nakajima, Hauser, Li & Pfeifer 2015), a network of photonic components (Vandoorne, Mechet, Van Vaerenbergh, Fiers, Morthier, Verstraeten, Schrauwen, Dambre & Bienstman 2014), and a network of mechanical oscillators (Coulombe, York & Sylvestre 2017), have been demonstrated as effective reservoirs, showcasing a certain degree of nonlinearity and memory for computational tasks. Similarly, fluidic circuits embrace the notion that physical systems can engage in computation, emphasizing the design of soft structures within fluid delivery systems to function as logic gates, eliminating the need for traditional pumps and valves (Rothemund, Ainla, Belding, Preston, Kurihara, Suo & Whitesides 2018). For instance, a bistable soft structure activated by fluidic pressure mimics the behavior of NOT, AND, and OR digital logic gates (Preston et al. 2019). Both reservoir computing and fluidic circuits exemplify computational processes grounded in physical structures, showcasing embodied intelligence. It becomes plausible to consider a transportation system as a physical reservoir, leveraging the flow of vehicles as a computational resource.

In the rest of this paper, we will explore the embodied intelligence in transportation system in Section 2, present the studies of performing computation with physical system such as reservoir in Section 3, and discuss the concept of considering road networks and vehicles as a computation unit in Section 4.

2. Embodied Intelligence Components in Transportation Systems

The evolution of transportation systems has played a pivotal role amid technological advancements, significantly influencing the design and functionality of contemporary transportation systems (Picone, Busanelli, Amoretti, Zanichelli & Ferrari 2015). Anticipating future challenges, the next generation of transportation systems is envisioned to seamlessly integrate connected vehicles, connected infrastructure, and increased automation with existing technologies (Coogan, Arcak & Belta 2017).

2.1. Computation in Transportation Systems

Bioinspired methods, a cornerstone in enhancing transportation system capabilities (Del Ser, Osaba, Sanchez-Medina & Fister 2019), draw inspiration from nature's intricate behavioral mechanisms and the adaptable processes of the brain (Holland 1975). Transportation systems integrate efficient, adaptive, and self-learning methods, encompassing fuzzy systems and artificial neural networks (Arbib 2003), (Engelbrecht 2007). For example, computational and distributed algorithm is a key element in efficient self-organizing strategies proposed by (Chu, Lam & Li 2020) that aims to dynamically alter road lane directions, contributing to the adaptability of autonomous vehicles. Moreover, it identifies optimal vehicle locations to cater to travel demand hotspots while supporting simultaneous vehicle-to-grid interactions demand a bio-inspired computation method such as genetic algorithm (Chu, Lam & Li 2022a).

2.2. Sensing Advancements

In the context of transportation with multiple vehicles, Crowd Sensing Intelligence (CSI) could be a useful sensing technique that introduces a paradigm where biological, digital, and robotic participants collaboratively contribute to sensing activities (Zhao, Hu, Zhu, Qiu, Chen, Jiao & Wang 2023). The envisioned progression of CSI from algorithmic to imaginative intelligence, guided by parallel intelligence (Ren, Jiang, Feng, Zhao, Liu & Yu 2022), involves digital participants operating in cyberspace, aiding in data collection and computation. Robotic participants, exemplified by intelligent vehicles, enhance sensing tasks with increased autonomy. Biological participants benefit from optimized productivity through the assistance offered by their digital and robotic counterparts (Teng, Hu, Deng, Li, Li, Ai, Yang, Li, Xuanyuan, Zhu et al. 2023), (Zhu, Wang, Zhao, Qiu, Liu, Chen & Wang 2022).

2.3. Design Integration

In the realm of design, the controller proposed in (Chu, Chen, Lam & Song 2023) is designed to orchestrate the coordination of transport requests, travel schedules, and charging/discharging for electric vehicles, optimizing utility across transportation and power distribution systems. Additionally, (Zardini, Milojevic, Censi & Frazzoli 2021) presents a structured co-design approach, leveraging monotone theory, to address embodied intelligence co-design problems in autonomous vehicle hardware and software stacks. This method showcases the potential to model complex autonomous vehicles with heterogeneous components, providing a comprehensive solution. Coordination strategies (Chu & Guo 2023) also leverage resources from multiple and heterogeneous transportation system users to maximize user satisfaction and long-term system profit while ensuring privacy.

Incorporating these embodied intelligence components, transportation systems are poised to navigate the complexities of future mobility challenges with enhanced efficiency, adaptability, and safety.

3. Embodied Computing

Reservoir computing and fluidic computing leverage the inherent characteristics of reservoirs and fluids for computational processes. In essence, both systems involve the dynamic response of a reservoir and fluid to inputs, leading to the generation of outputs shaped by these reactions. Specifically, a reservoir computer emphasizes the nonlinear and memory attributes of the reservoir to effectively emulate dynamic systems. On the other hand, a fluidic computer places significance on memoryless properties, such as those exhibited by a Markov chain, to execute logical operations. This distinction underscores the diverse computational approaches employed in reservoir and fluidic computing, where the former capitalizes on reservoir properties for dynamic system emulation, and the latter emphasizes memoryless characteristics for logical processing.

3.1. Reservoir Computing

Reservoir computing functions as an observer that extracts information from a dynamic reservoir. In this context, the reservoir's dynamism is crucial for manifesting intricate behaviors in response to varying inputs. The cornerstone of its computational prowess lies in the echo-state property, wherein the reservoir is transiently influenced by inputs, only retaining the memory of recent inputs while swiftly erasing the impact of inputs from an earlier time (Jaeger 2001b). For instance, when a tranquil reservoir receives a pulse as input, it activates, potentially oscillates briefly, and subsequently returns to a tranquil state. Notably, the reservoir retains the memory of the input during its oscillation but relinquishes that memory upon returning to a tranquil state. The echo-state property is indispensable for computational efficacy as it precludes perpetual oscillation, averting divergent output.

An additional crucial attribute of a reservoir computer is its memory, a facet partially encompassed by the previously mentioned echo-state property. However, this memory aspect specifically pertains to the retention of input information (Jaeger 2001a). A reservoir computer has to possess the capacity to remember inputs for a substantial duration (memory property) without persisting indefinitely (echo-state property). This characteristic proves instrumental in tasks requiring sequential reasoning, such as image recognition endeavors where the image is transformed into a digital sequence serving as inputs to the reservoir (Woods, Bürger & Teuscher 2015). The memory property is engendered by interconnections among nodes within the reservoir, facilitating information retention as it is passed from one node to another.

The third essential property is nonlinearity, signifying that the reservoir responds to inputs in a nonlinear fashion, showcasing diverse behaviors in response to different inputs (Maass et al. 2002). This property is instrumental in enabling the reservoir computer to replicate intricate systems, such as the Mackey-Glass chaotic system (Jaeger & Haas 2004). Leveraging this nonlinearity, the readout of the reservoir computer can adopt a straightforward linear regression approach while retaining the nonlinearity within the reservoir. This strategic approach significantly diminishes the requisite volume of data and time for training a reservoir computer, emphasizing the efficiency gained through the exploitation of nonlinear properties within the reservoir itself.

Viewing the traffic system as a dynamical system offers the potential to conceptualize it as a reservoir, contingent upon substantiating its echo-state property, memory, and nonlinearity. Nevertheless, adopting each vehicle as a node within the reservoir renders the system exceedingly intricate. Defining interactions among vehicles becomes challenging, and the complexity escalates exponentially with the increasing number of vehicles. Consequently, a methodological approach is imperative to simplify the traffic system effectively.

3.2. Traffic as Fluid

Researchers in fluid mechanics have identified parallels between traffic flow and airflow, noting similarities such as the resemblance between airflow shock waves and traffic jams (Lighthill & Whitham 1955). When considering a road, it becomes evident that the influx and efflux of vehicles adhere to a scalar conservation law, thereby giving rise to a nonlinear partial differential equation (LeVeque & Leveque 1992):

$$\frac{\partial k}{\partial t} + \frac{\partial ku}{\partial x} = 0, \tag{1}$$

where k is the density of vehicles, u is the vehicle speed, t is time, and x is position. This equation bears a striking resemblance to the conservation of mass equation in airflow, implying that phenomena such as shock waves in airflow can manifest in traffic flow, commonly referred to as traffic jams (Richards 1956). By drawing parallels between traffic flow and airflow, we can streamline the model of the traffic system into mathematical equations like the one represented by Eq.1. This formulation ensures the inclusion of nonlinearity, allowing us to treat the traffic system as a reservoir for computational purposes.

3.3. Fluidic Computing

Fluidic computing is centered on the construction of logic gates through fluidic circuits (Rothemund et al. 2018), providing a notable instance of leveraging fluid for computational processes, even though these circuits typically lack inherent memory. Through meticulous design of the soft fluid piping structure, a fluidic circuit can exhibit bistable behavior and effectively regulate fluidic pressure, emulating the functionality of NOT, AND, and OR gates (Preston et al. 2019). This innovative approach to piping structure design is not confined to fluidic circuits alone but can be extended to the development of a computer based on the principles of the traffic system.

In conclusion, the prospect of constructing a traffic-based computer emerges by simplifying the traffic system akin to airflow, involving the careful design of road network structures. Crucially, ensuring the incorporation of the echo-state property, memory, and nonlinearity within this simplified model is paramount to realizing a computational system inspired by traffic dynamics. This innovative approach holds promise for novel applications in computational science and underscores the adaptability of principles derived from fluid mechanics to diverse domains, including traffic-based computing.

4. Road Network Computing

In this section, we will first discuss the echo-state property in road networks. Subsequently, leveraging the echo-state property, we will illustrate how road networks, encompassing vehicular entities, may be construed as computation unit.

4.1. Information Flow in Road Networks

Consider a fleet of vehicles, denoted as $\mathcal{V} = \{v_1, \ldots, v_N\}$, entering a road at origin point o and traversing through a complex interconnected road network, eventually exiting the network at the sole destination point d, where N represents the total number of vehicles. Let v_1 be the first vehicle to pass through o at time $t_o^{v_1}$, and v_N be the last vehicle passing through o at time $t_o^{v_N}$. Similarly, they pass through d at time $t_d^{v_1}$ and $t_d^{v_N}$. The time required for vehicle v_n to traverse from o to d is defined as $T^{v_n}od$, and the time for the fleet of vehicles to pass through o and d is denoted as $T_o^{v_1 v_N}$ and $T^{v_1 v_N}d$, respectively. These quantities can be computed using the following equations:

$$T_{od}^{v_n} = t_d^{v_n} - t_o^{v_n}, \tag{2}$$

$$T_o^{v_1 v_N} = t_o^{v_N} - t_o^{v_1}, \tag{3}$$

$$T_d^{v_1 v_N} = t_d^{v_N} - t_d^{v_1}. \tag{4}$$

Intuitively, for a small number of vehicles N, it is relatively easy to maintain a constant speed such that there is a small variance in the travel time from one vehicle to another ($Var(T_{od}^{v_n})$ is small), and the overall travel times $T^{v_1 v_N}o$ and $T^{v_1 v_N}d$ are approximately equal, where $Var(\cdot)$ denotes the variance. However, with a large number of vehicles, traffic jams may occur, resulting in inconsistent vehicle platoons and longer travel times for later vehicles. This inconsistency could manifest as $T^{v_N}od \geq T^{v_1}od$ and $T_d^{v_1 v_N} \geq T_o^{v_1 v_N}$, reflecting nonlinear properties in the traffic flow and time relationship.

Travel time can be modeled using the Bureau of Public Roads (BPR) volume-delay function (Manual 1964), given by:

$$t^{BPR}(N) = t_f \left(1 + \alpha \left(\frac{N}{C}\right)^{\beta}\right), \tag{5}$$

where $t^{BPR}(N)$ is the estimated travel time, t_f is the free flow travel time, N is the traffic volume, C is the practical capacity, and α and β are the tuning parameters. This function indicates that as traffic volume increases beyond capacity ($N > C$), travel time also increases. Other models, like Greenshield's (Greenshields, Bibbins, Channing & Miller 1935) and the Highway Capacity Manual (Manual 2000), also depict the relationships among traffic speed, flow (number of vehicles per unit time), and density (number of vehicles per unit space).

The inverse relationship between traffic speed s and density k is expressed as:

$$s(k) = s_f \left(1 - \frac{k}{k_j}\right). \tag{6}$$

where s_f is free flow speed and k_j is the jam density. On the other hand, traffic flow ρ and density k relationship is modeled as a concave down parabola as flow is zero at both zero and jam density, while reaches the maximum flow in the in between zero and jam density. Therefore, the flow-density relation is given as

$$q(k) = s_f k \left(1 - \frac{k}{k_j}\right). \tag{7}$$

Similarly, the relation between traffic speed and flow is given by:

$$q(v) = k_j s \left(1 - \frac{s}{s_f}\right), \tag{8}$$

as speed can be at zero or free-flow speed when flow is zero, reaching the maximum flow between zero and free-flow speed.

These traffic states are interrelated, and as vehicles travel from o to d, the associated traffic information characterized by speed, flow, and density propagates. An interesting phenomenon in this propagation is the traffic wave, also known as shockwave traffic jam, stop waves, and phantom traffic jam (Sugiyama, Fukui, Kikuchi, Hasebe, Nakayama, Nishinari, Tadaki & Yukawa 2008). This occurrence involves disturbances in traffic flow propagating backward, resulting in localized slowdowns or stop-and-go traffic. Traffic waves can form without an apparent external cause, arising from the interactions between drivers and the inherent instability of traffic flow. This highlights that information transmission occurs not only through communication channels but also through the dynamic motions of vehicles.

Vehicles themselves can be considered as information carriers within a transportation system. In this context, information flow refers to the transmission of signals from input to output through the road network. The traffic flow or other measurable quantities in the road network can be analogous to voltage in an electric computer, forming the basis of digital information (0s and 1s).

4.2. Echo-State Property in Road Networks

As outlined in Section 3, the echo-state property (Yildiz, Jaeger & Kiebel 2012) characterizes a reservoir's ability to preserve and reflect input signals over a specific duration, facilitating the network's capture of temporal dependencies in the data. To investigate whether a road network aligns with the characteristics of a "reservoir", we analyze the echo-state property in the context of road networks.

Consider the road network as a discrete-time echo-state network with M^{in} input roads, M internal roads, and M^{out} output roads. Let the traffic flow q at input, internal, and output in time period t_1 to t_2 be the state of the echo-state network, denoted as $x^{in}_{t_1t_2} \in \mathbb{R}^{M^{in}\times 1}$, $x_{t_1t_2} \in \mathbb{R}^{M\times 1}$, and $x^{out}_{t_1t_2} \in \mathbb{R}^{M^{out}\times 1}$, respectively. The dynamics of traffic flow in the road network can be expressed as:

$$\begin{aligned} x_{t_1+1t_2+1} &= f\left(Wx_{t_1t_2} + W^{in}x^{in}_{t_1+1t_2+1} + W^{fb}x^{out}_{t_1t_2}\right), \\ x^{out}_{t_1t_2} &= g\left(W^{out}[x_{t_1t_2}; x^{in}_{t_1t_2}]\right), \end{aligned} \tag{9}$$

where $W^{in} \in \mathbb{R}^{M \times M^{in}}$, $W \in \mathbb{R}^{M \times M}$, $W^{fb} \in \mathbb{R}^{M \times M^{out}}$, and $W^{out} \in \mathbb{R}^{M^{out} \times (M+M^{in})}$ is the input, internal, feedback, and output weight matrices, respectively. Additionally, f and g denote activation functions.

Consider a simplified road network with a single road, i.e., $M^{in} = M = M^{out} = 1$, with designated input and output locations at o and d respectively. The output traffic flow from $t_d^{v_1}$ to $t_d^{v_N}$ equals the input traffic flow from $t_o^{v_1}$ to $t_o^{v_N}$, represented by $x^{out}_{t_d^{v_1}, t_d^{v_N}} = x^{in}_{t_o^{v_1}, t_o^{v_N}} = N$. We can obtain the weight matrices as:

$$W^{in} = 1, \quad W = 0, \quad W^{fb} = 0, \quad W^{out} = \begin{pmatrix} 1 & 0 \end{pmatrix}, \tag{10}$$

and f and g are identity functions. This configuration satisfies the echo-state property, whereby traffic flow persists in the network for a period of time from $t_o^{v_1}$ to $t_d^{v_N}$, and fades after time $t_d^{v_N}$. Therefore, a road network with specific structure exhibits the each-state property and can be conceptualized as a computational reservoir.

4.3. Road Network as a Computational Reservoir

After illustrating the echo-state property in road networks, we explore their potential as computational reservoirs. A specific road network structure can be conceptualized as an echo-state network, akin to a computational "reservoir." We exemplify the computational capabilities of road networks, focusing on fundamental logic operations.

Fig. 1 demonstrates the internal structure of the echo-state network represented by the road network. By segmenting the road into blocks, we can read the traffic information of each road segment to obtain the input, internal, and output states of the road network. For example, we can consider the traffic flow per road segment as a logical one if $x_{t_1+1t_2+1}$ is larger than a threshold value C_1, and zero if $x_{t_1+1t_2+1}$ is smaller then C_0, where $C_1 \geq C_0$.

A road network shown in Fig. 2 acts as an OR logic gate, jointly receiving traffic from input roads. When either or both input roads has high traffic, the output roads will have high traffic.

Another example is the XOR and AND road networks as shown in Fig. 3. For the illustration purpose, we use the horizontal dotted line at the outputs to represent a lengthy road segment. As for the output 2 road segment, it is a detour that vehicles normally will not enter if the traffic on output 1 road is light. Vehicle prefer to enter output 2 road when there is traffic jam on output 1 road. With the above structure, it can achieve the XOR and AND logic. For XOR logic, traffic flow in either input roads will trigger the logic on output 1. However, output 1 will be jammed if both input roads have high traffic flow, resulting a high density and low traffic flow (logic zero), as discussed in the flow-density relation. For the AND logic, there is traffic on output 2 (logic one) only when both input roads have high traffic (logic one). Since some of the traffic already diverted to output 1, we assume output 2 is less susceptible to congestion than output 1.

IOP Conf. Series: Materials Science and Engineering **1321** (2024) 012010 doi:10.1088/1757-899X/1321/1/012010

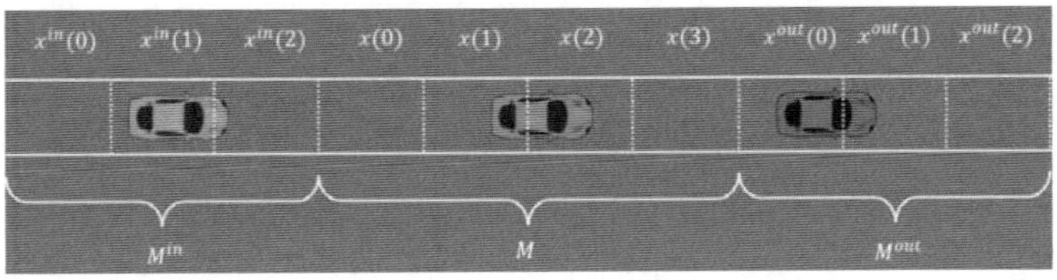

Figure 1. Input and read-out mechanisms of the road network computing.

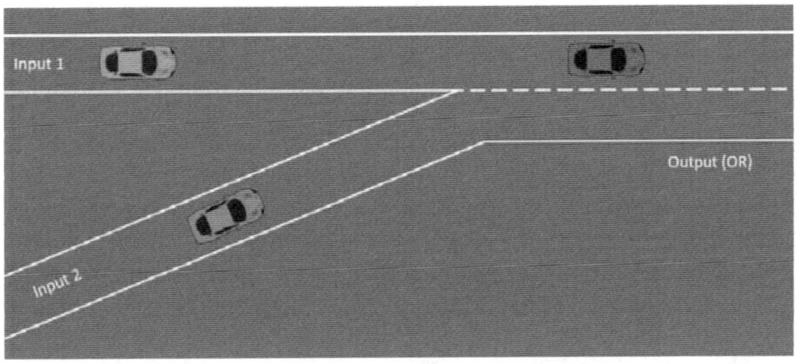

Figure 2. A road with two incoming road lanes exhibits OR logic.

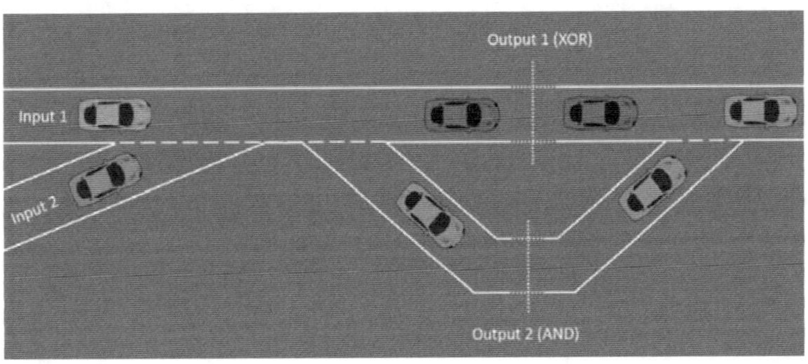

Figure 3. A road with two merging road lanes and a closed loop branch exhibits the XOR and AND logic.

These examples represent basic computational abilities within the network structure. Scaling up, multiple units can construct more powerful computational entities, such as a full adder using XOR and AND logic gates. In the urban city, much more complex road network are built on a city scale containing these fundamental logical components. The computational ability of such complex road network in city scale could be significant, and revolutionize the transportation systems with this new form of embodied intelligence.

5. Conclusion

Conceptualizing road networks and vehicles as computation units in the realm of embodied intelligence provides a novel perspective on enhancing transportation systems. This paper reviewed bio-inspired computation, reservoir computing, and fluidic computing, as fundamental theories supporting the novel perspective of road network computing. To demonstrate the effectiveness of road network computing, we analysed the echo-state property of road networks and illustrated sample road networks with OR, XOR, and AND logic capabilities.

By considering a transportation system as a computational reservoir, we can gain several significant insights and benefits. First, viewing the transportation network as a dynamic system allows for the identification of traffic flow disruptions and congestion patterns. This perspective can help pinpoint the origins of traffic jams and bottlenecks, facilitating targeted interventions to alleviate these issues. Second, treating the road network as a computational entity can optimize traffic flow by dynamically adjusting traffic signals, lane directions, and vehicle routing based on real-time data. This adaptive approach can lead to reduced travel times, lower emissions, and improved overall efficiency of the transportation system. Third, as a computational reservoir, the road network can perform real-time computations to potentially solve complex problems such as predicting traffic conditions, optimizing routing for emergency services, and managing vehicle-to-grid interactions. This capability can enhance the responsiveness and adaptability of the transportation system. This paper opens avenues for redefining transportation systems as not only physical infrastructures but also as intelligent computational entities. The consideration of road networks as computational reservoirs hints at the transformative possibilities, offering a new perspective on the future of transportation and the potential integration of embodied intelligence at a city scale.

Future research in this area holds the potential to further refine the integration of road networks and vehicles into a cohesive computational framework. Understanding how other road infrastructures and structures such as traffic lights and roundabouts play a role in road network computing could increase the possible road network computational units. The computing capability could ultimately lead to more sustainable, efficient, and functional transportation systems.

6. References

Akkaya, I., Andrychowicz, M., Chociej, M., Litwin, M., McGrew, B., Petron, A., Paino, A., Plappert, M., Powell, G., Ribas, R. et al. (2019). Solving rubik's cube with a robot hand, *arXiv preprint arXiv:1910.07113* .

Arbib, M. A. (2003). *The handbook of brain theory and neural networks*, MIT press.

Argall, B. D., Chernova, S., Veloso, M. & Browning, B. (2009). A survey of robot learning from demonstration, *Robotics and autonomous systems* **57**(5): 469–483.

Chu, K.-F., Chen, T., Lam, A. Y. & Song, Y. (2023). Collaborative routing and charging/discharging scheduling of electric autonomous vehicles in coupled power-traffic networks, *2023 IEEE 97th Vehicular Technology Conference (VTC2023-Spring)*, IEEE, pp. 1–6.

Chu, K.-F. & Guo, W. (2023). Privacy-preserving federated deep reinforcement learning for mobility-as-a-service, *IEEE Transactions on Intelligent Transportation Systems* .

Chu, K.-F., Lam, A. Y. & Li, V. O. (2020). Dynamic lane reversal routing and scheduling for connected and autonomous vehicles: Formulation and distributed algorithm, *IEEE Transactions on Intelligent Transportation Systems* **21**(6): 2557–2570.

Chu, K.-F., Lam, A. Y. & Li, V. O. (2022a). Joint rebalancing and vehicle-to-grid coordination for autonomous vehicle public transportation system, *IEEE Transactions on Intelligent Transportation Systems* **23**(7): 7156–7169.

Chu, K.-F., Lam, A. Y. & Li, V. O. (2022b). Traffic signal control using end-to-end off-policy deep reinforcement learning, *IEEE Transactions on Intelligent Transportation Systems* **23**(7): 7184–7195.

Coogan, S., Arcak, M. & Belta, C. (2017). Formal methods for control of traffic flow: Automated control synthesis from finite-state transition models, *IEEE Control Systems Magazine* **37**(2): 109–128.

Coulombe, J. C., York, M. C. & Sylvestre, J. (2017). Computing with networks of nonlinear mechanical oscillators, *PloS one* **12**(6): e0178663.

Del Ser, J., Osaba, E., Sanchez-Medina, J. J. & Fister, I. (2019). Bioinspired computational intelligence and transportation systems: A long road ahead, *IEEE Transactions on Intelligent Transportation Systems* **21**(2): 466–495.

Engelbrecht, A. P. (2007). *Computational intelligence: an introduction*, John Wiley & Sons.

Greenshields, B. D., Bibbins, J., Channing, W. & Miller, H. (1935). A study of traffic capacity, *Highway research board proceedings*, Vol. 14, Washington, DC, pp. 448–477.

Holland, J. (1975). An introductory analysis with applications to biology, control, and artificial intelligence, *Adaptation in Natural and Artificial Systems. First Edition, The University of Michigan, USA* .

Jaeger, H. (2001a). Short term memory in echo state networks.

Jaeger, H. (2001b). The "echo state" approach to analysing and training recurrent neural networks-with an erratum note, *Bonn, Germany: German National Research Center for Information Technology GMD Technical Report* **148**(34): 13.

Jaeger, H. & Haas, H. (2004). Harnessing nonlinearity: Predicting chaotic systems and saving energy in wireless communication, *science* **304**(5667): 78–80.

Khovanov, I. A. (2021). Stochastic approach for assessing the predictability of chaotic time series using reservoir computing, *Chaos: An Interdisciplinary Journal of Nonlinear Science* **31**(8).

LeVeque, R. J. & Leveque, R. J. (1992). *Numerical methods for conservation laws*, Vol. 214, Springer.

Lighthill, M. J. & Whitham, G. B. (1955). On kinematic waves ii. a theory of traffic flow on long crowded roads, *Proceedings of the royal society of london. series a. mathematical and physical sciences* **229**(1178): 317–345.

Maass, W., Natschläger, T. & Markram, H. (2002). Real-time computing without stable states: A new framework for neural computation based on perturbations, *Neural computation* **14**(11): 2531–2560.

Manual, H. C. (2000). Highway capacity manual, *Washington, DC* **2**(1).

Manual, T. A. (1964). Traffic assignment manual, *DC, USA* .

Nakajima, K., Hauser, H., Li, T. & Pfeifer, R. (2015). Information processing via physical soft body, *Scientific reports* **5**(1): 10487.

Picone, M., Busanelli, S., Amoretti, M., Zanichelli, F. & Ferrari, G. (2015). Advanced technologies for intelligent transportation systems.

Preston, D. J., Rothemund, P., Jiang, H. J., Nemitz, M. P., Rawson, J., Suo, Z. & Whitesides, G. M. (2019). Digital logic for soft devices, *Proceedings of the National Academy of Sciences* **116**(16): 7750–7759.

Ren, Y., Jiang, H., Feng, X., Zhao, Y., Liu, R. & Yu, H. (2022). Acp-based modeling of the parallel vehicular crowd sensing system: Framework, components and an application example, *IEEE Transactions on Intelligent Vehicles* **8**(2): 1536–1548.

Richards, P. I. (1956). Shock waves on the highway, *Operations research* **4**(1): 42–51.

Rothemund, P., Ainla, A., Belding, L., Preston, D. J., Kurihara, S., Suo, Z. & Whitesides, G. M. (2018). A soft, bistable valve for autonomous control of soft actuators, *Science Robotics* **3**(16): eaar7986.

Sugiyama, Y., Fukui, M., Kikuchi, M., Hasebe, K., Nakayama, A., Nishinari, K., Tadaki, S.-i. & Yukawa, S. (2008). Traffic jams without bottlenecks—experimental evidence for the physical mechanism of the formation of a jam, *New journal of physics* **10**(3): 033001.

Teng, S., Hu, X., Deng, P., Li, B., Li, Y., Ai, Y., Yang, D., Li, L., Xuanyuan, Z., Zhu, F. et al. (2023). Motion planning for autonomous driving: The state of the art and future perspectives, *IEEE Transactions on Intelligent Vehicles* .

Van Brummelen, J., O'Brien, M., Gruyer, D. & Najjaran, H. (2018). Autonomous vehicle perception: The technology of today and tomorrow, *Transportation research part C: emerging technologies* **89**: 384–406.

Vandoorne, K., Mechet, P., Van Vaerenbergh, T., Fiers, M., Morthier, G., Verstraeten, D., Schrauwen, B., Dambre, J. & Bienstman, P. (2014). Experimental demonstration of reservoir computing on a silicon photonics chip, *Nature communications* **5**(1): 3541.

Woods, W., Bürger, J. & Teuscher, C. (2015). Synaptic weight states in a locally competitive algorithm for neuromorphic memristive hardware, *IEEE Transactions on Nanotechnology* **14**(6): 945–953.

Yildiz, I. B., Jaeger, H. & Kiebel, S. J. (2012). Re-visiting the echo state property, *Neural networks* **35**: 1–9.

Zardini, G., Milojevic, D., Censi, A. & Frazzoli, E. (2021). Co-design of embodied intelligence: A structured approach, *2021 IEEE/RSJ International Conference on Intelligent Robots and Systems (IROS)*, IEEE, pp. 7536–7543.

Zhao, Y., Hu, C., Zhu, Z., Qiu, S., Chen, B., Jiao, P. & Wang, F.-Y. (2023). Crowd sensing intelligence for its: Participants, methods, and stages, *IEEE Transactions on Intelligent Vehicles* .

Zhu, Z., Wang, X., Zhao, Y., Qiu, S., Liu, Z., Chen, B. & Wang, F.-Y. (2022). Crowdsensing intelligence by decentralized autonomous vehicles organizations and operations, *IEEE Transactions on Intelligent Vehicles* **7**(4): 804–808.

IOP Conf. Series: Materials Science and Engineering **1321** (2024) 012011 doi:10.1088/1757-899X/1321/1/012011

Teaching Morphological Computation: A Case Study Using Paper Structures

Yi-Shiun Wu [1], **and Josie Hughes** [1,*]

[1]CREATE Lab, Institute of Mechanical Engineering, School of Engineering, EPFL, Lausanne, Switzerland

E-mail: `yi-shiun.wu@epfl.ch`, `josie.hughes@epfl.ch`

Abstract. Soft robotics is an interdisciplinary field that is constantly advancing, providing us with an improved understanding of materials, mechanisms, control systems, and design principles to enable the creation of innovative and adaptable robotic systems. Soft robotics leverages embodiment to attain novel capabilities or enable systems of increased complexity. However, to be able to design such increasingly intelligent systems, it is crucial to for engineers and researchers to understand how material and physical properties influence behavior. It is important that we educate individuals with the knowledge and skills necessary to tackle the complex system. We propose utilizing the V-shaped falling paper system as a teaching topic to delve into the role of morphological computation and design exploration. Parameterized V-shape papers exhibit diverse behaviors during the falling process, driven by their morphology. An easily deployable, cost-effective learning system was developed and employed to investigate participants' behaviors and learning outcomes using various learning objectives and methods. Participants recruited from university campus are expected to understand the V-shape falling paper system characteristic especially the link between the morphology and behaviors. Our findings indicate that participants' behaviors are influenced by the task objectives, while the learning methods employed can impact the learning outcomes. This research sheds light on the connection between task objectives and participants' behaviors as well as learning methods and learning outcomes, emphasizing the significance of morphological computation and design exploration in soft robotics education.

1. Introduction

Soft robotics is a growing field which is interdisciplinary, and exploits embodiment to achieve new capabilities ([1]). Embodied intelligence (EI), or computation through form, can provide new behaviors or simplify the required computation or decision making. It is particularly observable in soft or compliant system where the coupling between the body, brain and environment is strong ([2]). At a simplistic level, EI could be viewed as intelligence manifested in the body; furthermore, designing systems that exploit EI can have significant implications on the relationship between the physical body and the information or computation required ([3]). Interactions between soft structures and fluids, i.e., Fluid-Structure Interactions (FSI) show a high coupling between the environment and the soft structure. As such, these environmental interactions play a crucial role in shaping the behavior and capabilities of soft robotic systems. The dynamic interaction between soft robots and their surrounding fluid environment influences their locomotion, manipulation, and adaptability. As such, fluid-soft body interactions are an excellent means demonstrating EI.

To enable the design of improved capabilities in robots, we must teach and educate the next generation of roboticist as to how the body contributions to cognition, control and interactions ([4]). However, teaching EI is challenging due to the conceptual stubly and also the modeling complexity of physics and its interactions ([5]). We require methods and frameworks for teaching embodied intelligence, to understand how humans can best learn about embodiment and also how they approach the design systems which show embodied intelligence.

There are many different robotics platforms designed for education available, including hardware based systems (e.g. Lego Mindstorms, Thymio, Vex) ([6]) and also simulation based platforms ([7, 8]). These have been extremely successful in teaching many computational thinking concepts to a wide age range of students ([9]). However, due to the rigidity of these robots, their opportunity to explore the physical role of design through the lens of embodied intelligence and morphological computation is challenging. Although traditional mechanical engineering concepts can be taught, for example, electronics, mechanical design, computational thinking, and even art, the lack of complex interactions between body and environment means EI is not easily observable or exploitable. A number of soft robotics 'kits' have been introduced. This includes a soft robotics toolkit which provides a number of standardized designs for fabricating soft actuators and other structures ([10]). Whilst this introduces many fabrication concepts, it is more challenging to explore the role of design as the structure take time to fabricate to create. A soft robotic education kit utilizing sticks of hot melt adhesive has also been proposed which allows for rapid assembly of a variety of different structures and forms of the robot ([11]), promoting more design thinking and utilizing embodiment to aid the performance of the structure. However, this approach is still challenging for large scale education, assessment or benchmarking.

To address the limited platforms for teaching embodied intelligence, we propose the V-shaped paper system as a teaching tool for exploring the role of morphological computation and design exploration. This is a problem whereby V-shaped paper of different morphologies show different behavioral groups when falling, which results in different motion patterns and falling times introduced in the materials and methods section. Here, the interactions between the soft paper structure and the fluid drive the physical computation that occurs. Using this tool, we explore how the relationship between the falling paper design and its behavior can be explored and if physical interaction and sensory-based learning via students interacting with the paper is important to maximize their learning outcomes.

To address this question, we have created a design platform and procedure for optimization and exploration of morphological computation in a manner which is accessible, low cost and rapid. This leverages a Graphical User Interface (GUI) for selecting the design of the V-shaped paper, automated manufacturing with a laser cutter, robotic release mechanism and data capture using computer vision. Using this tool, we propose a number of different objectives for the individual in university to either perform exploration, or targeted optimization. By performing user trials with different learning objectives/methods and recording the outcomes, we explore the following questions:

- What objectives can be given to students to encourage humans to explore a morphological design space, and does this increase their learning outcome in identifying the relationship between shape and behavior?
- What approaches can be employed to optimize human learning outcomes when studying the morphology driven behavioral characteristics?

2. Materials and methods

In this section, we introduce the V-shaped falling paper problem, followed by the experimental design, task objectives, and workflow.

2.1. Falling Paper Problem

Our chosen focus is the design problem of the V-shaped falling paper, which involves a parameterized V-shaped paper exhibiting a range of different falling behaviors upon release, resulting in variable falling speeds. Introduced and studied in previous works ([12, 13]), it is an elegant example of morphological computation, with two design parameters controlling, or physically programming the falling behavior. In addition, the resulting design space is highly non-linear and stochastic in its nature. The study of falling paper is a wider research topic which has fascinated scientists for many years ([14, 15]).

Different morphologies will exhibit different behaviors which is a phenomena widely observed in fluid-structure interaction ([16]). The falling behaviors for the V-shaped paper have been previously categorized into four types ([12]). The first behavior, represented in Fig 1(a), is called **"plummeting,"** characterized by rapid descent without deformation. **"Undulant"** behavior (Fig 1(b)) is similar to plummeting but exhibits wave-like motion during the fall. When the V-shaped paper bends into a propeller shape and spins around the tip, it is classified as **"Helicopter"** behavior (Fig 1(c)). Another behavior, termed **"Asymmetric"** (Fig 1(d)), involves the V-shaped paper spinning around the tip with an angle without deformation. These distinct falling types have varying falling times, emphasizing the influence of behavior on the duration of descent. Moreover, the V-shaped falling paper is sensitive to environmental disturbances. The specific V-shaped paper design problem we consider is based upon paper of

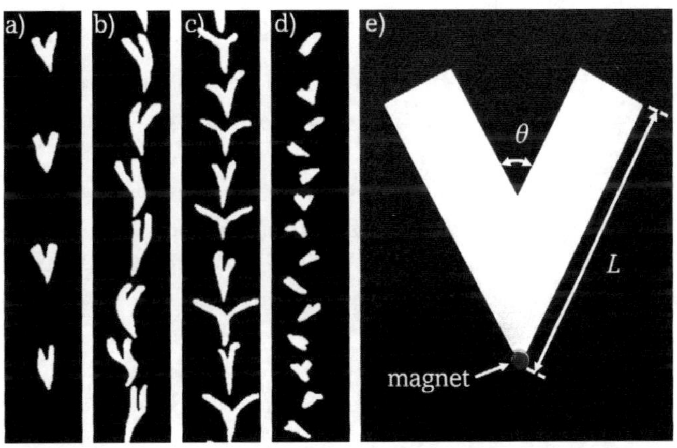

Figure 1. V-shaped falling paper behaviors and morphological parameters. a) Plummeting b) Undulant c) Helicopter d) Asymmetric e) V-shaped falling paper variables

mass 120 g/m^2, and a V-shaped paper where there are two design parameters, theta and length which can be varied between 30 to 70 degrees and 75 to 195 mm respectively. The width of the paper is fixed, and a fixed mass of 0.6 g is added to the paper. The mass is added using magnets on either side of the paper such that they can be easily added/removed. The design space (e.g., the fixed parameter and the variable ones) has been chosen such that all falling behaviors can be observed, and it is sufficiently large to navigate.

A grid search was conducted within the specific design space to measure the falling time and manually identify the behavioral types based on visual inspection. For each point, 10 experiments were conducted. Each behavior was assessed according to specific criteria defined in Fig 1. Although the system is stochastic, we can observe that a particular behavior is dominant in one design. The average time of the dominant behavior was taken and the results are depicted in Fig 2. The falling time has been interpolated for the sake of visualization and the behavior labels are given. This highlights the non-linear and complex nature of the design space and shows how the behavior maps to the falling speed, with the slowest speeds occurring with the asymmetric behavior, and the fastest with plummeting. The problem has been chosen due to the simplicity, multiplicity of states and also the accessible nature of the problem such that it is a teaching tool that could be used on a variety of ages of students.

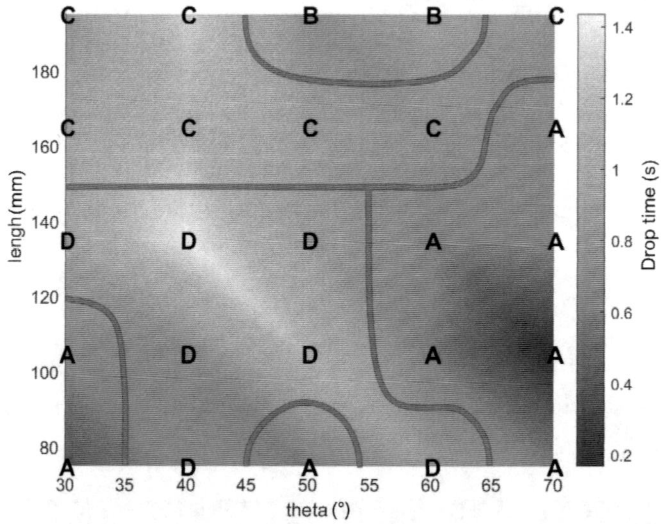

Figure 2. Falling time heat map interpolated from grid search and behaviors of the design space (A: Plummeting, B: Undulant, C: Helicopter, D: Asymmetric)

2.2. Experiment I: Investigation of human exploration methods

The goal of the first experiment is to investigate how humans explore, and how humans can be encouraged to explore and optimize within a specific design space which displays morphological computation. From this, we can then compare and contrast to computational optimization approaches. Thus, two different task objectives were chosen. There are two directions of the task objectives, exploratory and target-focused. The exploratory task requires the user to find as many as different kind of behaviors of the V-shaped falling paper while the target-focused task requires the user to find the fastest dropping design. We define these as Task A and Task B:

- **Task A**: find the most behaviors within the design space given 10 different trials.
- **Task B**: find the fastest dropping behavior within the design space given 8 trials.

For the first experiment, six participants were recruited from the university campus. The participants were randomly assigned into two groups. The male-to-female ratio is 5:1 and the ages are between 24 to 32 years old. None of the participants had previous knowledge of the V-shaped falling paper system nor was familiar with fluid-structure interaction (FSI). The two conditions were assigned in counterbalanced order. We consider two different scenarios. First where the participant performs Task A followed by Task B (i.e., exploratory and then target-focused), and also the reverse. This allows the effect of their prior knowledge gained through the previous task to be analyzed.

2.2.1. System setup and experiment workflow An automated robotic setup was created for the V-shaped paper experiment to allow user friendly design and fabrication. The experiment setup includes a computer running the V-shaped falling paper graphical user interface, a tabletop laser cutter, the gripper system and a webcam. The whole system is programmed in Matlab including GUI, control of hardware and computer vision. The participant uses the GUI to design, fabricate and execute the experiment. As shown in Fig 3, the participant designs the V-shaped falling paper on the GUI, then the laser cutter cuts out the V-shaped falling paper. The automated robotic system assists the experiment and participants use the data from the experiment to design the next iteration.

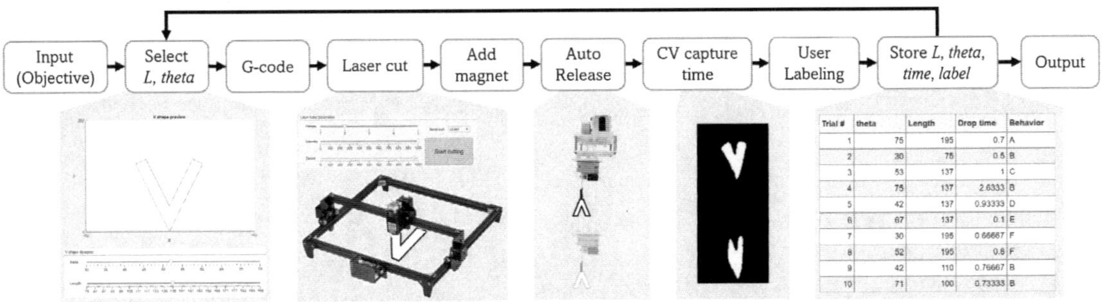

Figure 3. Flowchart of experiment procedure

The graphical interface shown in Fig 4 consisted of five sectors, V-shape designer, laser cutter control panel, gripper control panel, behavior recorder, and data display. The V-shape designer allows the user to adjust the angle and the length of the V-shape and preview the design. The user can use the laser cutter control panel to tune the cutting parameters, including laser power, moving speed, repeated passes, and initiate the cutting process. Once the design is ready, the user loads the A4-sized 120 g/m^2 DCP paper into the designated area then presses the start cutting button on the laser cutter control panel to generate G-code and send it directly to the laser cutter. Once the laser cutter received the G-code, it will automatically fabricate the V-shaped paper. The participant then added two magnets (Ø 5 mm, height 2 mm, 0.3g) to the tip of V-shaped paper. The user can manually control the gripper with the gripper control panel including open/close and lower/raise. By pressing the run experiment button, the system will automatically raise the V-shaped falling paper to the top and perform the dropping experiment. The user notes the behavior they observed in the behavior recorder and labels them in the data display sector. The data captured by the camera will be displayed in the data display area with history from previous experiments.

The experiment is assisted by an automatic gripper system. The participant first attaches the fabricated V-shaped paper to the gripper. The run experiment command sends from GUI makes the gripper automatically raise the V-shaped paper to 3 meters above the ground. Once the V-shaped paper is in position, the camera starts capturing frames and the gripper releases the V-shaped paper. The falling time is identified using computer vision method. The captured frames are processed by using background subtraction to isolate and identify the V-shaped falling paper. The frames including the V-shaped falling paper are counted then used for calculating the falling time.

The length, theta, falling time and behaviors labeled by the participant are stored and displayed in the data display sector. The participant was able to see the data from previous experiments and use them as references for future iterations.

2.2.2. Task Process Each session included one participant and one facilitator. The facilitator was one of the researchers. The facilitator's role was to provide clear task instructions, assist with any technical issues, and ensure a smooth overall experience for the participant. The participant was given a step-by-step tutorial on how to create the V-shaped falling paper and

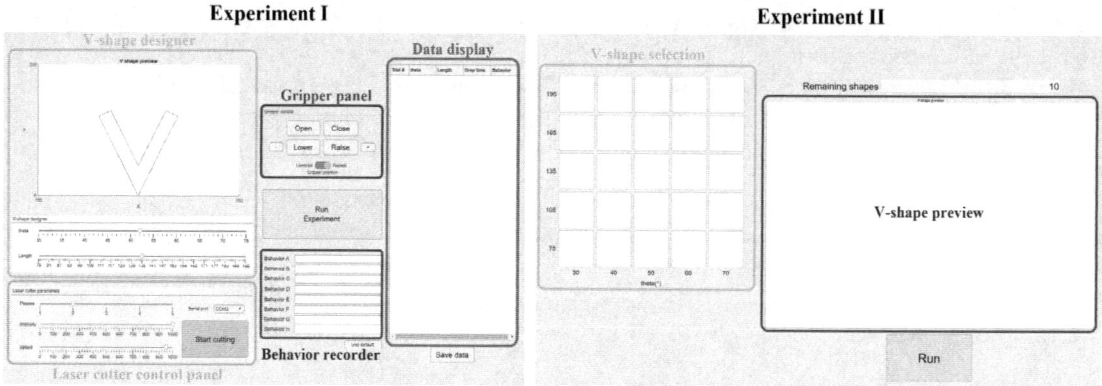

Figure 4. Graphical user interfaces of V-shaped falling paper experiment system

execute the experiment. The tutorial introduce the participant to the experimental setup and experiment procedure. The participant fills out individual pre-test and post-test questionnaires for both tasks. The facilitator gave task instructions before the pre-test questionnaire was filled. The pre-test questionnaire ask the participants about their planned approach for the task and expected effect of the morphology to the behaviors or falling time. The post-test questionnaire ask about the actual approach they used during activities and their observation of the link between morphology and the falling behavior or time. The entire session duration, from greeting the participant to thanking them for their participation, typically ranged from 45 minutes to 1 hour, depending on the participant's pace and questions. There was a 15-minute break between the two activities.

2.2.3. Bayesian Optimization Comparison & Modeling To evaluate the effectiveness of the exploration and also the internal model of the design space that the participant may develop, we compare human exploration to a black-box optimization method. Bayesian optimization(BO) is a black box optimization approach that seeks to minimize an objective function ([17]). Bayesian optimization provides a systematic and efficient framework for optimizing parameters or exploring a design space. It utilizes probabilistic models to guide the search process and adaptively selects new sample points to explore, based on previous observations. Using a surrogate model, test points are selected that seek to maximize the reward and minimize uncertainty in the model. Previously, Bayesian Optimization has been used as a means of explaining human active search and also as a benchmark or comparison ([18]). Bayesian optimization has also been explored as ways of imitating or learning how humans optimize or explore processes ([19]). In its simplest form, BO can be parameterized by one single ratio, the exploration and exploitation ratio; in choosing the next evaluation point what is the ratio between exploiting current knowledge or exploring new areas with a potential for high reward.

Using the experimental data for falling time and behavioral labeling (Fig 2) and interpolating we can simulate different Bayesian optimization by obtaining the fitness function from this interpolated data. This allows for the performance of Bayesian optimization with different exploration ratios to be computed to allow for comparison to human behavior for different objectives, and also to see if Bayesian optimization can be used to explain human active search behaviors. Bayesian optimization is performed using the Matlab Statistics and Machine Learning Toolbox, and varying the number of seeds and the exploration ratio. Surrogate models (i.e., creating a Gaussian process model) generated using participants' data allow us to examine the helpfulness of prior knowledge of doing target focused task. We compare the surrogate model from the participant as a mean of simulating their internal model of the design space and the surrogate model after the optimization task.

2.3. Experiment II: Comparison between multi-sensory and traditional learning methods
The second experiment aimed to test the hypothesis that hands-on and multi-sensory activities would enhance the learning outcomes related to morphology-driven behavioral characteristics. By providing participants with opportunities to physically interact with the V-shaped falling paper and engage their senses, the researchers expected that the hands-on and multi-sensory approach would lead to a deeper understanding of how morphology influences behavior.

The second experiment has 10 participants recruited from the university campus. All the participants were male and the ages are between 25 to 35 years old. None of the participants had previous knowledge of the V-shaped falling paper system nor was familiar with fluid-structure interaction (FSI). The participants were randomly assigned into two groups. The participants in the experiment group were able to access the physical V-shaped paper within the design space and do real experiments, while the control group participants used the GUI on the computer displaying the V-shaped paper in real scale and watched the pre-recorded videos.

2.3.1. System setup and experiment workflow The experiment setup includes a computer running the V-shaped falling paper graphical user interface for the control group and a simple release mechanism to drop the V-shaped paper for the experiment group. The participants chose the shape they would like to explore and received the chosen V-shaped in different ways depending on which group they were assigned. The participants in the control group were shown the real scale V-shaped paper on the GUI. The participants in the experiment group received the physical V-shaped paper. The participants in both group run the experiment under their conditions and observe the behavior after they have given a prediction of the chosen shape.

A GUI (right GUI in Fig 4) is designed for the control group participants. The design selection area allows participants to choose the V-shaped paper they would like to explore. The preview area shows the selected V-shaped paper in real scale. In order to preserve the stochastic characteristic of the system, there are six video recordings for each design and will be randomly displayed to the participants.

Pre-defined V-shaped papers were prepared for the experiment group. The V-shaped papers used in the second experiment are with the same parameters in first experiment except they are prepared by the facilitator before the experiment.

2.3.2. Post-test exam The post-test exam consists of three sections. The first section is the theoretical test to evaluate the participants' understanding of the V-shaped paper system, where the participants were given the parameter same as the design space and asked them to select the behaviors. The second section tests their transfer knowledge. The task is similar to the first section except the shape given were in different parameters including width, length and angles. The experiment group was given the cut V-shaped paper for each question and was able to interact with the paper. The control group was given the shapes printed on A4 papers instead of getting the V-shaped paper thus only visual information was given. The third section of the exam is guided questions asking the participants about the characteristic of the system they observed including behavior distribution, falling speed relationship and stochasticity of the system.

2.3.3. Task Process Each session included one participant and one facilitator. The facilitator was one of the researchers. The facilitator's role was to provide clear task instructions and ensure the task procedures were followed. Each participant were ask to explore ten out of 25 shapes of the design space. For the experiment group, the participants chose the shape they would like to explore and the facilitator handed over the physical paper of the chosen V-shaped. The participants were asked to interact with the paper on hand without dropping it and give a prediction of the behavior. After they were satisfied with their experience, the V-shaped papers were put into the release mechanism for the dropping experiment. For each design, the participants were asked to drop it at least twice with an option to do a third experiment if needed. The control group was asked to do similar procedure as the experiment group. Instead of having the V-shaped paper on hands, their choice will be displayed on the computer screen. They were also asked to give a prediction of the behavior. Once they have given a prediction, they can start the simulation process. The participants were given a sheet to record their observations and where they have explored.

3. Results and Discussion

3.1. Experiment I

3.1.1. Study Results The exploration patterns for each participant are shown in Fig 5. The participants' exploration points are widely spread for the exploratory task and more concentrated for the target-focused task. A more systematic approach can be observed from some participants (P2, P3 and P4) for the exploratory task. All the participants explored the four different behaviors in the exploratory task. The search pattern on target-focused task clustered on one area which shows more exploitation rather than exploration.

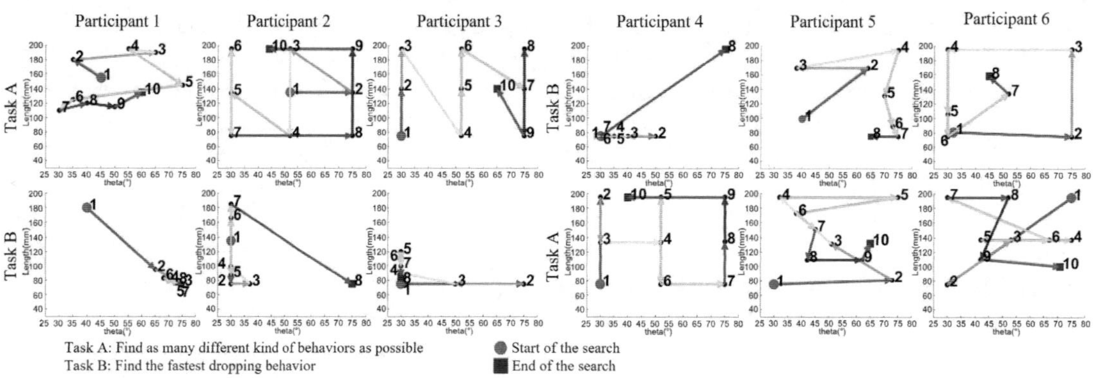

Figure 5. Exploration patterns of participants. Visited points sequence during the experiment were labeled by numbers. Participants 1 to 3 performed task A followed by task B, whereas participants 4 to 6 followed the reverse sequence, completing task B before task A.

When participants undertake the exploration task before the target-focused task (P1 to P3), the participants' search patterns for the target-focused task are very clustered. They have used their prior knowledge as reference to find the fastest dropping behaviors. While the participants (P4 to P6) do not have prior knowledge before the target-focused task, they do some exploration in the beginning and then exploitation. Comparing the search patterns of the target-focused task of P4 to P6, the first choice has a great impact. The first iteration of P4 was already close to free fall which led the participant to search in the near area instead of exploring in the first few iterations. For some (P2 and P4), we see exploration at the end after exploitation. They commented to the facilitator that they believe they have found the fastest dropping behavior in the area and wanted to see if there was a chance to find a faster design within the given trials.

The Euclidean distance between steps was used to evaluate the changes between each iterations. Fig 6 shows that the step size is greater in the exploratory task and with no discernible trends. However, the step size of the target-focused task reduces with iterations. Participants with prior knowledge from the exploratory task lead to a smaller step size earlier than those without. Thus, participants with prior exploration were searching in a more precise manner.

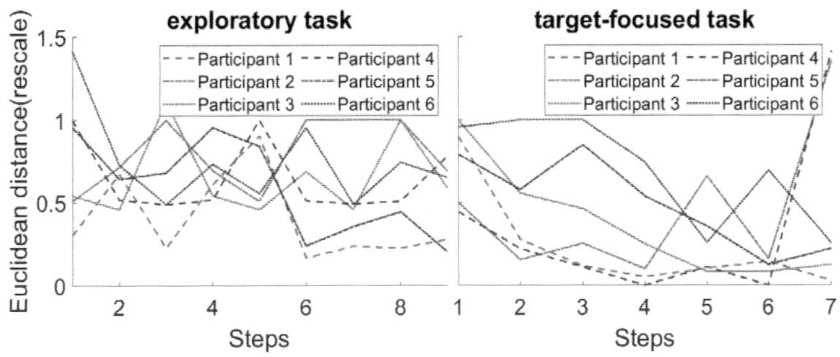

Figure 6. Euclidean distance of participants' design iterations. The differences of length and theta were both normalized between 0 and 1.

3.1.2. Comparison between human and Bayesian Optimization To compare both the performance and methodology of human optimization and search of the morphology space, we utilize Bayesian optimization to perform the same process, with a goal of minimizing falling time, using the interpolated falling time results (Fig 2) to determine the fitness function. We perform the Bayesian optimization with different exploration ratios and the results are shown in Fig 7. Comparing the search patterns from Bayesian optimizer and humans, human behaviors are similar to those with exploration ratios of 0.5 and 0.8.

A visualization of the difference between human and Bayesian optimization search paths is also shown in Fig 7. The Euclidean distance is smaller when the exploration ratio is smaller. We can see that the Euclidean distance of the human search path is between exploration ratio 0.5 and 0.8, which supports the previous conclusion that the human behaviors are similar between the exploration ratio of 0.5 and 0.8.

The surrogate model constructed from the participants' experiment data in Fig 8 suggests that participants likely used the model they learned from the exploratory task for the target-focused task. During the target-focused task, the participants leverage the model from the exploratory

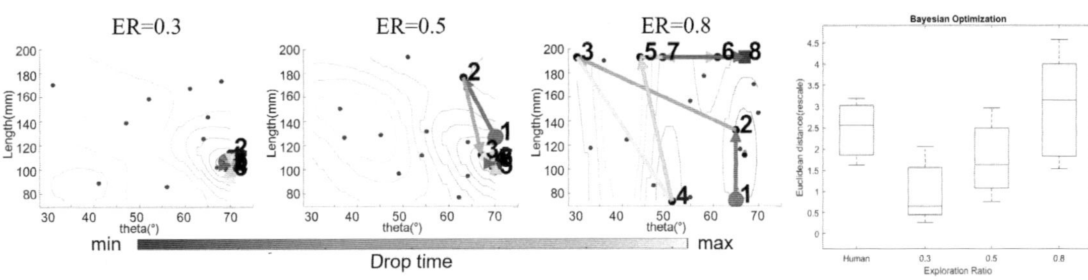

Figure 7. Example search patterns of Bayesian optimizer with different exploration ratio and Euclidean distance boxplot of human compare to Bayesian optimizer. Bayesian optimization was carried out multiple times for each exploration ratio, and the presented example provides a representation of different exploration ratio. The boxplot displays the distribution of the total Euclidean distance for each participant or optimization.

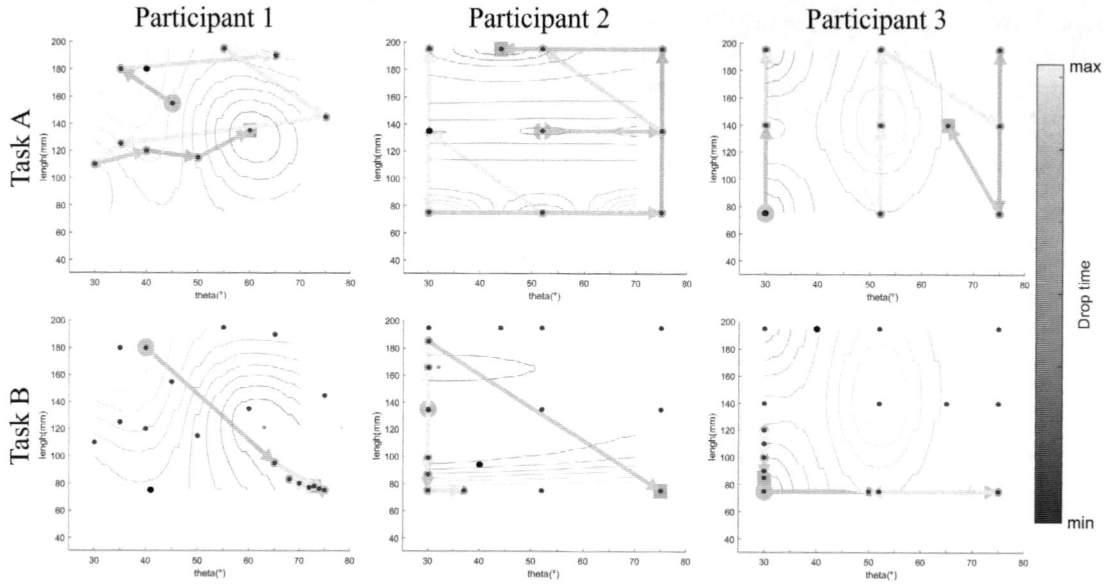

Figure 8. Reconstruction of objective functions using participants' experiment data. Participants' visited points in Task A were used to reconstruct the objective function on the upper row and all points in Task A and Task B were used for the lower row.

task to choose the starting points where they saw previous success. Consequently, the model (P1 and P3) established using data including both the exploratory and target-focused tasks is similar to the one only using the exploratory task data. The uncertainty reduced in certain areas (close to the minimum) reflects the Bayesian nature of searching. Except for P2 where the participant continued to search a relatively large area. In the target-focused task, some exploration can still be observed (high risk, high reward), which can reflect the Bayesian approach of expectation maximization. For the exploration task, humans have previous prior knowledge in addition to the 'internal model' they build which assisted them in initial exploration. For the target-focused task, the optimization approach is better exploit the model created from exploitation. Prior knowledge is not as useful for optimization, when the correct approximate area has been discovered.

3.2. Experiment II

3.2.1. Study result The participants' test performance of the post-test is visualized in Fig 9. The results indicate that the participants in the experiment group who had access to the hands-on and multi-sensory activity of interacting with the physical V-shaped falling paper, performed better in both theoretical and application tests compared to the control group. The result suggests that the multi-sensory activity indeed improved the learning outcomes for embodied intelligence education.

The physical interaction with the V-shaped paper allowed the participants in the experiment group to have a more immersive and engaging learning experience. By engaging multiple senses, such as touch and movement, they may have developed a deeper understanding of the concepts being taught. This hands-on approach might have facilitated better comprehension, retention, and application of the knowledge compared to the control group relying solely on visual cues.

These findings support the hypothesis that hands-on and multi-sensory activities can enhance learning outcomes for embodied intelligence education. The physical engagement and

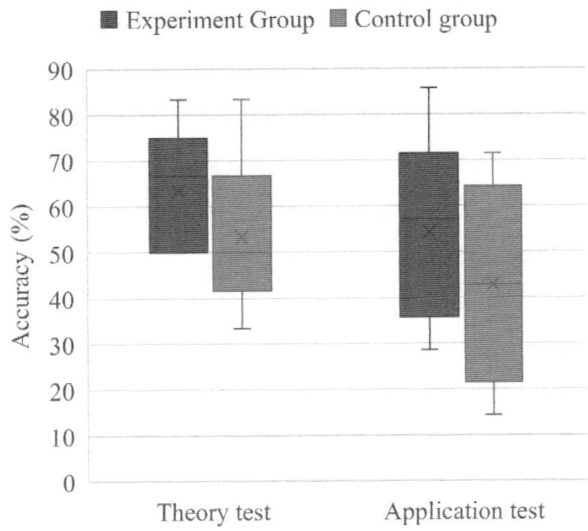

Figure 9. Exam result comparison between experiment group and control group

multi-sensory stimulation provided an enriched learning experience that facilitated a deeper understanding of the subject matter and better application of knowledge.

The ranking of participants' performance in the experiment (shown in Fig 10) indicates a clear dominance of the experiment group, with only one participant from that group performing below the control group participants. This suggests that the hands-on and multi-sensory activity provided to the experiment group had a significant positive impact on their learning outcomes.

Additionally, when observing the selected exploration points, another interesting finding emerged. Participants who utilized an adaptive search distribution tended to have better performance compared to those who employed grid search patterns. This implies that adaptability in exploring the design space, rather than adhering to a rigid grid-based approach, contributed to improved learning outcomes.

The adaptive search distribution approach allows participants to dynamically adjust their exploration strategies based on the feedback and information they receive during the learning process. This flexibility enables them to efficiently explore and discover optimal solutions within the design space. On the other hand, participants utilizing grid search patterns follow a predetermined grid-based approach, which may limit their ability to explore and identify the most effective solutions.

Participants in both groups were able to capture the characteristics of the V-shaped system. They reported understanding the stochastic nature of the system and successfully linked the falling speed with the observed behaviors. The participants' illustrations of the system's behaviors in the design space aligned with their observations during the experiment. Some of the illustrations closely resembled the grid search pattern depicted in Fig 2. Despite the differences in the learning experiences between the experiment and control groups, both groups were able to comprehend and articulate the behavior of the V-shaped system.

4. Conclusion

In this paper, we presented the infrastructure to enable paper 'V' shapes to be a platform for both educating students about the role of morphological computation, and also to understand how individuals optimize morphology. We investigated the difference on learning behaviors between two learning objective directions, exploratory and target-focused. The search space and step size are wider for the exploratory task. The participants who completed the exploratory task performed better on the target-focused task compared to those who started directly with the

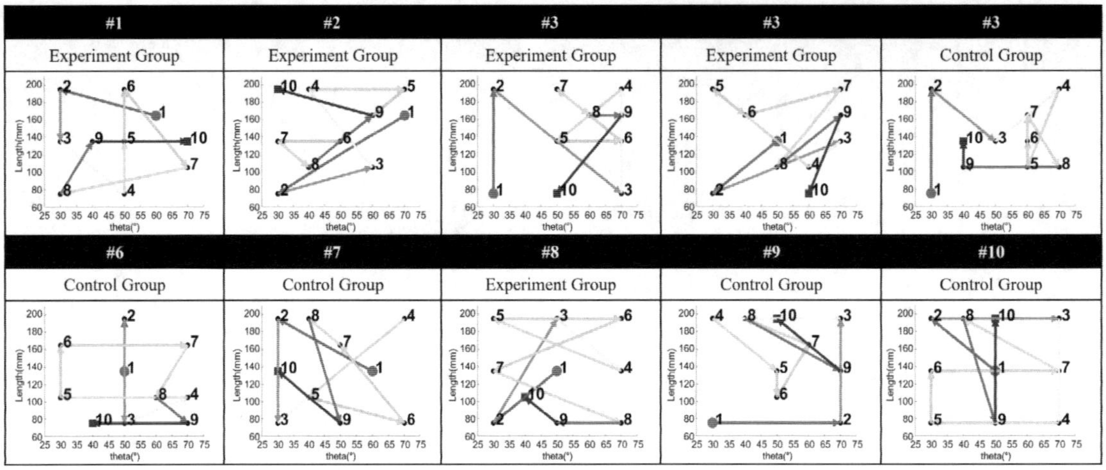

Figure 10. Participants' exploration patterns with order of the ranking of their performance.

target-focused task.

Results from the second experiment highlight the effectiveness of hands-on and multi-sensory activities in improving learning outcomes for embodied intelligence education. They also underscore the benefits of incorporating adaptive exploration strategies to enhance performance and problem-solving abilities. These insights have significant implications for educational practices, emphasizing the value of providing students with interactive and adaptable learning experiences.

The sample size of the current experiment is very limited and it is anticipated to have a larger scale of user study. The focuses remain using the current system to assist humans in exploring morphological computation and optimization, in addition, investigating what different tasks would be benefit to human on morphology computational thinking and understanding.

Acknowledgments
We are grateful to the support from the Jacob's Foundation which made this work possible.

References
[1] Sangbae Kim, Cecilia Laschi, and Barry Trimmer. "Soft robotics: a bioinspired evolution in robotics". In: *Trends in biotechnology* 31.5 (2013), pp. 287–294.

[2] Rolf Pfeifer and Gabriel Gómez. "Morphological computation–connecting brain, body, and environment". In: *Creating brain-like intelligence*. Springer, 2009, pp. 66–83.

[3] Rudolf M Füchslin et al. "Morphological computation and morphological control: steps toward a formal theory and applications". In: *Artificial life* 19.1 (2013), pp. 9–34.

[4] Vincent C Müller and Matej Hoffmann. "What is morphological computation? On how the body contributes to cognition and control". In: *Artificial life* 23.1 (2017), pp. 1–24.

[5] Francesco Stella and Josie Hughes. "The science of soft robot design: A review of motivations, methods and enabling technologies". In: *Frontiers in Robotics and AI* 9 (2022).

[6] Omar Mubin et al. "A review of the applicability of robots in education". In: *Journal of Technology in Education and Learning* 1.209-0015 (2013), p. 13.

[7] Stefano Carpin et al. "USARSim: a robot simulator for research and education". In: *Proceedings 2007 IEEE International Conference on Robotics and Automation*. IEEE. 2007, pp. 1400–1405.

[8] Josie Hughes. "Robotic rescue simulation for computing teaching in the UK: A case study". In: *2016 IEEE Global Engineering Education Conference (EDUCON)*. IEEE. 2016, pp. 1051–1055.

[9] Andri Ioannou and Eria Makridou. "Exploring the potentials of educational robotics in the development of computational thinking: A summary of current research and practical proposal for future work". In: *Education and Information Technologies* 23.6 (2018), pp. 2531–2544.

[10] Dónal P Holland et al. "The soft robotics toolkit: Shared resources for research and design". In: *Soft Robotics* 1.3 (2014), pp. 224–230.

[11] Xiaoxiang Yu et al. "Soft robotics education". In: *Soft Robotics* 1.3 (2014), pp. 202–212.

[12] Toby Howison, Josie Hughes, and Fumiya Iida. "Morphologically programming the interactions of V-shaped falling papers". In: *ALIFE 2020: The 2020 Conference on Artificial Life*. MIT Press. 2020, pp. 359–366.

[13] Toby Howison et al. "Physics driven behavioural clustering of free-falling paper shapes". In: *PloS one* 14.6 (2019), e0217997.

[14] Yoshihiro Tanabe and Kunihiko Kaneko. "Behavior of a falling paper". In: *Physical Review Letters* 73.10 (1994), p. 1372.

[15] Umberto Pesavento and Z Jane Wang. "Falling paper: Navier-Stokes solutions, model of fluid forces, and center of mass elevation". In: *Physical review letters* 93.14 (2004), p. 144501.

[16] Tayyaba Bano et al. "Investigation of Fluid-Structure Interaction Induced Bending for Elastic Flaps in a Cross Flow". In: *Applied Sciences* 10.18 (2020). ISSN: 2076-3417. DOI: 10.3390/app10186177. URL: https://www.mdpi.com/2076-3417/10/18/6177.

[17] Peter I Frazier. "Bayesian optimization". In: *Recent advances in optimization and modeling of contemporary problems*. Informs, 2018, pp. 255–278.

[18] Ali Borji and Laurent Itti. "Bayesian optimization explains human active search". In: *Advances in neural information processing systems* 26 (2013).

[19] Nathan Sandholtz et al. "Inverse Bayesian Optimization: Learning Human Acquisition Functions in an Exploration vs Exploitation Search Task". In: *arXiv preprint arXiv:2104.09237* (2021).

Evaluation of The Stochasticity in The Collective Bernoulli-ball System as A Physical Reservoir Computer

Fan Ye, Arsen Abdulali, and Fumiya Iida

Bio-Inspired Robotics Lab, Department of Engineering, University of Cambridge, Trumpington Street, Cambridge CB2 1PZ, United Kingdom

E-mail: fy264@cam.ac.uk, aa2335@cam.ac.uk, fi224@cam.ac.uk

Abstract. Reservoir computing is a type of machine learning approach that uses a fixed, randomly generated dynamic system called a reservoir to process input data for prediction and classification tasks. The performance of a reservoir computer is highly dependent on the properties of its physical reservoir, particularly its stochasticity. Currently, there is no practical guide for selecting an appropriate reservoir to accomplish desired tasks. In this paper, we utilized the collective Bernoulli-ball system as the reservoir and investigated the effect of stochasticity on its performance in a non-linear prediction task. Additionally, we examined other properties of the reservoir computer, such as the input delay, input gain, n-samples average, and prediction target delay. This paper reveals how stochasticity affects the prediction performance of a reservoir and provides some possible methods of adjusting the parameters mentioned above to mitigate the drawbacks of stochasticity. Our findings show that in low-noise cases the reservoir output follows the trend of the target, but sometimes jumps to a significant offset. For high-level noise, the system remains stable, but the trend information in the output is compressed since noises cover the effect of input, which finally makes the reservoir computer useless for the task.

1. Introduction

Artificial intelligence has brought about a revolution in data processing. By mimicking the neural network of human brains, it excels in various tasks such as computer vision [28], function approximation [9], speech recognition [1], and robot control [2]. However, there is growing concern over the efficiency of artificial intelligence [7]. Even for seemingly simple tasks, neural networks still require a substantial amount of data and consume significant time and energy [22, 14]. To address this issue, researchers have explored leveraging physical systems' properties for computation [6]. For instance, one research proposed to use a low-power resonance system to realize the learning process [13]. This method of learning through a physical system is called reservoir computing [24, 17].

Reservoir computing was initially developed as a technique to address temporal problems using linear post-processing methods [23]. It comprises an input layer, a reservoir (either a pre-defined neural network or a physical model), and an output layer, which uses linear regression for training. This training process is less expensive than training a new neural network. Two prototypes of reservoir computers are the echo state network [9] and the liquid state machine [15]. The echo state network incorporates a recurrent neural network in its reservoir to handle periodic sequence learning and switchable point attractor learning tasks. The liquid state machine employs computer-simulated neural microcircuits in its reservoir to classify spoken words.

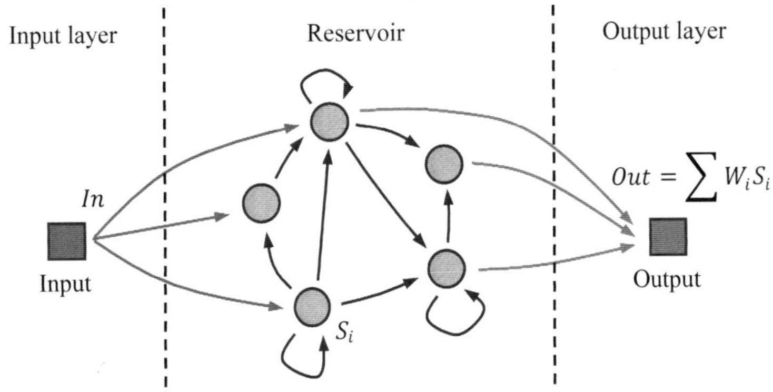

Figure 1. The prototype of reservoir computer. The input layer provides inputs In to the reservoir; the reservoir (a 5-node echo state network) exhibits behaviors influenced by the input, given by node state $S_i = \mathcal{F}(In, S), i = 1, \cdots, 5$; the output layer generates output by a linear combination of reservoir state $Out = \sum W_i S_i$. For other reservoirs, e.g., soft silicone arms [18], they use intrinsic dynamics to update S.

Various types of systems have been applied to reservoir computing to investigate their properties and performances, including pre-defined neural networks [8, 10, 11, 3], photonics chips [27], soft silicone arms [18, 20, 19], tapered whiskers [31, 32], and even dynamic memristors [4]. While there have been successful examples of reservoir computers and some guidelines [14] for their application, there are no practical rules for selecting an appropriate physical system as the reservoir to accomplish desired tasks. Previous research has shown that the high-dimensional nonlinear dynamics of physical reservoirs empower the reservoir computer [17], but such physical systems in nature always incorporate a different level of stochasticity, which affects the stability of the reservoir. Therefore, it is necessary to accomplish two key tasks: determining the degree of stochasticity that is tolerable for a physical reservoir, and identifying the method to enhance the performance of a reservoir computer with a certain level of stochasticity.

To assess the impact of stochasticity on the reservoir computer, we selected a collective Bernoulli-ball system as the reservoir, as depicted in Fig. 2. This system consists of a nozzle that releases an upward vertical airflow, allowing four balls to levitate. Each individual ball is balanced vertically by its weight and the drag force of the airflow [25], and horizontally by a side force towards the horizontal center, governed by Bernoulli's principle. The collective behavior of the balls exhibits collaboration and competition due to their interactions with each other [30]. In our previous work [29], we developed a simulator that incorporates the balancing forces of the flow field [16, 5], the interaction forces among the collective balls [12], and the stochastic forces caused by turbulence [21, 26]. As a physical reservoir, the collective Bernoulli-ball system consists of abundant behavior patterns for computation and stochasticity for investigation. Furthermore, its adjustable parameters like noise strength make it a perfect prototype to investigate how the properties of the reservoir influence the performance of the reservoir computer.

In this paper, we applied the collective Bernoulli-ball simulator as the reservoir and assessed its performance in a non-linear prediction task using the Mackey-Glass chaotic time series. We

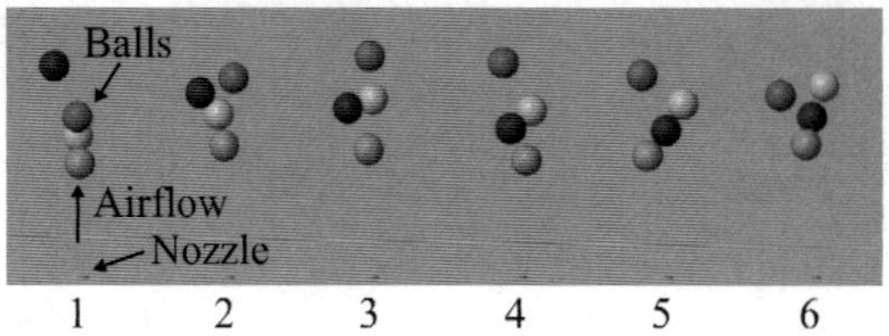

Figure 2. The Collective Bernoulli-Ball System. The nozzle releases vertical airflow upward, with balls floating inside the airflow. The interactions between the balls and the airflow give rise to intricate collective behaviors. The chaos of this system comes from the complex interaction between Bernoulli-balls, while the stochasticity is due to the turbulence in the airflow.

explored various properties of the reservoir computer, such as the variance of Gaussian noise, the input delay, the input gain, the n-samples average, and the prediction target delay. The performance of the reservoir computer was evaluated using the mean prediction error and its standard deviation. Finally, we analyzed the relationship between the properties of the reservoir computer and its performance in non-linear prediction tasks to determine the acceptable level of stochasticity in a reservoir and methods for enhancing the performance of the reservoir computer.

2. Methods

In this section, we introduce A.) the method to introduce a collective Bernoulli-ball system to a reservoir computer, B.) the nonlinear prediction task that examines the reservoir performance, and C.) the parameters that we investigated in the reservoir computer. We select the collective Bernoulli-ball system as the physical reservoir due to its abundant behavior patterns for computation and stochastic nature for performance analysis.

2.1. Bernoulli-ball Reservoir Computer

The Bernoulli-ball reservoir computer comprises three components, as depicted in Fig. 3. The first component is the input layer, which accepts input data and converts it into a flow speed signal. The second component is the reservoir, which is a collective Bernoulli-ball system where a vertical airflow lifts four balls, and they interact with each other. The nozzle receives the stairstep signal of flow speed from the input layer, and the balls react to the varying flow speed by exhibiting different behaviors. The third component is the output layer, which receives the time trajectories of the positions of the balls from the reservoir. The linear regression process in the output layer samples the trajectory data and transfers it into the output.

2.2. Task of Nonlinear Prediction

We use a nonlinear prediction task to evaluate the computational performance of this system. The input is a Mackey-Glass chaotic time series, which models the variation in the relative quantity of mature cells in the blood. The equation is:

$$\frac{dx}{dt} = \beta \frac{x_\tau}{1 + x_\tau^n} - \gamma x, \qquad \beta, \tau, n, \gamma > 0, \tag{1}$$

where β, τ, n, γ are constants and x_τ is the value of x at time $(t-\tau)$. Under a certain combination of parameters, this system displays a range of periodic chaotic dynamics. In our task, the parameters (β, τ, n, γ) are set to be $(0.2, 17, 10, 0.1)$, and the initial condition is $(x_0 = 1)$. The target is the input with a prediction target delay, see the output layer of Fig. 3.

We select one part of the Mackey-Glass chaotic time series as the training dataset to train the reservoir computer to predict this nonlinear trajectory. After training, we select another

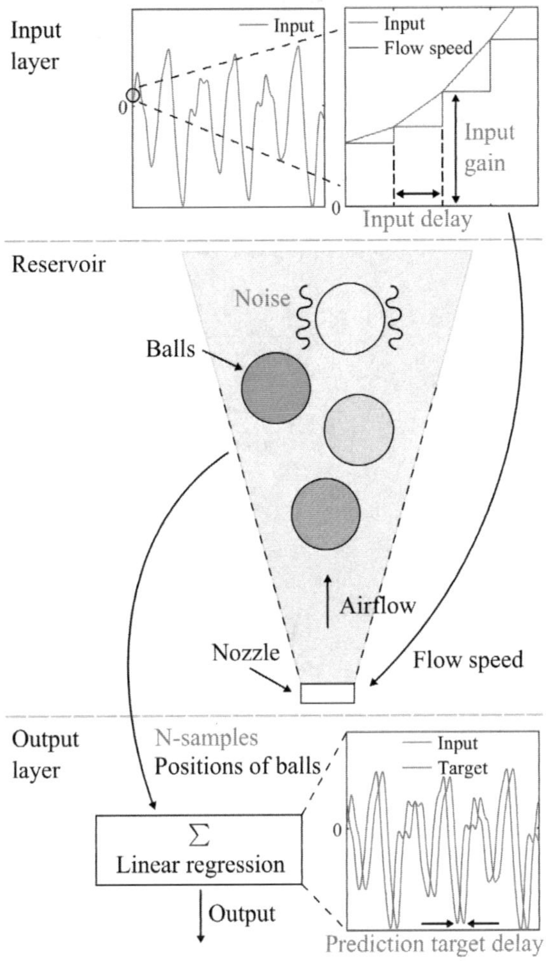

Figure 3. Schematic diagram of the reservoir computer for nonlinear prediction task based on the collective Bernoulli-ball system. The approximation quality is affected by the following parameters: input delay, input gain, noise variance, the n-samples (the sampling method when collecting the reservoir data for linear regression), and the prediction target delay.

part of the nonlinear trajectory as the testing dataset and compute the prediction error between the testing output and the target to evaluate the performance of the reservoir computer. The prediction error is calculated as the mean squared error

$$e = \frac{1}{N}\sum_{i=1}^{N}(Y_i - \hat{Y}_i)^2, \tag{2}$$

where e is the prediction error, N is the number of sample points, Y is the testing output, and $\hat{Y}$ is the target. Note that the first 10% of testing output is ignored when calculating the prediction error to reduce the effect of the initial condition.

2.3. Investigated Parameters

Our study aims to explore how five parameters affect the performance of the reservoir computer, as depicted in Fig. 3. These parameters include the noise strength, input delay, input gain, n-samples average, and prediction target delay. The noise strength factor (nsf) quantifies the degree of randomness that affects the movement of balls in the reservoir and ranges from 0 to 1, where $nsf = 0$ means no noise while $nsf = 1$ indicates the same noise strength as in real-world experiments, which was determined in our previous work about building the Bernoulli-ball simulator [29]. The input delay factor (idf) reflects the time delay of input sent to the reservoir and varies from 1 to 50. For instance, $idf = 1$ means the reservoir receives a new input every frame it computes, while $idf = 50$ means the reservoir receives a new input every 50 frames it computes. The input gain factor (igf) determines the scale when the input change is transformed into the change of airflow speed and ranges from 0.5 to 1.5, where $igf = 1$ indicates that the flow speed changes the same amount as the input in m/s. The n-samples average (n_s) corresponds to the number of samples that the linear regression uses to collect data from the reservoir. When $n_s = last$, the linear regression retrieves the last frame of the reservoir before receiving a new input. When $n_s = half$, it computes the average of the second half of frames of the reservoir before receiving a new input. The prediction target delay (ptd) indicates the frame delay between the target and input, which varies from 1 to 10. Table. 1 summarizes the default values and limits of the investigated parameters.

Table 1. Investigated Parameters.

Parameter	Limits	Default
noise strength factor (nsf)	0∼1	1
input delay factor (idf)	1∼50	50
input gain factor (igf)	0.5∼1.5	1.1
n-samples average (n_s)	last, half	last
prediction target delay (ptd)	1, 10	1

1 When investigating the effect of one parameter, all other parameters are set to default.

As the reservoir computer incorporates stochasticity, the performance may vary based on the random seeds of the Gaussian noise. Therefore, we conduct 50 training and prediction processes for each parameter configuration and compute the mean and standard deviation of the error to assess the prediction performance.

3. Results

The computational process is illustrated in Fig. 4. The input layer shows that the flow speed is represented by stairsteps of the input. The reservoir is depicted by the positions of four balls that are affected by the variations in the flow speed. Trajectories that lie above $0.2(m)$ show the vertical z-axis positions, while those that lie below $0.2(m)$ depict the horizontal xy-axis positions. After the flow speed changes, the second half of the positions are sampled and passed to the linear regression process. The output layer is depicted by a comparison between the predicted output and the target. The Bernoulli-ball reservoir computer predicts the nonlinear trajectory with a reasonably low error.

3.1. Noise Strength

The interaction between the balls and the airflow generates turbulence that affects the motion of nearby balls in a random manner. This randomness is measured by the variance of a Gaussian-distributed random signal in the simulator. Previous research has shown that noise contributes to the complexity of the collective behaviors among balls [30]. In this study, we investigate the impact of noise on the performance of nonlinear prediction. Fig. 5 illustrates that when there

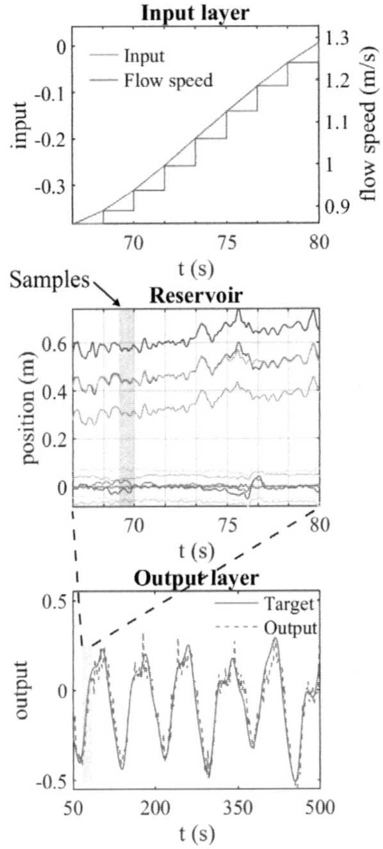

Figure 4. The result of one Bernoulli reservoir computation. The parameters are (nsf = 0.1, $idf = 50$, $igf = 1.1$, $n = half$, $ptd = 1$), and the prediction error is ($e = 0.0040$).

was no noise ($nsf = 0$), the mean error was very low ($e = 0.0020$) with a standard deviation of 0 ($std = 0$). In contrast, when there was low noise ($nsf = 0.1$), the mean error was significantly higher ($e = 0.3081$) and had a high standard deviation ($std = 0.9949$). When there was relatively high noise ($nsf \in [0.2, 1]$), the mean error was moderate ($e \approx 0.055$) with an almost moderate standard deviation ($std \approx 0.008$). Notably, when the noise strength was similar to that of the real-world experiment ($nsf = 1$), the mean error was moderate ($e = 0.0570$) and had a medium standard deviation ($std = 0.0109$).

3.2. Input Delay

The input delay factor (idf) determines the duration for which the reservoir responds to the input. Specifically, the reservoir receives the input with a time delay of ($dt = idf/30\ s$). As depicted in Fig. 6, a high mean error ($e = 0.1040$) and high standard deviation ($std = 0.0612$) were observed when $idf = 1$, indicating that a shorter input delay negatively affects the prediction performance. On the other hand, when idf ranged from 5 to 50, a medium mean error ($e \approx 0.055$) and almost medium standard deviation ($std \approx 0.07$) were observed. Particularly, a medium mean error ($e = 0.0552$) and standard deviation ($std = 0.0074$) were observed when $idf = 50$, indicating that a longer input delay may contribute to better prediction performance.

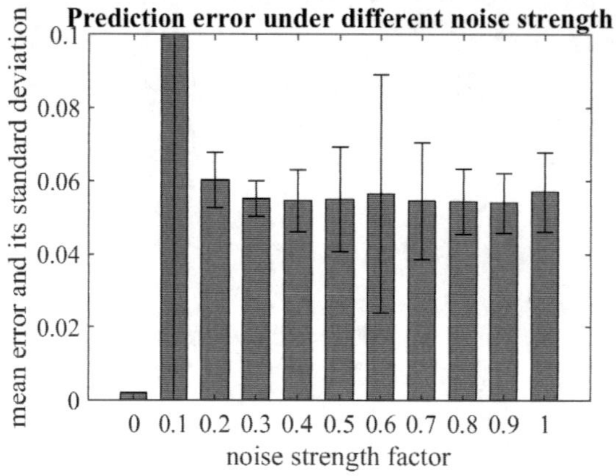

Figure 5. Prediction error of reservoir computing under different variances of Gaussian noise. The bar chart shows the mean error and its standard deviation.

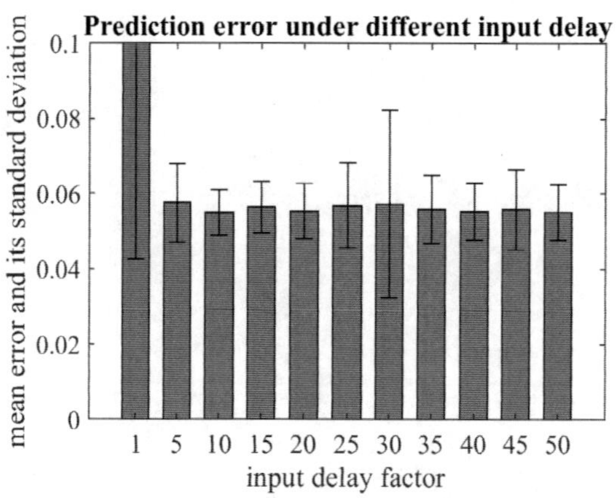

Figure 6. Prediction error of reservoir computing under different input delays.

3.3. Input Gain

The input gain factor (igf) determines the transfer of the input to the airflow speed of the reservoir as ($v = v_0 + input * igf$), where ($v_0 = 1.2801$) is the default flow speed that simulates the real-world experiment [29]. Fig. 7 shows that the input gain factor did not significantly affect the prediction performance when varying from 0.5 to 1.5. The mean error was consistently medium ($e \approx 0.055$), and its standard deviation ranged from low to medium ($std \in [0.0021, 0.0152]$). Notably, when $igf = 1.1$, the prediction error was medium ($e = 0.0570$), and its standard deviation was also medium ($std = 0.0109$).

3.4. N-samples Average

As the collective Bernoulli-ball system involves stochasticity, increasing the number of samples and averaging them during linear regression could be a promising method to reduce prediction uncertainty. However, according to Fig. 8, the half-sampling approach did not improve the performance of nonlinear prediction compared to the last-sampling method. Both methods yielded similar prediction error and standard deviation, except for when $nsf = 0.1$, where the last-sampling method ($e = 0.3081$, $std = 0.9949$) performed better than the half-sampling

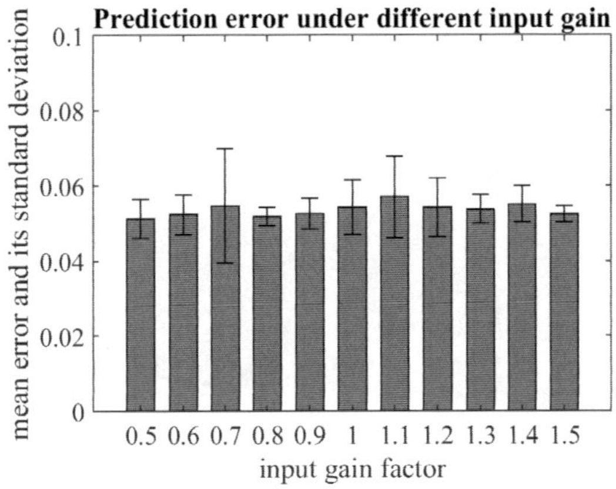

Figure 7. Prediction error of reservoir computing under different input gain.

method ($e = 0.4211$, $std = 1.1797$). Notably, for $nsf = 1$, the half-sampling method achieved moderate performance with $e = 0.0571$ and $std = 0.0208$.

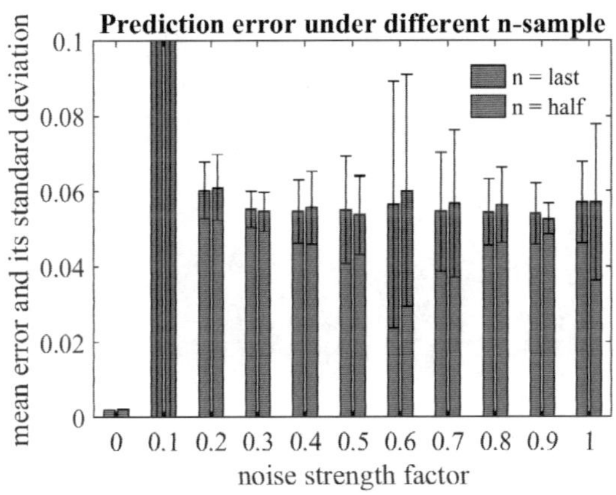

Figure 8. Prediction error of reservoir computing under different variances of Gaussian noise to compare the last-sampling method and the half-sampling method.

3.5. Prediction Target Delay

The difficulty of the task can also affect the performance of non-linear prediction. A longer prediction target delay means that the reservoir computer must predict the non-linear trajectory further into the future. Fig. 9 shows that when ($nsf = 0$), the prediction error significantly increased from ($delay = 1$, $e = 0.0021$) to ($delay = 10$, $e = 0.0436$). When ($nsf = 0.1$), the prediction error and its standard deviation significantly decreased from ($delay = 1$, $e = 0.4211$, $std = 1.1797$) to ($delay = 10$, $e = 0.1465$, $std = 0.1917$). When ($nsf \in [0.2, 1]$), the prediction error and its standard deviation generally decreased slightly from ($delay = 1$) to ($delay = 10$). In particular, when ($nsf = 1$), the case with ($delay = 10$, $e = 0.0506$, $std = 0.0116$) had better performance than the case with ($delay = 1$, $e = 0.0571$, $std = 0.0208$).

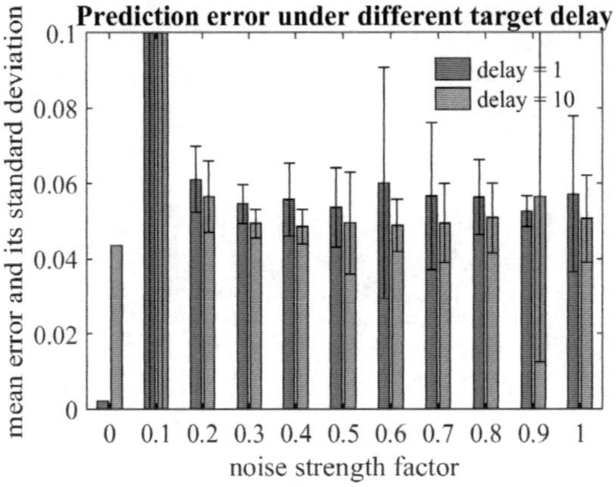

Figure 9. Prediction error of reservoir computing under different variance of Gaussian noise to compare different prediction target delay.

3.6. Joint Effect of All Parameters

We have investigated the effects of five parameters in the previous sections. However, these parameters might have joint effects on the performance of the Bernoulli-ball reservoir computer. Fig. 10 shows how these five parameters jointly contribute to the prediction error and its standard deviation. In general, when ($nsf = 0$), the prediction error and its standard deviation were low. When ($nsf \approx 0.1$), the prediction error and its standard deviation were extremely high. When ($nsf \in [0.2, 1]$), the prediction error and its standard deviation were medium. Increasing the input delay factor resulted in a decrease in the prediction error and its standard deviation. The input gain factor did not have a significant effect on the prediction error and its standard deviation.

Some interesting joint effects are found in Fig. 10. When the noise strength was high ($nsf \in [0.5, 1]$), decreasing the input gain factor slightly reduced the prediction error and its standard deviation. In the absence of noise ($nsf = 0$), increasing the prediction target delay significantly increased the prediction error and its standard deviation. However, when the noise strength was relatively higher ($nsf \in [0.2, 1]$), increasing the prediction target delay slightly decreased the prediction error and its standard deviation.

3.7. Effect of Noise in The Reservoir

As mentioned in Fig. 5 and Fig. 10, the mean error and standard deviation increased as the noise strength factor increased from $nsf = 0$ to $nsf = 0.1$, and decreased as nsf increased from $nsf = 0.1$ to $nsf = 1$. This indicates that the effect of noise might work differently in the low-noise reservoir ($nsf = 0.1$) and high-noise reservoir ($nsf \in [0.2, 1]$). Therefore, Fig. 11 shows the internal states of the reservoir under different nsf, which can help explain the role of noise in a non-linear prediction task.

Fig. 11A shows that when $nsf = 0$, the output of the reservoir matched the target at low error. When $nsf = 0.1$, most of the outputs followed the trend of the target but occasionally jumped to a big offset. When $nsf = 0.2$, the outputs were gathered in the range of the target ($[-0.5, 0.5]$) but it was difficult for them to follow the trend. When $nsf = 1$, the outputs were gathered around the mean of the target (0) with low amplitude (≈ 0.1). Fig. 11B and C show that when $nsf = 0$, the trajectory of the yellow ball was determined and stable. When $nsf = 0.1$, the trajectory mainly converged to three discrete states: the high, middle, and low states at heights around ($0.7, 0.6, 0.4\ m$). Although every time the ball started at a certain

IOP Conf. Series: Materials Science and Engineering **1321** (2024) 012012
doi:10.1088/1757-899X/1321/1/012012

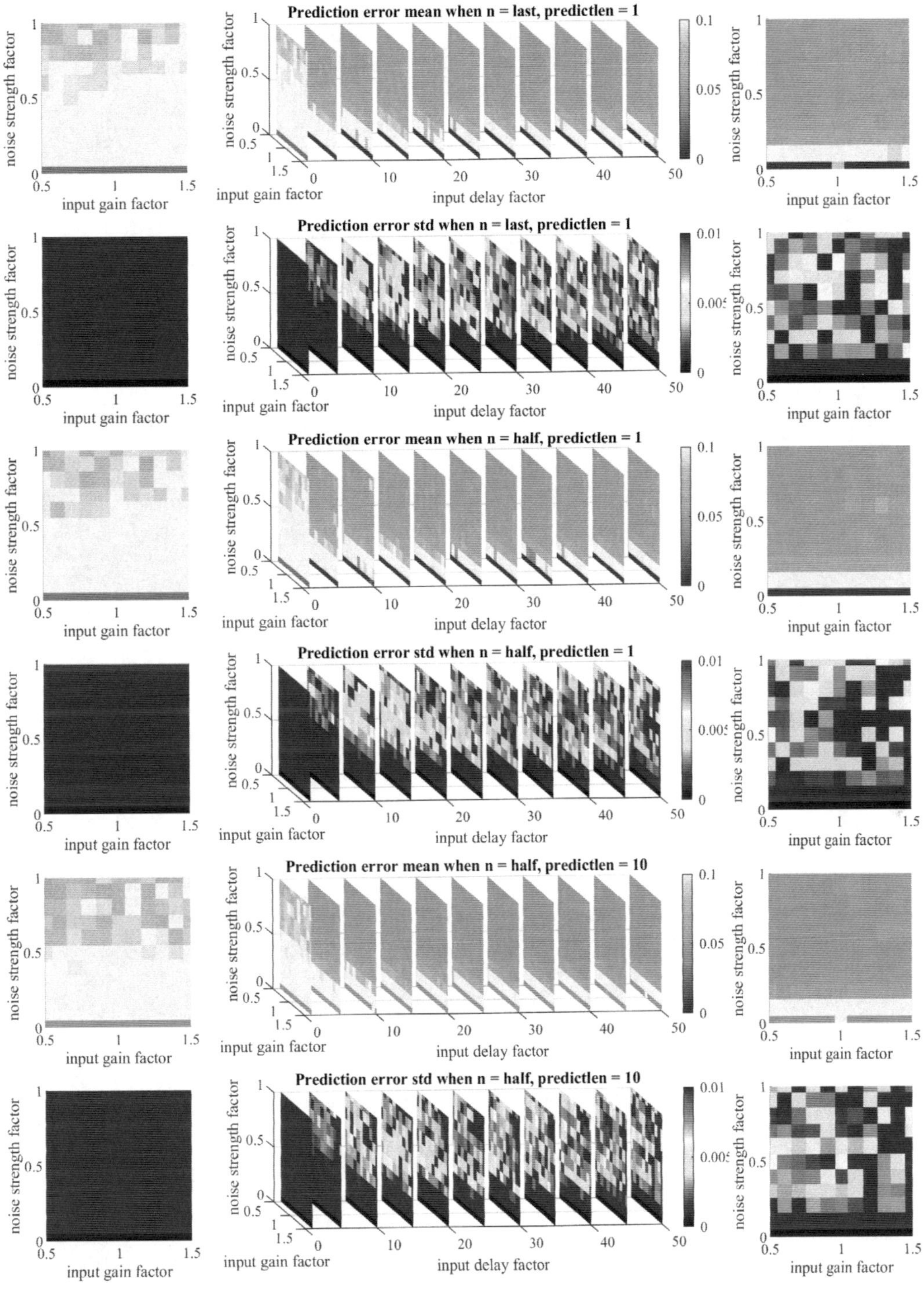

Figure 10. Prediction error of reservoir computing under different properties. The color on the frames shows the mean error and its standard deviation. The *predictlen* term refers to the prediction target delay.

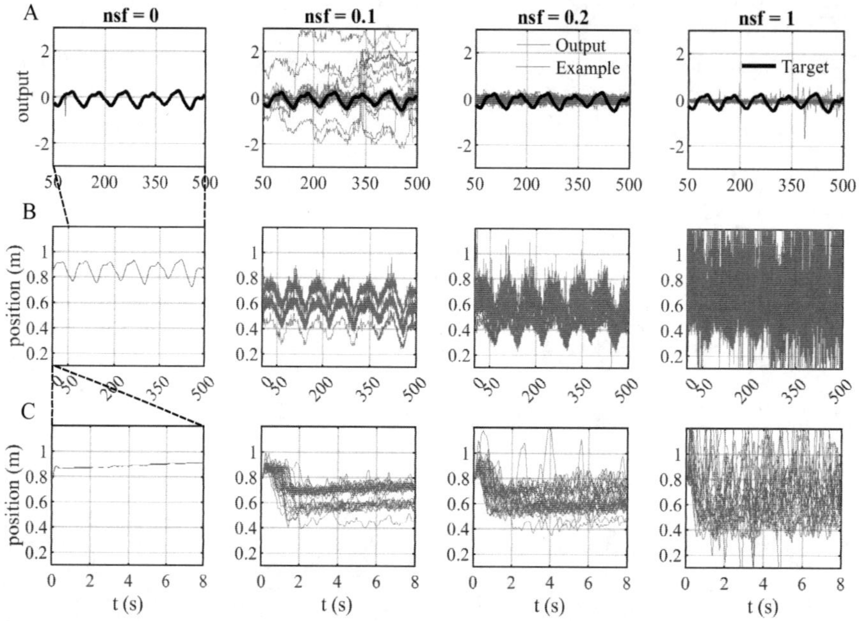

Figure 11. Internal states of the reservoir under different noise strength factors (nsf) while other parameters are the same ($idf = 50$, $igf = 1$, $n = half$, $ptd = 1$). The output of 50 times of prediction compared to the target when $t \in [50, 300]s$ (A). The z-axis position of the yellow ball in the reservoir when $t \in [0, 300]s$ (B) and $t \in [0, 8]s$ (C). The example of one prediction is colored in red.

height ($0.8\ m$), the discrete states developed over a short period ($t \in [0, 2]s$), and afterwards the trajectory of the ball followed these states. When $nsf = 0.2$, the variance of the three states increased, making them hard to distinguish. When $nsf = 1$, the trajectories became completely random, and no obvious state could be distinguished from the plot. One computational example is colored in red in Fig. 11, revealing that the effect of noise can switch the state of the ball when $nsf = 0.1, 0.2$ in Fig. 11B and ultimately lead to the probable offset of the output when $nsf = 0.1$ in Fig. 11A.

4. Discussions and Conclusions

This paper presents a study on a reservoir computer that utilizes a collective Bernoulli-ball system and its performance in non-linear prediction with various systematic parameters. The primary focus of this research is to investigate how these five parameters affect the non-linear prediction error and its standard deviation, both individually and in combination.

The effect of the noise strength factor reveals how stochasticity influences the reasoning performance of the reservoir. As shown in Fig. 5, the system predicts the non-linear trajectory with a low mean error in a no-noise case. When it comes to a low-noise case, the mean error and standard deviation increase significantly, while in a high-noise case, the mean error and standard deviation reduce to a medium level. Fig. 11 shows that, compared to high-noise cases, a low-noise case follows the trend of the target better. The occasional offset in the output increases the mean prediction error and standard deviation, and this offset emerges since the Bernoulli-ball system is a complex system that is highly sensitive to small perturbations. This offset does not show up in high-noise cases since the noise gets common in the trajectories that even covers up the information of the input, and the linear regression process compresses the trajectory signal to reduce the effect of noise and make this system stable. Therefore, the amplitude of the output gets lower than that of the target in a high-noise case, as shown in Fig. 11. However, this signal compression process makes the reservoir useless in prediction tasks under a high-noise case, since it cannot correctly reveal the trend of the target. In conclusion, when using a complex physical system as a reservoir, it is always good to reduce the level of stochasticity in the reservoir, but we should be aware of the occasional offset produced by small perturbations in a low-noise case. One probable solution might be computing the same task a number of times to take the modal of the output. If the noise is too high, the stochasticity may cover up the properties of the reservoir and make it fail in tasks.

Some other discoveries are also worth discussing. First, as shown in Fig. 6, input delay matters in our reservoir computer: we need to match the input delay with the timescale of our reservoir. When the input delay is too short, the reservoir may not be able to react to changes in the input, resulting in poor performance. When the input delay is long enough for the reservoir, continuously increasing the input delay will not help improve performance, but on the other hand, consumes more computation time. Fig. 10 shows that the most efficient time delay is ($idf = 10$): the prediction error, its standard deviation, and the time consumption are all relatively low. Second, although the input gain does not solely show a significant effect on prediction performance in Fig. 7, it slightly influences the prediction performance when the noise strength increases, as shown in Fig. 10. This indicates that we should match the input gain and noise strength for better performance: when the noise strength is high, the input gain should be lower to reduce the prediction error and its standard deviation. Lastly, the change of sampling method does not have a significant effect on prediction performance, as shown in Fig. 8 and Fig. 10. Perhaps the linear regression section in the output layer already has a similar function, so the effect of sampling methods is reduced in this case.

The non-linear prediction task is only one example of the evaluation of the reservoir computer. In the future, we will apply the collective Bernoulli-ball reservoir computer to different tasks to evaluate its performance, such as filtering signal noise. Additionally, since the Bernoulli-ball system incorporates noise (turbulence) in its dynamics, we will investigate other physical reservoirs that have negligible noise but still exhibit complex behaviors under different inputs.

ACKNOWLEDGMENT

This work was supported by a Huawei HiSilicon Scholarship.

5. Reference

[1] Ossama Abdel-Hamid, Abdel-rahman Mohamed, Hui Jiang, Li Deng, Gerald Penn, and Dong Yu. Convolutional neural networks for speech recognition. *IEEE/ACM Transactions on audio, speech, and language processing*, 22(10):1533–1545, 2014.

[2] Elijah Almanzor, Fan Ye, Jialei Shi, Thomas George Thuruthel, Helge A Wurdemann, and Fumiya Iida. Static shape control of soft continuum robots using deep visual inverse kinematic models. *IEEE Transactions on Robotics*, 2023.

[3] Eric Antonelo, Benjamin Schrauwen, and Dirk Stroobandt. Modeling multiple autonomous robot behaviors and behavior switching with a single reservoir computing network. In *2008 IEEE International Conference on Systems, Man and Cybernetics*, pages 1843–1848. IEEE.

[4] Chao Du, Fuxi Cai, Mohammed A Zidan, Wen Ma, Seung Hwan Lee, and Wei D Lu. Reservoir computing using dynamic memristors for temporal information processing. *Nature communications*, 8(1):1–10, 2017.

[5] Rory LC Flemmer and CL Banks. On the drag coefficient of a sphere. *Powder Technology*, 48(3):217–221, 1986.

[6] Nir Fresco. *Physical computation and cognitive science.* Springer, 2014.

[7] Josie Hughes, Arsen Abdulali, Ryman Hashem, and Fumiya Iida. Embodied artificial intelligence: Enabling the next intelligence revolution. In *IOP Conference Series: Materials Science and Engineering*, volume 1261, page 012001. IOP Publishing.

[8] Herbert Jaeger. Short term memory in echo state networks. 2001.

[9] Herbert Jaeger. The "echo state" approach to analysing and training recurrent neural networks-with an erratum note. *Bonn, Germany: German National Research Center for Information Technology GMD Technical Report*, 148(34):13, 2001.

[10] Herbert Jaeger. Tutorial on training recurrent neural networks, covering bppt, rtrl, ekf and the" echo state network" approach. 2002.

[11] Herbert Jaeger and Harald Haas. Harnessing nonlinearity: Predicting chaotic systems and saving energy in wireless communication. *science*, 304(5667):78–80, 2004.

[12] KC Lee. Aerodynamic interaction between two spheres at reynolds numbers around 104. *Aeronautical Quarterly*, 30(1):371–385, 1979.

[13] Z. Q. Liao, Z. Y. Wang, H. Yamahara, and H. Tabata. Low-power-consumption physical reservoir computing model based on overdamped bistable stochastic resonance system. *Neurocomputing*, 468:137–147, 2022. Liao, Zhiqiang Wang, Zeyu Yamahara, Hiroyasu Tabata, Hitoshi Liao, Zhiqiang/0000-0002-4030-1364 1872-8286.

[14] Mantas Lukoševičius. *A practical guide to applying echo state networks*, pages 659–686. Springer, 2012.

[15] Wolfgang Maass, Thomas Natschläger, and Henry Markram. Real-time computing without stable states: A new framework for neural computation based on perturbations. *Neural computation*, 14(11):2531–2560, 2002.

[16] RH Magarvey and Roy L Bishop. Transition ranges for three-dimensional wakes. *Canadian Journal of Physics*, 39(10):1418–1422, 1961.

[17] Kohei Nakajima. Physical reservoir computing—an introductory perspective. *Japanese Journal of Applied Physics*, 59(6):060501, 2020.

[18] Kohei Nakajima, Helmut Hauser, Rongjie Kang, Emanuele Guglielmino, Darwin G Caldwell, and Rolf Pfeifer. A soft body as a reservoir: case studies in a dynamic model of octopus-inspired soft robotic arm. *Frontiers in computational neuroscience*, 7:91, 2013.

[19] Kohei Nakajima, Helmut Hauser, Tao Li, and Rolf Pfeifer. Information processing via physical soft body. *Scientific reports*, 5(1):1–11, 2015.

[20] Kohei Nakajima, Tao Li, Helmut Hauser, and Rolf Pfeifer. Exploiting short-term memory in soft body dynamics as a computational resource. *Journal of The Royal Society Interface*, 11(100):20140437, 2014.

[21] Delphine Ormières and Michel Provansal. Transition to turbulence in the wake of a sphere. *Physical review letters*, 83(1):80, 1999.

[22] David E Rumelhart, Geoffrey E Hinton, and Ronald J Williams. Learning internal representations by error propagation, 1985.

[23] Benjamin Schrauwen, David Verstraeten, and Jan Van Campenhout. An overview of reservoir computing: theory, applications and implementations. In *Proceedings of the 15th european symposium on artificial neural networks. p. 471-482 2007*, pages 471–482.

[24] Gouhei Tanaka, Toshiyuki Yamane, Jean Benoit Héroux, Ryosho Nakane, Naoki Kanazawa, Seiji Takeda, Hidetoshi Numata, Daiju Nakano, and Akira Hirose. Recent advances in physical reservoir computing: A review. *Neural Networks*, 115:100–123, 2019.

[25] Sadatoshi Taneda. Experimental investigation of the wake behind a sphere at low reynolds numbers. *Journal of the Physical Society of Japan*, 11(10):1104–1108, 1956.

[26] Himanshu Tyagi, Rui Liu, David S-K Ting, and Clifton R Johnston. Measurement of wake properties of a sphere in freestream turbulence. *Experimental thermal and fluid science*, 30(6):587–604, 2006.

[27] Kristof Vandoorne, Pauline Mechet, Thomas Van Vaerenbergh, Martin Fiers, Geert Morthier, David Verstraeten, Benjamin Schrauwen, Joni Dambre, and Peter Bienstman. Experimental demonstration of reservoir computing on a silicon photonics chip. *Nature communications*, 5(1):1–6, 2014.

[28] Athanasios Voulodimos, Nikolaos Doulamis, Anastasios Doulamis, and Eftychios Protopapadakis. Deep learning for computer vision: A brief review. *Computational intelligence and neuroscience*, 2018, 2018.

[29] Fan Ye, Arsen Abdulali, and Fumiya Iida. Simulation of collective bernoulli-ball system for characterizing dynamic self-stability. In *Annual Conference Towards Autonomous Robotic Systems*, pages 367–378. Springer.

[30] Fan Ye, Arsen Abdulali, and Fumiya Iida. State-transfer modeling collective behavior of multi-ball bernoulli system based on local interaction forces. *Frontiers in Robotics and AI*, 9:980586, 2022.

[31] Zhenhua Yu, Uswatta Liyanage Shehara Perera, Helmut Hauser, Peter RN Childs, and Thrishantha Nanayakkara. A tapered whisker-based physical reservoir computing system for mobile robot terrain identification in unstructured environments. *IEEE Robotics and Automation Letters*, 2022.

[32] Zhenhua Yu, SM Hadi Sadati, Helmut Hauser, Peter RN Childs, and Thrishantha Nanayakkara. A semi-supervised reservoir computing system based on tapered whisker for mobile robot terrain identification and roughness estimation. *IEEE Robotics and Automation Letters*, 7(2):5655–5662, 2022.

IOP Conf. Series: Materials Science and Engineering **1321** (2024) 012013 doi:10.1088/1757-899X/1321/1/012013

Continual Developmental Neurosimulation Using Embodied Computational Agents

B Alicea[1,2], R Chakrabarty[3], S Dvoretskii[4], AV Gopiiswaminathan[5], A Lim[1], and J Parent[1]

[1] Orthogonal Research and Education Laboratory, Champaign-Urbana, IL USA
[2] University of Illinois, Urbana-Champaign, USA
[3] Independent Researcher, Paris, France
[4] Technical University of Munich, Munich, Germany
[5] Manipal Institute of Technology, Manipal, India
E-mail: bradly.alicea@outlook.com

Abstract. There is much to learn through synthesis of Developmental Biology, Cognitive Science and Computational Modeling. Our path forward is to present a design for developmentally-inspired learning agents based on Braitenberg Vehicles. Using these agents to exemplify the embodied nature of computational autonomy, we move closer to modeling embodied experience and morphogenetic growth as components of cognitive developmental capacity. We consider biological and cognitive development which influence the generation of adult phenotypes and the contingency of available developmental pathways. Continual developmental neurosimulation allows us to consider the role of developmental trajectories in bridging the related phenomena of nervous system morphogenesis, developmental learning, and plasticity. Being closely tied to continual learning, our approach is tightly integrated with developmental embodiment, and can be implemented using a type of agent called developmental Braitenberg Vehicles (dBVs). dBVs begin their lives as a set of undefined structures that transform into agent-based systems including a body, sensors, effectors, and nervous system. This phenotype is characterized in terms of developmental timing: with distinct morphogenetic, critical, and acquisition (developmental learning) periods. These properties produce emergent connectivity with shifting weights and adaptive network topologies in the context of developmental processes and spatially-dependent embodiment in training a dBVs neuronal network. We further propose that network morphogenesis can be accomplished using a genetic algorithmic approach, while developmental learning can be implemented using a number of computational methodologies. This approach provides a framework for adaptive agent behavior that might result from a developmental approach: namely by exploiting critical periods or growth and acquisition, an explicitly embodied network architecture, and a distinction between the assembly of neuronal networks and active learning on these networks. In conclusion, we will consider agent learning and development at different timescales, from very short (<100ms) intervals to long-term evolution. The development, evolution, and learning in an embodied agent-based approach is key to an integrative view of biologically-inspired intelligence.

1. Introduction

The process of biological development provides many novel lessons for embodied approaches to computational agents. Development serves as an inspiration for artificially intelligent processes such as the serial acquisition of linguistic information [1]. This can be summarized as a compositional approach to development. While serial acquisition can capture the linguistic and psychological aspects of information acquisition in a naively intelligent system, it is largely insufficient to properly characterize the developmental process that creates an embodied nervous system. Even when developmental processes are made explicit in the algorithm [2], they often focus on the psychological (particularly constructivist) nature of development [3, 4]. One alternative involves the exploration of developmental transformations in growth and connectivity, particularly as they relate to an agent's body and neuronal network topology [5]. A neuronal network is different from a conventional neural network in that sensory input results in motor output with both linear and nonlinear components. This connectivity matrix depends on edges connecting the network's individual processing units, and so can change both in terms of size and reconfigurability over the course of development. In short, dBV neuronal networks are analogous to small animal connectomes, where small circuits can regulate simple behaviors. Elman pointed out in 1993 [6] that in the context of learning language, the connectionist approach fails except in cases where the network is subject to a growth process where the topology starts small then expands. Yet even in this case, such studies do not explicitly leverage embodiment as a fundamental aspect of the developmental process. Here, we use developmental Braitenberg Vehicles (dBVs) [7] to understand the developmental process via two different frames of analysis: 1) the level of network topology and 2) the impact of both innate or genetically-defined factors (the initial condition for any phenotypic or behavioral trait) and external environment on agent phenotype. Genetically-defined factors are indeed directly influenced by the external environment through plasticity. A plastic response provides the capacity of an agent to change or exhibit variation among its parts when responding to an environmental stimulus.

Braitenberg Vehicles [8] serve as toy models of an embodied nervous system with a simple input/output relationship between sensors and effectors. In a Braitenberg Vehicle, the relationship between sensor and effector is a white box relationship [9]. A minimal set of connections are made, and conduct a sensorimotor signal that can either be linearly mapped from sensory to effector or experience contralateral cross-talk. This occurs when sensors from one side of the vehicle carry a signal to an effector on the opposite side. dBVs (see Figure 1) refine this basic model in three ways: a generative model of a multicellular agent with defined intercellular communication channels, an increase in representational complexity over developmental time, and allow for multiple hidden layers between sensor and effector [10]. These two points can be collectively understood as connectogenesis [11] in the biological context, and refers to the generativity and emergence of this process. Indeed, these properties often capture the flexible and sub-optimal wiring patterns in biological brains [12]. Thirdly, learning is enabled by different types of acquisitions called morphogenetic (structural) and behavioral (functional), which themselves are further enabled by a substrate determined by innate capabilities. Developmental neurosimulation offers a balanced approach to learning: growth of the internal network, plasticity from an initial set of innately-defined conditions, and morphogenesis (the process of growth and transformation of the agent's neuronal network and body) of the agent's body all enable a naturalistic agent containing a developing morphology and plastic neuronal network.

We will utilize the dBV approach to explore the nature of developmental learning and the heterogeneous timing of information acquisition. The first aspect of this can be summarized as *developmental freedom*, which summarizes the effect of network reconfigurations given connectivity constraints of the neuronal network and plasticity at the network level. A system exhibiting developmental freedom demonstrates the ability to produce alternative network configurations given connectivity constraints of the neuronal network and plasticity at the network level. A greater amount of freedom results from an increased amount of reconfigurability and ability to produce nonlinear outputs, with spatial embodiment [13] providing context for learning. In Continual Developmental Neurosimulation, *spatial embodiment* is a developmental antecedent of learning, but also provides a substrate for learning. Alternate arrangements of sensors, effectors, neuronal network connectivity, and phenotypic geometry results in different outcomes from similar sets of innate capabilities and environmental experiences. The third aspect is better known as the *critical period* of development, or an enhanced period of learning enabled by a heightened nervous system plasticity. While the critical period itself is determined through genetic mechanisms, its effect on learning is environmentally variable. Depending on when the critical period occurs relative to sensory acquisition, the agent can learn a vast amount or miss out on learning altogether. Collectively, developmental freedom, spatial embodiment, and critical period regulation contribute to a form of artificial learning with explicit developmental mechanisms.

1.1 *Deep Learning Approaches to Continual Learning*

Continual learning has been explored in the world of deep learning in ways that are distinct from our approach. In deep learning-related bio-inspired approaches, the focus is on implementing continual learning as a means to avoid catastrophic forgetting. Lassig et.al [14] provide three main strategies for implementing continual learning in a computational agent: methods for reviewing data from old tasks while engaging in new tasks (replay), constraining current learning in order to protect previous learning (regularization), and selective malleability of weights and neurons in the network (architecture). For deep learning approaches, the focus is on learning specific tasks, which are the impetus of adaptation, memory maintenance, and generalization. In dBVs, acquisition via embodied context (shape and relative size of body parts) for specific tasks are connected to potential task-independent behavioral responses [15].

Replay and regularization rely on implementing a form of the stability-plasticity tradeoff [16], or the need to maintain stable memories while also enabling adaptive changes in synaptic weights. The stability-plasticity tradeoff also plays a role in age-related learning constraints [17], which we will revisit later in this paper. This suggests a generalized homeostatic mechanism, which is responsible for behavioral robustness and associated resiliency [18]. The ability to capture adequate context, particularly in the form of multisensory integration [15, 19], is key to enabling agent generalization. Implementing a suitable architecture relies on modeling plasticity and neurogenesis [20], which can enable reconfiguration and repair.

Critically, conventional deep learning methods do not focus directly on input-output relations. This is in line with a focus on discrete tasks: the connection between inputs shaping an internal representation and a wide range of output behaviors over different stages of an agent's interactions are irrelevant. Task structure is of primary importance: temporal structure and relation to other tasks impacts performance [21]. Of more immediate interest is the replay of specific experiences, which work on the level of dynamic stability and recall ability [22]. More important to continual learning is the encoding of task-specific architectural features that generalize across multiple tasks [23]. To do this, methods such as Deep Feedback Control [14] and Synaptic Intelligence [24] are implemented to maintain robustness to task learning of various types.

In [15], the authors propose several biological mechanisms that support lifelong learning in computational agents. We will focus on the following: metaplasticity, cognition outside the brain, reconfigurable organisms, and multisensory integration. We also present additional mechanisms that are important to enabling continual cognition and behavior, chief among them a connection between reconfigurability and embodiment. The developmental neurosimulation approach also offers an embodied dimension to this literature, which provides new challenges and opportunities.

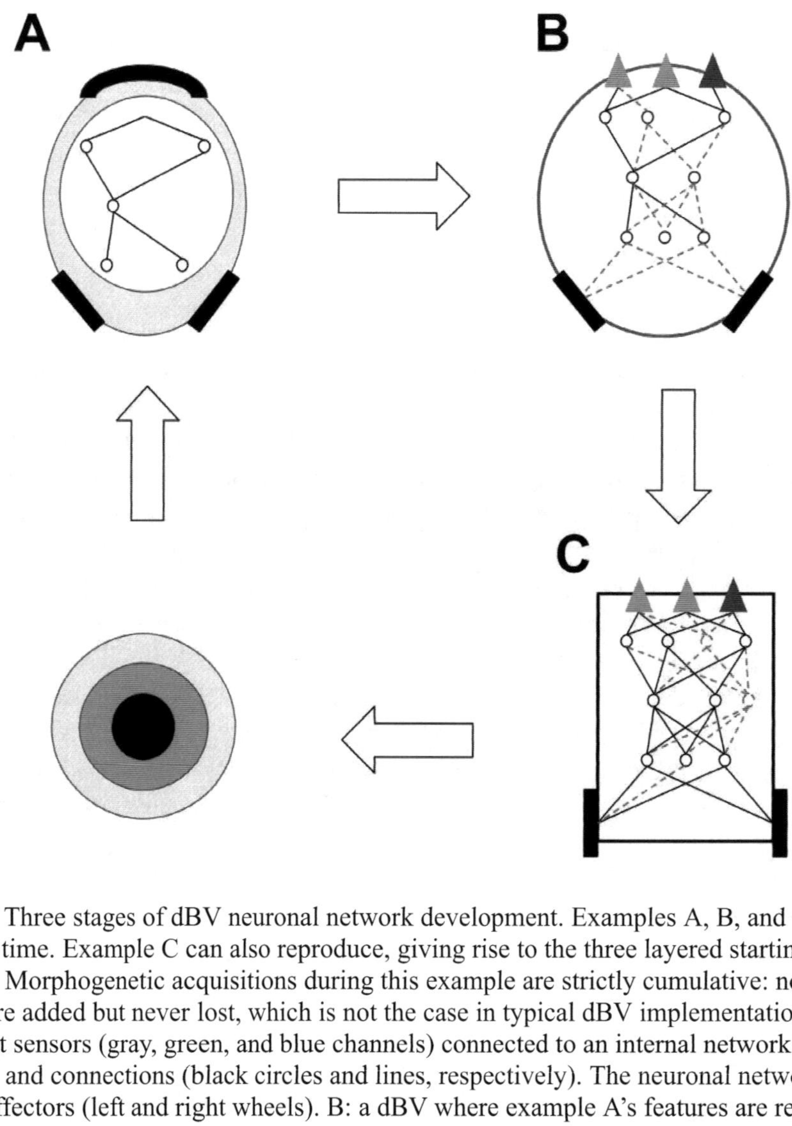

Figure 1. Three stages of dBV neuronal network development. Examples A, B, and C occur sequentially in time. Example C can also reproduce, giving rise to the three layered starting form at the lower left. Morphogenetic acquisitions during this example are strictly cumulative: nodes and connections are added but never lost, which is not the case in typical dBV implementation. A: a dBV with three light sensors (gray, green, and blue channels) connected to an internal network with a small set of nodes and connections (black circles and lines, respectively). The neuronal network output drives two effectors (left and right wheels). B: a dBV where example A's features are retained, but with the birth of additional nodes and connections (red dotted circles and lines, respectively). C: a dBV where features of example B are retained, but with the birth of additional nodes shown in example B.

1.2 *Development, Learning, and Developmental Freedom*

In dBVs, development and learning are nominally independent: development is an anatomical phenomenon, while learning is a representational one. Taken collectively, the process of a dynamic neuronal network defines the scope of developmental freedom, or the degrees of freedom enabled by the architecture. Developmental freedom is the tendency for information and representations in particular to take advantage of a stable underlying topology. This can be understood in relation to contingency. Developmental contingency restricts the developmental neuronal network to an increasingly limited set of possible trajectories. Contingency results in earlier developmental events (cell birth or the establishment of connections) to constrain the range of possibilities for future developmental events (formation of a network motif). Furthermore, contingency locks the neuronal network into an increasingly smaller number of pathways over developmental time, and results in path dependency. Once the network configuration becomes formalized, formation of a stable but robust circuit benefits from developmental freedom.

In Figures 1 and 4, we introduce an embryogenetic model of Braitenberg vehicle development. The standard Braitenberg vehicle is shown to develop from an undifferentiated initial state (analogous to an egg), denoted by a single spherical object. The next stage of development involves either diploblastic (two-layered) or triploblastic (three-layered) concentric layering. Layered structures (blastic structures) establish the basis for various functional components of the agent body. For the three-layered structure shown in Figure 1, the inner, middle, and outer layer become raw material for sensors/effectors, body shell, and the neuronal network chamber, respectively. As the dBV matures, these layers separate and migrate to their adult locations. For example, a single layer (sensors/effectors shown in black on Figure 1) can split into four parts, two migrating towards the vehicle front (sensors) and two migrating towards the back (effectors). An agent that starts from an undifferentiated initial state with three layers experiences morphogenesis to reach its mature form as a Braitenberg vehicle. Components for each layer allow for compositional phenotypes, which can be mutated and recombined during development. This can be determined via the genomic encoding, shown in Figure 2.

To discuss developmental freedom in the context of Braitenberg Vehicle nervous systems, we must first consider how nervous systems connect together during development. The time window model of Lim and Kaiser [25] suggests that this process is a highly parallel process where the state of connectivity at any one developmental time point can have far-reaching effects across an agent's life-history. A model of contingency (see Figure 1) allows us to think about the mapping between sensor and effector in terms of branching developmental stages, while each time point serves to determine the trajectory of future growth and development. Recent work [26] has also brought into question that neuronal network assembly occurs exclusively via shortest paths, resulting in a sprawling constellation of alternative network communication models. In developing nervous systems, complexity accumulates through the duplication and divergence model [27, 28], wherein mappings between sensor and effector can be reused and duplicated as sensors, effectors, and nervous system modules are duplicated and deployed to yield more complex phenotypes. This expansion of connectivity between sensor and effector provides a more flexible network than a standard neuronal network, and is enabled by spatially-explicit maps between environmental features and topology in the internal network. We propose that the combination of connectogenesis, artificial morphogenesis, and direct sensory processing enable neurosimulation in ways that simply taking weights and nodes in layers of a disembodied network cannot.

1.2.1 Developmental Freedom as an Emergent Phenomenon

As the neuronal network is expressed, it serves as a substrate for information processing similar to that of morphogenetic and behavioral substrates at the periphery. However, rather than being exposed to environmental information or serving as the scaffolding for other traits, the neuronal network serves as a substrate for information transfer between the sensor and effector. There are a number of pathways from sensor to effector, which grow exponentially as the neuronal network grows in size. To this end, growth and transformation of the phenotype also contribute to developmental freedom, particularly in terms of where and how sensory inputs are established. Changes in sensory input, spatial arrangement of the agent phenotype, and phenotypic symmetry can also affect the underlying substrate. Developmental freedom is the tendency of information processing to take new pathways along the substrate. While not autonomous itself, developmental freedom is advantageous, and can be related to ontogenetic strategies often observed during development [11, 29]. This ability to find new paths comes from the stability of a network that is no longer growing. For example, exploratory behaviors cannot stably emerge on networks that are constantly reconfiguring. The best and shortest pathways are constantly changing, undercutting the efficiency of information processing. Similarly, during connectogenesis, the potential for pruning connections due to lost inputs or the emergence of feedback loops can fundamentally alter information processing itself. Thus freedom on the substrate enables learning and stable behaviors to emerge.

1.3 *Spatially-dependent Embodiment*

There are a number of interesting connections between dBVs and their embodiment. The first is how embodiment allows the agent to sample the incoming data in a stratified fashion: the relative position and orientation of sensors allows the agent to compare points in a differential manner. Differential comparison results from spatialized sensor arrays, which is analogous to biological sensory organs. This allows for different frames of reference to be distinguished, and is distinct from the input layer of a neuronal network. Each sensor observes the world at a specific angular perspective relative to the vehicle's midline. This constant interaction with an environment of sensory cues allows the agent to keep concepts about the world continuously up to date, and provides a contextual fabric for the internal representation [10]. Variation in bilateral symmetry may yield some interesting effects. One effect is that complex movement outputs can be amplified and modulated. Asymmetrical sensory systems can also be advantageous in learning more about the vehicle's environment. A dBVs overall bodily structure allows it to extract more information from a single data point than in a disembodied neural network. Changes in the relationship between sensor and effector provides context for simple behaviors.

1.4 *Network Reconfiguration and the Critical Period*

Another attribute of development is the unfolding of time-dependent processes. One powerful developmental principle we can draw from is the existence of a critical period [30-32]. The critical period allows for enhanced sensitivity at specific points in developmental time [33]: acquisition of sensory information and phenotypic plasticity during this time have an outsized effect on what our agent looks like and how it behaves. The critical period is also an important time for learning from sensory inputs, and loss of function during this time can radically reconfigure the neural architecture [34, 35]. In the development of touch, delays in development as compared to the normal developmental trajectory can lead to a loss of sensory processing ability [36].

Hensch [30] proposes two potential mechanisms that define enhanced neural reorganization during the developmental critical period in support of learning. First is a functional competition between inputs that might be thought of as developmental allocation. Second is a role for electrical activity in the structural consolidation of selected pathways. This has a number of consequences on the later stages of development and ultimately the adult phenotype [37]. Learning and anatomical change also seem to occur at different time scales [38], but learning in the service of behavior might be dependent on the length of time afforded to developmental growth [39].

According to this view, while both development and learning are mechanisms that induce neural and behavioral plasticity, development is a generative phenotype largely influenced by experience-expectant mechanisms [40]. As the dBV interacts with its environment, these interactions shape the manner in which the network is generated. Learning that occurs independent of large-scale changes in the network is the product of experience-dependent influence. Changes in the timing of this transition from a neuronal network largely influenced by experience-expectancy to one primarily influenced by experience-dependency is key to understanding how developmental processes shape information acquisition and the supervision of learning.

2. Methodological Perspectives

2.1 *Artificial Genetics and Environment*

Development is captured using a generative process of constructing nervous system nodes and weighted connections, while learning is defined by a process of connectionist plasticity. In this paper, we have instantiated these mechanisms using Genetic Algorithms and Hebbian Learning mechanisms [10] on top of an architecture for developmental innate capabilities. The generalized influence of sensory information is captured using the concept of Gibsonian Information [41]. In the proposed architecture for innateness, a genetic algorithm is used to demonstrate the link between encodings representing innate capabilities and expressed substrates for various acquisitions.

2.1.1 *BraGenBrain encoding*

The first version of the genetic encoding is utilized by the BraGenBrain implementation demonstrated in [7]. Sensors (inputs) and effectors (outputs) are connected by an artificial neuronal network, the components of which are generated by an algorithm. In an object-oriented framework, vehicles are a class, instances of which can be generated by a factory function. All generated neuronal networks are evaluated by a fitness function, characterized by force (non-inertial movement), rotation (movement trajectory curvature), and displacement (sliding movement). These also combine into a speed vector, which are linear components of a behavioral signal. Additional parameters in the BraGenBrain approach are effect strength and a specific gravity constant. BraGenBrain uses four genetic operators to determine various aspects of evolutionary change: initial population size (default value of 10), world size (with a default size of 800x800 pixels), a mutation rate (default frequency of 0.05), and crossover (random reproduction with a default frequency of 0.05). Phenotypes in new generations of vehicles are biased by natural selection (fitness) or neutral drift.

Each developmental neuronal network is represented by a directed acyclic graph (DAG) with a unique ordering. Nodes are added at a rate determined by the genomic encoding, particularly innate timing mechanisms (specified for each agent lineage, see Figure 6). Each neuron's output is normalized as synaptic weights (a total of 255), which sum up to one. A matrix graph representation is used to store connectivity information. The neuronal network is ordered from input (root) to output (tip). Root nodes of the neuronal network (proximal to the sensors) correspond to sensory inputs. Leaves of the network (in the direction of the effectors) correspond to internal movement circuits and their associated representations. Tip nodes (proximal to the effectors) correspond to movement outputs. The neuronal network architecture is structured by a developmentally-inspired three-step evolutionary algorithm. This is distinct from canalization function-driven morphogenesis, and occurs within the inner blastic layer (neuronal network chamber) among newly-born agents. Initially, a mutation occurs in the standard binary encoding at a specified probability. Vehicles are then evaluated for the next generation using the past generation's speed vectors as a criterion. Crossover allows for variation in the architecture: neuronal network components are inherited from two randomly-selected agents. This produces a hybrid offspring with a nominally high fitness.

2.1.2 *Critical Period encoding*

The second type of genetic encoding is specific to dBV agents utilizing critical periods to enable development [42]. The onset of a critical period is triggered by a binary switch, leading to alternate expressions of the plasticity encoding. This mimics a regulatory switch that can be turned on and off based on regulatory control mechanisms. Each agent also has a genetic encoding of sensory organ acuity consisting of discrete register entries representing sensor-specific acquisition (with the potential for developing differential sensory acuity) during the critical period. Direct sensory transformations provide proto-representations that extract different components of information (e.g. low-pass and high-pass). These transformations can be captured by both simple logic gates as well as more complicated functions related to reversible logic gates and canalization processes. Canalization refers to a specific developmental trajectory resulting from a series of differentiation events. Differentiation allows a layer component to acquire a more specific form or function. For example, a germ layer can divide into two spatially discrete components that later go on to produce four sensors and two effectors. Differentiation events occur in a serial fashion, with changes being acquired one at a time as shown in Figure 6. The process of canalization then is analogous to a system of channels in a river system: subsequent differentiation events lead to increasingly specific trajectories, and increasingly specialized (and constrained) developmental trajectories. The other component is environmental acquisition itself, which we can approximate using Gibsonian Information. GI mediates the output of each sensory array, which serves to process and integrate direct sensory information. In combination with the canalization function, a subset of values dominate the output at different points in time. This is generally applicable to developmental contingency, where sensory units in an array are acquired at different points in time.

2.2 *Architecture for Developmental Innateness*

In developmental neurosimulation, an innate capability is the substrate of a morphogenetic or behavioral acquisition encoded in an artificial chromosome. An artificial chromosome is a vector that contains a genomic encoding, consisting of instructions that define the innate response and agent architecture. Our genomic encoding is represented on a binary string which represents the structure and function of traits related to the body (morphology) and innate behavioral primitives. While the mapping from encoding to substrate can vary, it typically follows a correspondence principle: features of the substrate are represented by a certain number of bits. In a dBV, we can use the morphogenetic substrate of light sensors as an example. A simple suite of light sensors have the following features: spatial coordinates for feature placement, multiple subsampling cells, aperture width of the sensor, angle of acquisition, and maintenance cells that provide real-time feedback. The entirety of the light sensor morphogenetic substrate occupies a certain number of bits in the encoding, while multiple sensors can result from duplication of these bits. The substrate of specific behavioral acquisitions requires additional bits and a separate encoding for features such as information bandwidth and visual angle. All parts of the encoding must be in place at the beginning of development, while the morphogenetic substrate must be in place to express the behavioral substrate (Figure 2). The corresponding encodings can be mutated or recombined in ways that modify function. Plasticity encodings allow for changes in encoding expression in real-time, often in response to environmental information (see Figure 2).

The origins of innateness in our computational agents is implemented through evolutionary computing techniques. One example of an architecture comes from the BraGenBrain platform [7]. Briefly, features of a phenotype are encoded on a linear artificial chromosome. These features are directly related to the size of the agent's neuronal network. A population of agents is generated, and each agent is subject to recombination and mutation. Over time, this population converges to a single set of navigation behaviors. While the BraGenBrain model works in implementation, the developmental aspects of this model is limited: for example, there is no canalization nor expression of artificial chromosomal encodings. Extending the BraGenBrain encoding strategy to a general architecture for dBVs (as well as other instances of developmental neurosimulation), we can refine this model in a number of important ways. Generally, dBVs are clonal (self-reproduction) that are subject to recombination and mutation. As these genomic encodings (contained on an artificial chromosome) and their associated phenotypes (the shape and transformational properties of an agent body) grow in complexity, it requires both the coordination of innate capabilities and the selective timing of innate capability expression.

2.2.1 *Coordination of Innate Capabilities*

Expression of the genomic encoding into multiple realizations of an agent is based on the biological observations of [43], in which cell behaviors and expressed genes are regulated through weak regulatory linkage. Weak regulatory linkage characterizes genetically-defined units linked by shared regulatory components. There tend to be multiple ways to activate a weak regulatory relationship, making our genomic encodings robust to a plastic response. Weak regulatory linkage is linked to the concept of evolvability, or the degree to which an organism or lineage is prone to evolutionary innovation, and affects embodiment by coordinating development of a phenotype through common function and location. When regulatory linkage is weak, traits that are linked together in a non-deterministic fashion. This allows sensors on the left and right side of the agent's phenotype to develop in parallel, or other symmetries to be enforced. The sensors not only become functional in an embodied context, but also experience a common critical period that allows them to synchronize their acquisition of sensory information. A binary encoding (shown in Figure 3B) can be mapped to newly-born neurons in the vehicle's internal model (Figure 3A), and experience variable expression due to experience over developmental time. These artificial chromosome encodings have two types of switches: identity and regulatory. Each one maps to a particular gene order (m) and level (n) combination.

This arrangement of cell identities in the binary encoding mimics the arrangement of *Hox* genes, which maintain body segmentation and anatomical order in animal development. Rearranging the gene order can provide compositional variety and novel wiring arrangements. Artificial chromosome encodings can also serve as instructions that modify the arrangement of phenotypic components. Timing of critical periods can be adjusted across a population or between populations according to the expression of the genomic encodings. Components of the agent phenotype can be continually reconnected according to their activation strength across different stages of learning (see Figure 3C). In this example, regulatory elements are activated to trigger the birth of components (neuronal network cells or body components), then express an activation strength fluctuating according to an encoding-specific distribution. Once a cell is wired to the sensor, however, the environment can modulate this activation strength. The internal network's collective behavior determines the canalization function and specialization of a vehicle's behavior. Innate capabilities serve to guide behavioral training. Adaptive behaviors and behavioral variety results as the neuronal network changes in time.

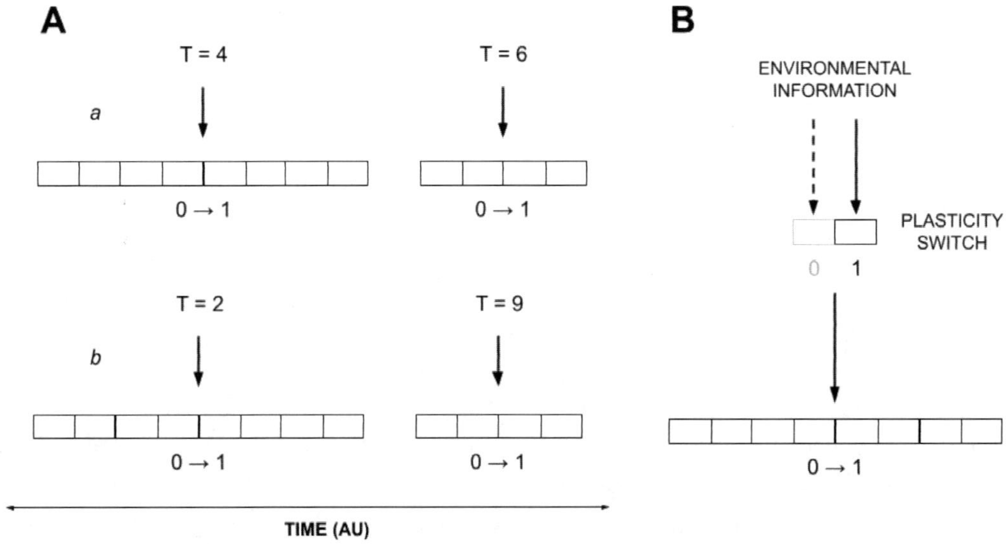

Figure 2. An example of an encoding via artificial chromosome (A), and an alternate form called the plasticity encoding (B). A shows two scenarios for the morphogenetic and behavioral substrates. *a* shows the expression of these traits at sequential points in time, and *b* shows changes in timing both earlier and later than the initial condition.

2.2.2 *Heterochrony and the expression of encodings*

Encodings are expressed at specific times in the developmental process, and can be optimized for specific computational agents. However, the timing and sequence of expression can be moved forward or backward in processes that resemble dynamical and sequence heterochrony, respectively [44]. Heterochrony is a genetically-determined mechanism by which organisms (or in this case, agents) experience relative changes in the timing of component differentiation, rate of growth, or the general duration of development. For example, changes in the position of effectors relative to body length results in consequences relative to motor control. As there tends to be variation across organisms, heterochrony can provide differences in the size and shape of the mature dBV. Heterochrony is a Figure 2A shows the role of sequence and dynamical heterochrony in terms of both the ordering of morphogenetic and behavioral substrate encodings on a virtual chromosome, and changes in timing of expression of these traits in the agent phenotype (T = time of appearance, arbitrary time unit). While sequence heterochrony (the ordering of developmental events) is important for maintaining a viable mature form [45], dynamical heterochrony is also important for extending the growth trajectory or limiting the size of certain features. In Figure 2A*a*, morphogenetic and behavioral traits are expressed closer in time, while in Figure 2A*b*, the morphogenetic substrate is expressed earlier in developmental time and the behavioral substrate is expressed later in time.

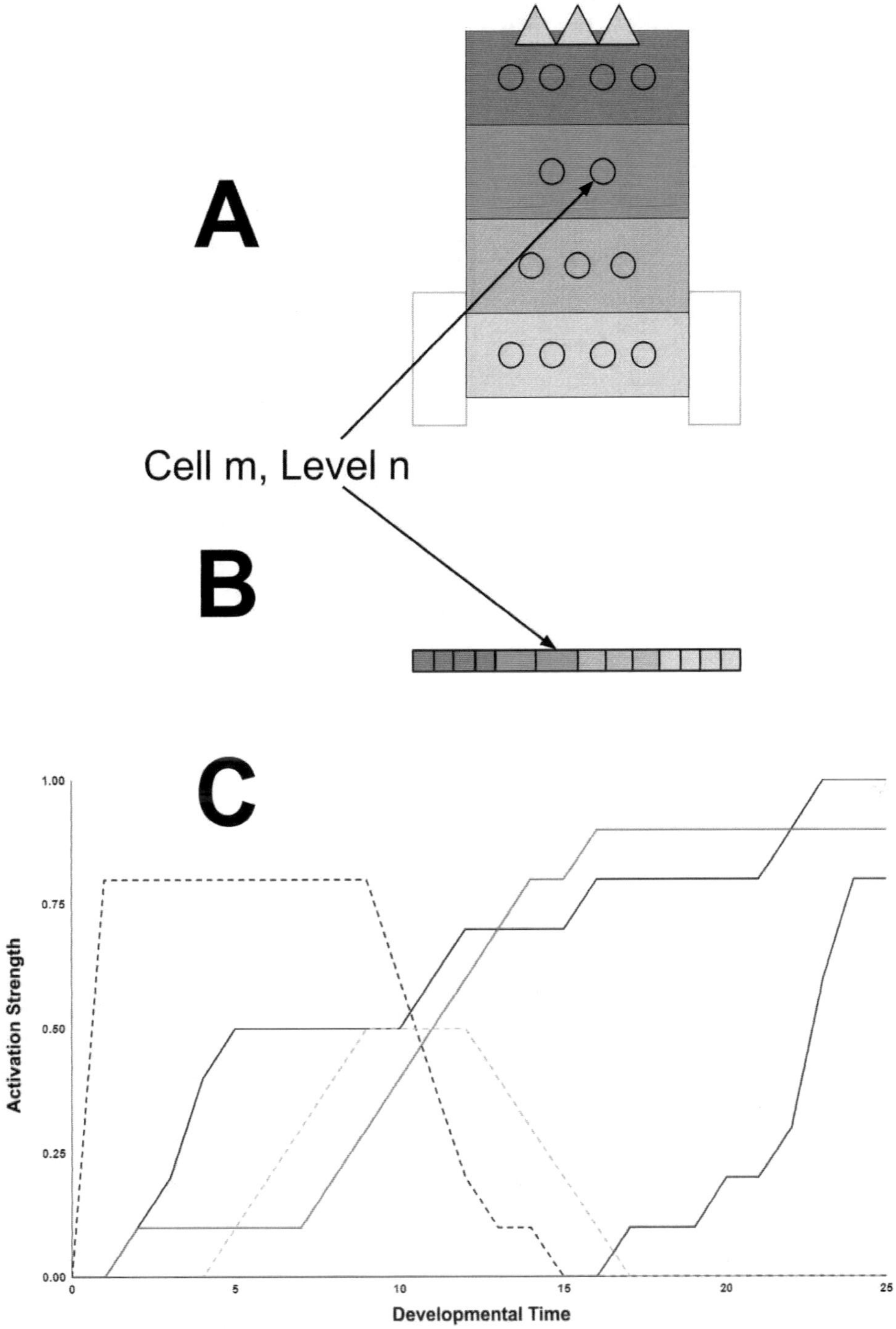

Figure 3. Demonstration of how morphogenesis and developmental time proceed in a dBV. A: a dBV with four levels of internal nodes (interneurons between the sensor and effect). B: a linear binary encoding that maps to each cell and level combination. C: the behavior of five neurons as they are born (or die) and acquire (or lose) an activation strength over developmental time. Black and gray dotted

lines are cells that are born and then die off during development. Black, red, and blue solid lines show birth and increasingly strong activation over developmental time.

2.3 *Agent Embryogenesis*

One key aspect of continual developmental neurosimulation is the development of agent phenotypes from innate capabilities for learning and behavior. The source of innate capabilities are not only provided through artificial chromosome encodings, but in singular, undifferentiated precursor phenotypes that both reproduce and exhibit morphogenesis. Agent phenotypes (bodies, sensors, and internal networks) emerge according to a series of transformations shown in Figure 4. The transformations proceed from a single undifferentiated cell, transitioning to a three-layered embryonic form, which then results in a mature phenotype. These tissue layers (hereinafter referred to as types) are determined by mature function, and are ordered by their relative size in the final adult form. For example, if more sensors or effectors are needed, the black type will be bigger and located as the central ring in a three-layered developmental phenotype.

The three developmental stages shown in Figure 4 results from a set of transformation conditions, encoded in our artificial chromosome encodings and applied to the developing phenotypic applied at each stage. The difference between stages 1 and 2 is defined by two differentiation events and a sorting of types by order of size. Creating initial differentiation events not only allows the agent to attain more complex phenotypes, but also raw materials for later transformations. Each blastic layer shown in Figure 4 becomes a type that is restricted to a range of potential phenotypic components. As per Figure 4, the light gray type is restricted to forming the outer part of the body. The dark gray type is restricted to forming neuronal network and actuation components (e.g. muscles and joints), and the black type is restricted to sensor and effectors. This allows for a balance between developmental specificity and reconfigurability.

2.4 *Positional Information and Connectivity in dBVs*

Our architecture for developmental innateness is focused on the specification of both the dBV phenotype and neuronal network. Positional information is a more general mechanism behind the *Hox* gene analogy for the arrangement of cell identities in the binary encoding discussed in section 2.2.1. Embryonic development of the agent body is in part defined by the positional information of each cell and tissue relative to the whole body. In a biological context, cells and tissues acquire positional (coordinate-based) information based on their location relative to coordinating signals [46]. This allows us to build a body and neuronal network without an explicit blueprint. As a result, these positions become part of an addressable system of interactions which give rise to spatial segregation, movements of components from the same blastic layer, and global pattern formation.

In Figure 3, transformation conditions are accompanied by movements of layer components rather than cells or tissues. The black layer divides and extrudes to form a single frontal sensor and two rear side wheels. This requires positional information based on functional requirements and the relative position of the type in terms of germ layer organization. Positional information can also be characterized in terms of bits. Coordinating signals can be summarized as discrete molecular states that in biological systems differ between cells of a distinct tissue type [47], but in our case are specified by innate factors. Differentiation into body parts represents a gain in information content. The addressable cell location can also be represented as information content (bits), where a location is embedded in a local concentration gradient of patterning signals [48]. This can be represented as a stochastic gradient, which when combined with considering how events unfold in developmental stages (Figure 4) opens the door for applying various gradient descent models [49] to our model of dBV phenotypes.

Neuronal network growth and plasticity proceeds through *connectivity-activation encoding*, in which the matrix *W(i,j)* of potential connections between internal nodes is shaped in a selective manner over developmental time. Expansion of the matrix represents an increase of cell number in the network, and the rate of expansion from the input-output initialization can either be deterministic or stochastic. All non-zero values represent active connections between all cells existing at a given time step. Updating *W(i,j)* during the morphogenetic growth phase of development is governed by a discrete genetic algorithm and encoding featuring recombination, mutation, and selection (for further technical details about BraGenBrain, see [7]).

Regardless of the exact method of generating variation in cell number expansion and active connections, we still must evaluate a large number of candidate network topologies. Thus, our genetic algorithm utilizes a fitness function that is determined by movement of the vehicle relative to target stimuli. The fitness function evaluating a given dBV's *W(i,j)* matrix is based on stimulus-to-vehicle distance and force generated by vehicles in response to a given stimulus. Even though the introduction of variety randomizes matrix elements over time, selection nevertheless favors the emergence of networks that enable recognizable behaviors [8].

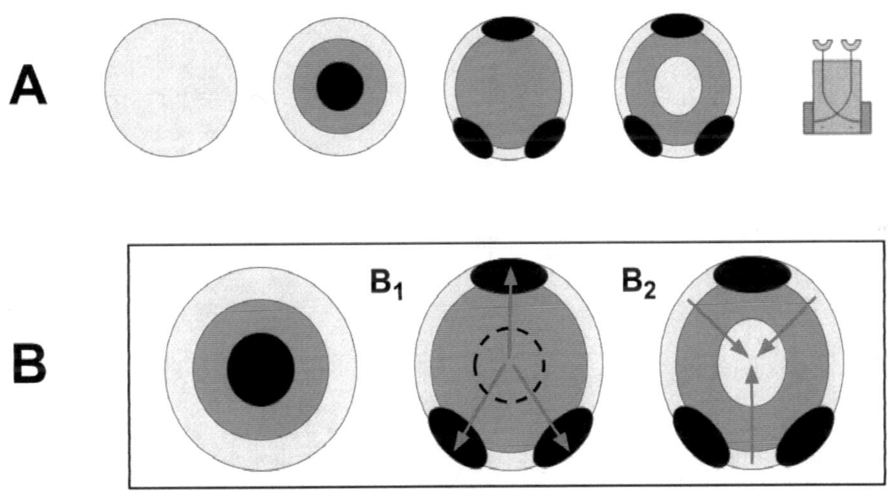

Figure 4. The embryonic development of a dBV from a single phenotype to a mature vehicle. A: progression of development from an undifferentiated initial state to a three layered phenotype, to a phenotype with proto-sensors/effectors, to a phenotype with a proto-neuronal network, to a mature vehicle. B: the three type phenotype, phenotype with proto-sensors/effectors (B_1), and phenotype with a proto-neuronal network (B_2) with red arrows showing the migration of types to other regions of the phenotype. In B_1, the inner type (black) migrates to the edge of the phenotype, forming one proto-sensor and two proto-effectors. B_2 shows the formation of a proto-neuronal network as partial migration of the outer phenotype (white) inward to the former location of the inner type (black).

2.5 *Mechanisms of developmental plasticity and learning*

Using a genetic algorithm as a training mechanism also generates a variety of topological orderings of matrix *W(i,j)* through mutation and recombination. For larger dBV neuronal networks, an encoding of coordinated projection identity can be used to more closely mimic a biological nervous system [50]. In the scheme introduced here, the critical period corresponds to a period of accelerated evolution in which the mutation rate is increased. This increases the number of potential configurations, but using a fitness criterion ensures that all of them are capable of learning. At the end of the critical period, the topological mutation rate is set at near zero (or turned off completely). In previous work, it has been demonstrated that associative learning can occur on a stable (post-critical period) *W(i,j)* matrix [7, 10] using a Generalized Hebbian Algorithm (GHA). Associative learning in dBVs is instantiated by building associations between sources of multiple senses simultaneously, and is thus the essence of developmental freedom. In [7, 10], olfactory and gustatory stimuli are distributed independently in a multidimensional space. The associative learning process acts to change the connection weights in matrix *W(i,j)* according to the co-occurrence of the two stimuli. In this way, spatially-explicit multimodal associations can be acquired.

3. Broader Methodological Concepts

3.1 *Gibsonian Information and Naturalistic Behavior*

dBVs experience sensory flow as they move around their environment with respect to environmental stimuli. The dynamic nature of the reference frame itself involves active direct perception [51] and can be detected using an approach called Gibsonian Information [41]. The latter might be identified using quantitative measures that produce lawlike relationships [52]. Aside from taking advantage of changes in timing, our developmental system also exploits relative information, or the continuous difference between multiple sensors. The current sensory state allows for the detection of coherent motion and spatial differences in a complex scene more generally. Angular differences between sensors in different locations across the front and sides of the vehicle capture the intensity or shape of a stimulus. dBV receives continuous sensory input from this stimulus. One way is to integrate the information using the GHA, while another way is to compare incoming sensory information to movement of the vehicle. This latter operation is a recurrent mechanism, and can be fully utilized by implementing other types of learning algorithms.

3.2 *Differentiation Trees and the Evolution of dBV Lineages*

Aside from biological and cultural evolution, we can also use phylogenetic techniques to understand the developmental origins of specific morphogenetic acquisitions. We use a differentiation tree-inspired model [53, 54] to represent the potential for developmental variation amongst dBVs. In biology, differentiation trees provide a means to understand waves of expansion and contraction that contribute to the emergence of new tissues in the embryo (Figure 5). For dBV agents, expansion and contraction waves represent phenotypic expansion (gain) and phenotypic contraction (loss) that unfold to the right and to the left, respectively. This provides us with basic conditions for growth and morphological changes that translate into different conditions for embodiment.

A differentiation tree represents all possible developmental pathways through which a population of dBVs can proceed. At each branching point, a transformation may occur, leading to developmental phenotype contingent upon its past developmental trajectory. Each colored portion of the differentiation tree in Figure 5 begins and ends at a differentiation event. For example, the red branches of the differentiation tree define a differentiation event (originating at a blue node) that transforms all possible configurations of the phenotype until the next viable differentiation event (shown in green). In this example, there is an non-viable differentiation event marked in red, which leads to a transformation that can make the dBV non-functional (the red branch in the middle of the tree leading to no differentiation event). This represents a transformation in the phenotype. These transformations can have an effect on future neuronal network development, particularly if the transformations have an effect on either sensors or effectors. For the purely developmental example in Figure 5, contraction waves represent restricting the phenotype and results in a lower growth rate. A version of the differentiation tree adapted to developmental neurosimulation is shown in Figure 6, with the branching process proceeding downwards instead of upwards. In Figure 6, this translates into a loss of agent parts, representing progressive degeneracy the farther left in the tree we move. By contrast, expansion waves in Figure 5 yield an expansion of the phenotype and a higher growth rate. Figure 6 adapts this to a gain in phenotypic parts and an overall increase in phenotypic complexity.

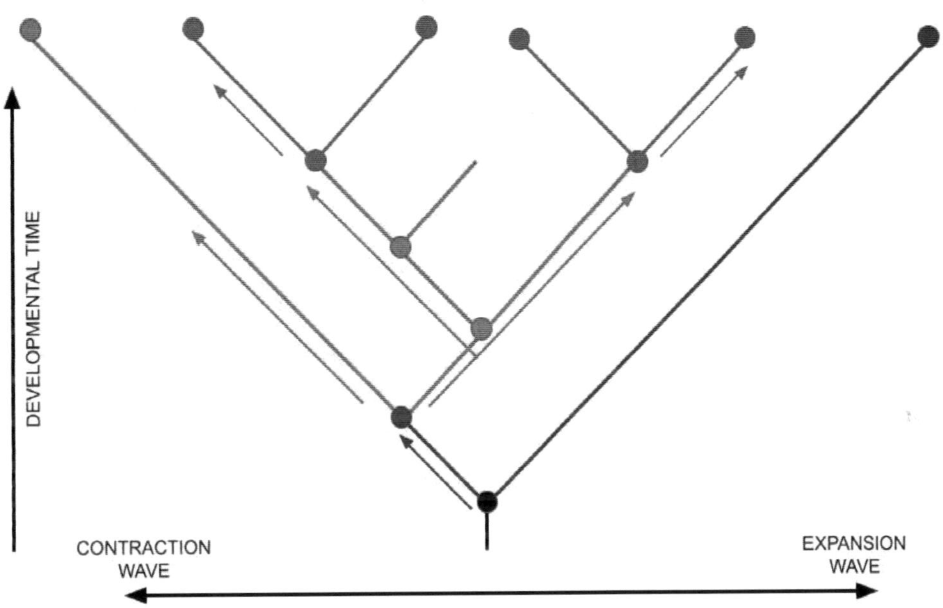

Figure 5. A differentiation tree: a branching model that represents the differentiation of tissue types in the embryo. This example will map to the dBV context with modifications as necessary. Each color (black, blue, red, green) is a differentiation event that proceeds from another (black → blue → red → green). These events result in one contraction wave (produced restricted phenotypic modules with a lower growth rate) and one expansion wave (produces expanded tissues with a higher growth rate).

Building a differentiation tree for agent development is useful for composing BV phenotypes in development. In agents, contraction and expansion waves represent organizational rearrangements and growth of the ancestral phenotype. In Figure 6, we can see the emergence of several dBV lineages representing variations on basic innovations. A single branch of the differentiation tree results in two vehicle morphologies. Some of these are viable, and some are not. In the Figure 6 example, we first see axial expansion of square body elements, followed by the emergence of triangular motifs and various rearrangements and ultimately proliferations of the basic dBV phenotype.

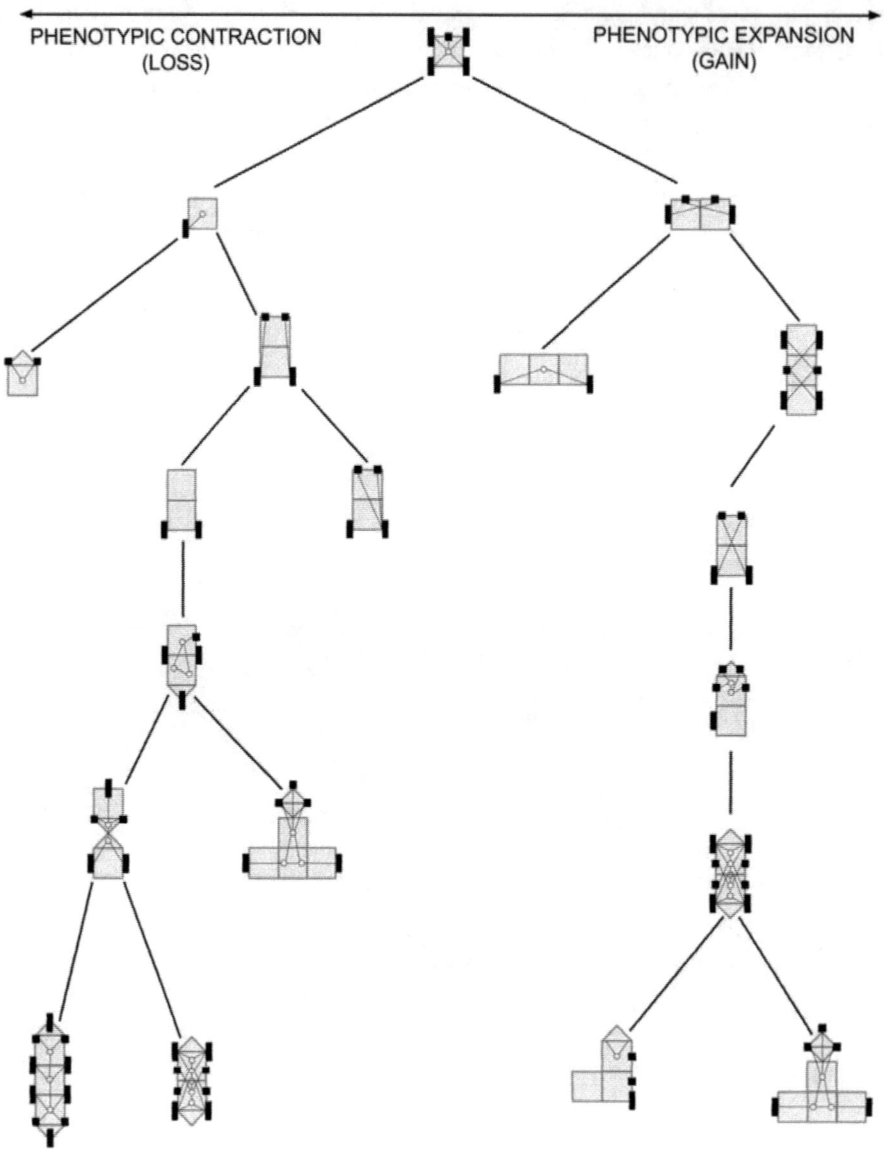

Figure 6. Developmental composition of BVs using a binary differentiation tree.

3.3 *Potential for Reinforcement Learning*

In general, dBVs are a special case of a reinforcement learning model. Yet there are differences between developmental and reinforcement learning. The first major difference is that feedback through reinforcement is not made explicit during the developmental process. While a dBV receives environmental feedback, this is often decoupled from the expression of an internal set of processes. The dual process of generating a neuronal network and learning on the network could be formulated as a reinforcement learning problem. We argue that this constant interaction with the environment scaled with time plays a significant role in the development process of the dBVs. Our dBV model can be contrasted with the dual world/self model of [55]. In our case, we can benefit from reinforcement learning through disentangling the intrinsic and learned representations which contribute to the developmental process of adult behaviors.

Rather than the Genetic Algorithm/GHA implementation, implementation of a model-free Reinforcement Learning paradigm is possible, where the continuous sensorimotor space grows temporally; this approach is more continuous across developmental time and work while the agent is learning a policy. In such a case, learning could be optimized as a function of morphogenetic growth. This might improve performance in the adult phenotype. As for the effects of critical period timing, maximizing the consequences of sensory feedback during this time would allow us to take advantage of the correspondingly intense rate of morphogenetic growth.

Developmental Braitenberg vehicles are models of embodied nervous systems that navigate and constantly interact with their environment. To implement an RL model of dBV learning, we can appeal to the principle of parallel, loosely coupled processes. This principle states that intelligence emerges from the agents' interactions with the environment. These interactions are based on loosely coupled processes run in parallel and connected directly to the sensory-motor apparatus of the agent [56]. However this continuous sensorimotor space which is inflated with time along with the large continuous action spaces makes it difficult to learn good strategies (train optimal policies) to perform a task, some approaches have been proposed to counter this problem- for instance, adding developmental layers to the neuronal networks this could help the agent to learn better policies, embedding information about the source task in the reward function [57] and also applying other curriculum learning [58] techniques which we believe requires further consideration.

4. Analysis

4.1 *Contingency Analysis*

Aside from being characterized by generativity and plasticity, development is also characterized by developmental contingency [59, 60]. Consider a neuronal network resulting from an innately-defined relationship between sensors and effectors. Each time step introduces components such as an intermediate node or the formation of a new connection between two nodes. At the end of development, a complex nervous system results. We hypothesize that a neuronal network where no circuit components can be removed without functional consequences results in a representational capacity quite unlike a traditional BV. From the initial input-output mapping, every component generated by the developmental process essentially locks the network into a subset of possible future topologies. Topological evolution is an internal embodiment of path dependency. From a design perspective, our choice of critical period timing has an effect on what the vehicle can learn and how it behaves.

4.2 *Different Scenarios for the Critical Period and Developmental Freedom*

Figure 7 demonstrates how dBVs respond to changes in the timing of the critical period and the resulting onset of developmental freedom. Consider two scenarios: early onset of the critical period and late onset of the critical period. The evaluation of early and late onset is defined by when the addition of nervous system elements occurs (early or late). Early critical period onset means that relatively few nervous system connections are generated after the end of the morphogenetic period, occurring mostly within this period. By contrast, late onset allows for a larger proportion of nervous system connections to be added after termination of the morphogenetic phase. Early and late developmental freedom are a direct consequence of critical period plasticity timing, but the timing of this process also has unique signatures on the adult neuronal network.

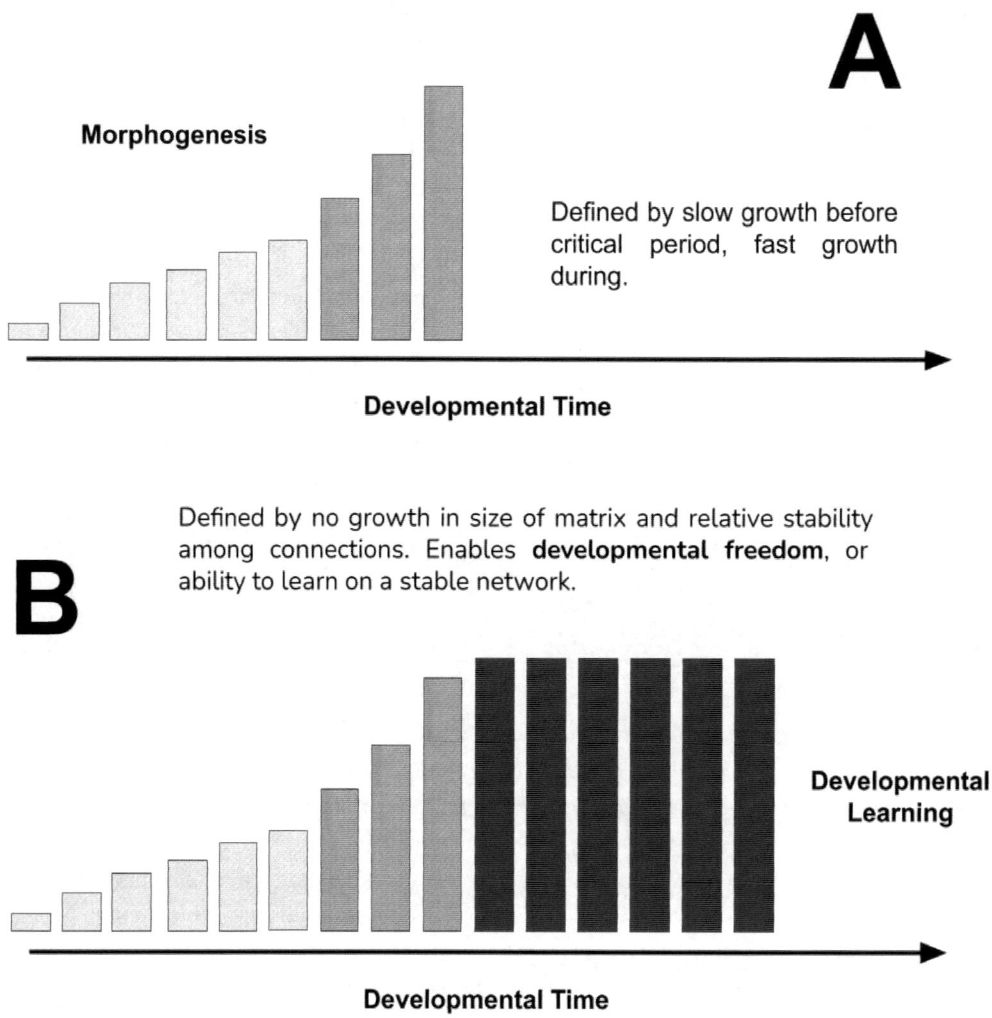

Figure 7. Diagram showing the relationship between the morphogenetic period (light and dark gray), the critical period (dark gray), and developmental learning period (black). Developmental freedom is defined by the black squares, while the shifts in the critical period that define our hypotheticals occur along the axis of developmental time.

4.2.1 Early Critical Period results in Early Onset of Developmental Freedom

When the critical period occurs early, the adult neuronal network tends to be more homogeneous, exhibiting the same types of configurations and motifs. In this scenario, there is less time for preliminary or auxiliary nervous subsystems or behavioral traits to develop. There are also fewer varieties of experience prior to maturity, but since the neural topology is more constrained, the capacity for developmental freedom is homogeneous across individual dBVs. The experience of an early critical period results in a high degree of overall developmental freedom. An early critical period allows for faster ascendance to a fully mature (adult) suite of behaviors suite and capabilities, and thus more exploration of network weights on a stable topology. This enables more refined behaviors and associations to emerge and for shared experiences between individual dBVs to be more common.

4.2.2 *Late Critical Period results in Late Onset of Developmental Freedom*

When the critical period occurs later, the adult neuronal network can exhibit a larger number of possible configurations across a population of dBVs. Delaying the onset of the critical period affords more time for exploration. This results in a slower ascendance to the mature behavioral suite, and provides the adult phenotype with a greater number of alternate network topologies, which can lead to divergent capacity for developmental freedom across individual dBVs. By contrast, dBVs that experience a later critical period results in a low degree of overall developmental freedom. Due to a later critical period, there is a slower ascendance to adult behaviors and capabilities. The consequence of this is less exploration of weights on a stable network topology, which results in more stereotyped behaviors and associations. However, due to divergent network topologies, we should see a heterogeneity of function, and less shared experience when comparing individual dBVs.

Our conception of an agent neuronal network goes beyond what is typically found in a Braitenberg Vehicle in two ways: continual fluctuation in the number of internal processing units and connections, and formation of recursive structures such as internal processing loops and sensorimotor feedback. This forces the system to adapt through the adoption of adaptive capacity rather than specializing towards specific optimal outcomes [61]. In dBVs, morphogenetic events result in two types of feedforward elements: singular connections from A to B, and composite connections from A and C to B (or from A to B and C). While the singular connection type provides a linear feedthrough of the sensorimotor signal, composite connections provide nonlinearities such as convolution, synthetic effects such as cross-talk, and even the potential for interference that can degrade the sensorimotor signal.

4.3 *Canalization Function*

We propose that the Genetic Algorithmic and Hebbian Learning models can be unified using a developmental loss function based on a canalization function [62, 63]. A canalization function is a binary switch that when triggered by the canalyzing selection of a non-uniform environment [64], can shift the developmental state to a new trajectory [65]. This is useful for modeling branching processes that describe restricted regions of developmental possibility space for specific neuronal network topologies [66]. In this case, the canalization function enforces branching after a certain number of morphogenetic events despite slightly different explorations of developmental possibility space [67]. The alternate morphologies featured in Figure 6 can lead to developmentally non-viable vehicles, or present novel means to process a vehicle's sensorimotor signal.

4.4 *Agent Developmental Cognition at Different Scales*

Figure 8 demonstrates how continual learning is not only a property of a single time scale (e.g. encoding and recall of a remembered stimulus), but a property of the entire lifespan as well as intergenerational cultural and biological inheritance. In Figure 8A, we see a timeline of different cognitive events alongside their time course. This comparison is to place different types of events in context. We expect that agents will not need to experience these events directly in order to experience them. Rather, they develop the conditions suitable to experiencing these events.

Figure 8B shows the same timescale, but with different types of behaviors manifest at different durations of time. When these behaviors occur at different time scales (e.g. single stimulus presentation, lifespan), they are said to be compositional: long-term memory consolidation builds upon neuronal activity, as historical events build upon short coffee breaks. The precursors of compositionality are enabled by a canalization function that implements development of the phenotype and neuronal network in timed stages. In this way, learning is structured by multiple developmentally-inspired mechanisms. This is particularly useful in the phenomenological sense [68], particularly in terms of producing an agent's response at multiple time scales. In connecting phenomenology with multiple timescales, a hierarchical model has been proposed by [69].

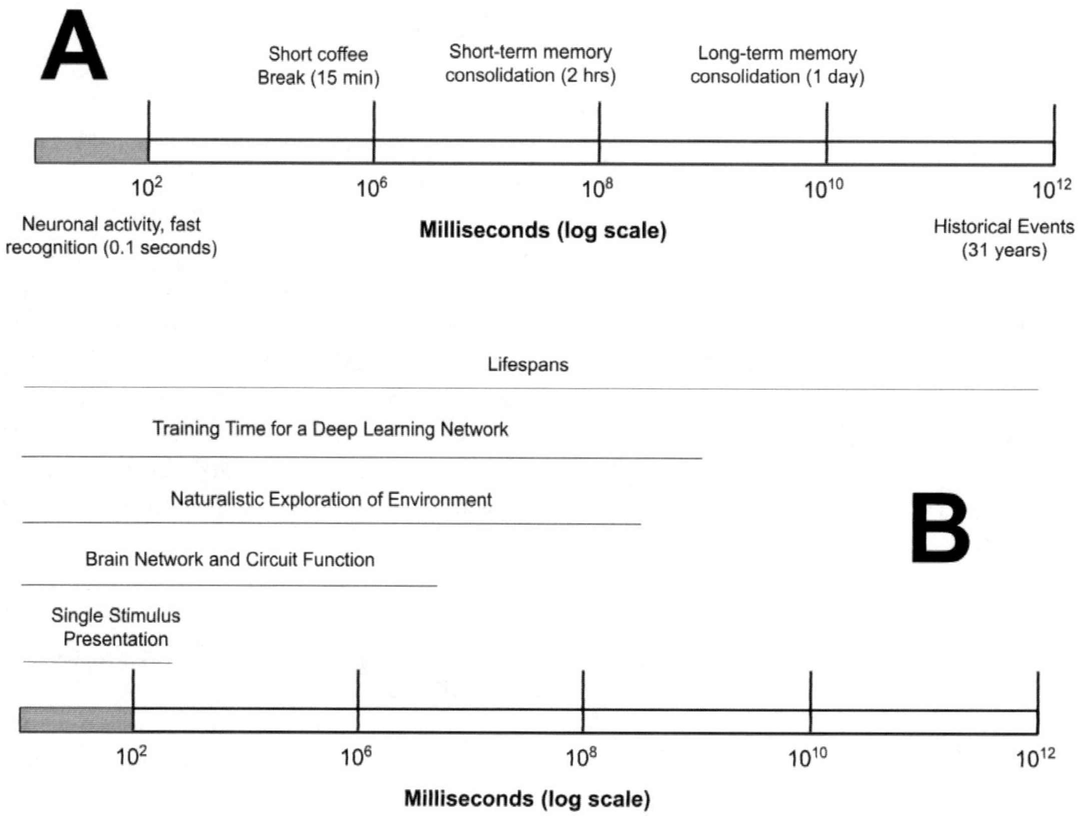

Figure 8. A labeled timeline of agent development as a continual process, from short timescales of cognition to multi-generational inheritance and interactions. A: log-millisecond (ms) timeline of naturalistic cognition with select events and their approximate duration (seconds, minutes, hours, days, and years). B: a log-millisecond (ms) timeline with different aspects of agent biology, artificial neuronal networks, and interaction with the environment and their approximate duration.

4.4.1 Timescales of development and learning

Integrating embodied agents with a developmentally bio-inspired approach to neurosimulation provides a unique opportunity to study these multiple timescales. In a population of agents, we observe some agents in egg form, others in developmental mode, and still others in reproductive mode. This provides multiple life stages and generations against a background of environmental information and cultural practices. At some life stages, this background can serve as both directly and indirectly influential to learning. For example, neuronal activity and fast recognition is important for training during expression of innate capabilities (morphogenetic and behavioral substrate). Experiencing a coffee break and consolidating short-term memory is critical for building on top of the agent neuronal network. Historical events are relevant to intergenerational learning and strategies for reproduction. Replicating the complexities of organismal biology and its behavioral milieu will improve our chances of simulating intelligent behaviors far beyond task-constrained and human-centric examples.

4.4.2 Meta-brain dBV modeling

A meta-brained dBV can be used to model the multiple modes of agent cognition as shown in Figure 9. The meta-brain demonstrates two modes of cognition: the low-representation state is indicative of smaller timescales of neural activity and pairwise connectivity, while the high-representation state is indicative of longer timescales at which event-related phenomena are integrated. One way to conceptualize this is in the form of neuronal avalanches [70, 71]. Neuronal avalanches are large-scale responses that result from the accumulation of small changes. The sandpile model [72] describes this as large-scale displacements of sand in response to the stochastic grain-wise growth of a sandpile. Sandpile displacements are the product of nonlinear responses to additive growth. While this has been demonstrated to be a property of neuronal dynamics, it is also much more complex than an input-output (sensor-effector) relationship. In the agent's neuronal network, events that respond from short-timescale activity accumulate and get passed to the high-representation state. In this layer, accumulation of activity can yield large-scale changes in the output behavior [73]. For example, a 15-minute interval of naturalistic behavior leads not to a linear repetition of simple outputs, but multiple transitions and multiplicative combinations of output behavior.

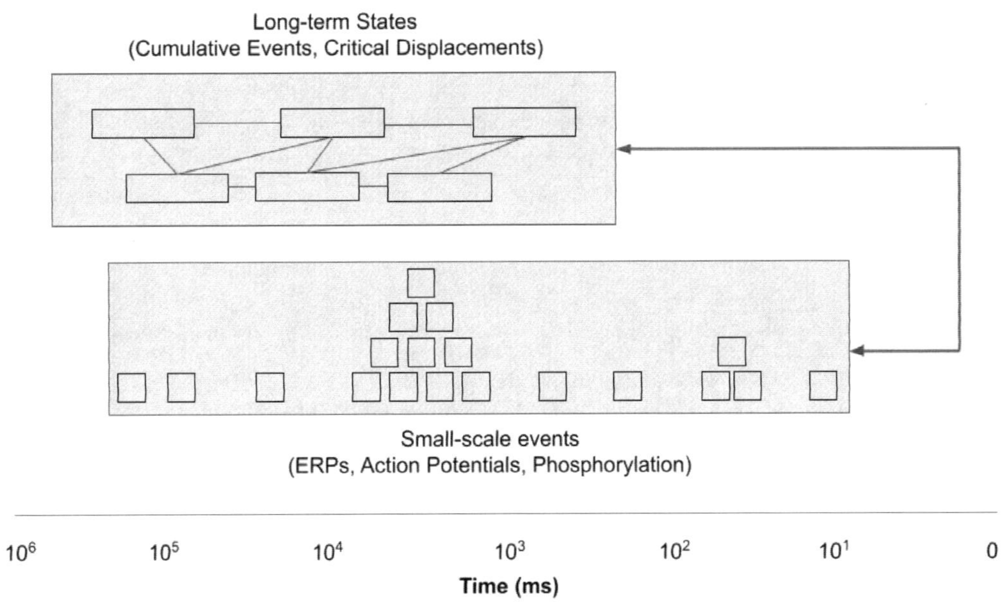

Figure 9. Subcognitive networks (corresponding to small-scale events) and cognitive networks (corresponding to large-scale events) modeled using a meta-brains configuration. A log-millisecond (ms) timeline of naturalistic cognition with various events and their approximate duration is used to characterize their place in time.

4.4.3 Development of layers as enabling global response.

The advantage of the meta-brain model is to demonstrate how small-scale events in individual neuronal network units can be translated into global states. Meta-brains allow for these incompatible descriptions to be summarized by leveraging the different representational complexity of laminar organization. Laminar complexity allows for compartmentalization of information as it is processed internally: single units can serve as localized and specialized processors, while brain-wide states that interact with the surrounding somatic layers can also play a role in information processing. This can also provide a link between more complex behaviors. For example, steering behaviors that respond to 3-D curvature can be approximated using taxis in traditional Braitenberg Vehicles [74]. This can be modeled with representations using non-linear control systems or ODEs, but this specifies reactive behaviors without making the connection between neural microcircuits and global states. Walter et.al [75] propose that neural microcircuits are fundamental units of processing in artificial neuronal networks, and replicate these in artificial systems using heterogeneous network topologies. In addition, we can use multi-layered representations to model social behaviors, which involve both neurobiological mechanisms and behavioral interactions [76]. As an enabler of developmental generativity, the subcognitive/global state meta-brain improves upon Braitenberg's original typology of simple behavioral states [8].

5. Discussion

5.1 *Broad Overview*

In this paper, we have presented a broad overview of the developmental neurosimulation approach to embodied cognition, focusing on developmental Braitenberg vehicles (dBVs). Taking an embodied developmental approach to agent simulation is superior to both canonical approaches to embodied intelligence as well as simply recapitulating developmental psychology. Approaches that simply account for development as a growing neuronal network is similarly insufficient. A sufficient developmental approach takes into account what Lux et.al [77] and Lickliter [78] summarize as embodiment and embeddedness. In both cases, embodiment refers to advances beyond the psychology of the agent [79]. As we argue with respect to phenomenology, understanding the world involves multiple timescales and multiple mechanisms at each timescale. Establishing the dBV in space also involves both a body-centric reference frame and a goal-directed frame. However, that goal-directed frame is simply relevant to immediate behaviors, not necessarily its more ultimate origins [80, 81]. For example, the energetic imperatives or various developmental transformations of the agent might conflict with the priorities of goal-directed behavior. This leads us to considering embeddedness as a factor in developmental neurosimulation. Agents of more than trivial complexity must deal with relations between physical, biological, and cultural phenomena. As these multifaceted objectives often do not point in the same direction [82], understanding the relational nature of these different contributors to the emergence of behaviors in development [83].

In [10], we have previously argued that dBV architectures are low-representation networks that require other models to compute context and other, more sophisticated representations. Yet it may be that aspects of the dBVs embodied nature may afford dBVs a more robust comprehension of the transactions made with their environment over developmental time [84]. We have also argued that developmental freedom, spatial embodiment, and critical period regulation play a role in learning that is missed using other types of artificial learning. The dBV approach may also be a means to balance the processing previously unseen information (unsupervised developmental learning) with the need for a set of innate capabilities. Aside from our previous example of canalizing selection, in which the environment is the perturbing factor, we can also use sensory deprivation to analyze the potential of dBV neuronal networks across development and in its mature state. Sensory deprivation allows for environmental inheritance to be disrupted in a way that affects propagation of the sensorimotor signal more generally. One example of sensory deprivation is the removal of sensors that provide a coherent sensorimotor signal. Blocking the sensory inputs on one side of the vehicle will lead to behaviors that are biased to the non-occluded side. Like morphological asymmetries, sensory deprivation does not affect the morphogenetic process, but does affect how sensorimotor signals are conducted. A more indirect example of sensory deprivation with respect to dBVs is a stray cat that does not have the proper socialization experience.

5.2 *Reassessing Developmental Neurosimulation as Continual Learning*

We can also put developmental neurosimulation in the context of continual learning methodology by thinking about dBVs as agents. The first methodological issue is *concept drift*. Concept drift is where various changes (ranging from gradual to sudden) in the agent's environment occur with respect to their representational capacity [85]. Statistically, this translates into how the distribution of events changes over time [86]. This is similar to the Sandpile model example shown in Figure 9. While the sandpile model allows for new conceptual frameworks to be attained in a non-linear manner, it too can also lead to potentially catastrophic shifts in function. Yet it might also be that naturalistic behavior can smooth out the transitions between concept classes by providing context in an embodied manner. Therefore, combining embodiment with a non-linear account of a non-stationary world might reduce the chance of catastrophic forgetting [87].

The other methodologies of interest are *class-incremental learning* and *forward transfer*. Class-incremental learning assumes that the agent's environment is non-stationary, and disallows for classes to be reused across experiential contexts [88]. In [89], a single context variable is used to compensate for shifts in accumulated knowledge. We believe that our approach tackles this same problem in multiple ways, including utilizing criticality to manage transitions and asymmetries in the neuronal network and morphology. Indeed, the incorporation of anatomical segregation [90] and detailed synaptic processing models [91] can resolve some drawbacks of non-stationarity. In forward learning, transferring what is learned in the past to tasks that are encountered in the future must overcome model loss characterized as catastrophic forgetting [92]. In both cases, plasticity and regeneration on a late developmental network can play a role consistent with these strategies.

5.3 *Lessons from Embodied Cognitive Morphogenesis*

There are a number of lessons from a related approach called Embodied Cognitive Morphogenesis [93], which is based on the developmental dynamics and emerging sensory systems of biological embryos. The first lesson is that computational agent development is governed by multiple overlapping networks. While Figure 1 demonstrates the emerging agent neuronal network, other networks such as the relationships between different artificial chromosome encodings, sensory structures, or morphogenetic differentiation overlap. Another lesson is that as an agent's neuronal network begins to grow in terms of internal processing units and connections, it becomes increasingly embedded with morphogenetic and behavioral traits that define the agent's phenotype. Thus, information processing becomes embedded in a more active spatiotemporal context. The advantage of developmental embodiment over static embodiment involves symmetry-breaking events that can be summarized by tree representations (canalization function and differentiation tree) at multiple scales.

In our overlapping networks, symmetry-breaking also introduces a phenotypic information eddy (circular flows from laminar ones) that can enable further structure and feedback loops.

As these structural and functional networks come online, information processing is progressively embedded within a phenotype [94]. Importantly, symmetry-breaking events provide anatomical context for the emergence of cognition through the phenotype (body) and nervous system (connectogenesis). Figure 6 demonstrates how these developmental processes can result in phenotypic symmetry-breaking events, which in turn leads to differential sensory processing and potentially even generating behavioral states from small-scale phenotypic features. Both rooted (directed graphs) and unrooted (networks) tree structures are ubiquitous in the developmental neurosimulation approach. Embodied Cognitive Morphogenesis offers several additional formulations of tree structure derivatives, the basic lesson is that embodied cognition can be explicitly tied to the contingencies produced by specific developmental trajectories. Related to this are the transformational aspects of developmental phenotypes, which also intersect with the expansion of neuronal networks across development. In the case of morphogenetic substrates, different patterns of differentiation of computational agent germ layers into distinct features can potentially result in radically different neuronal network architectures.

5.4 *Expressivity and Interpretability of Developmental Neurosimulation*

As bio-inspired algorithmic systems, dBVs require not only high expressivity but also extensive interpretability. Expressivity is crucial for capturing naturalistic behaviors that unfold over continuous time at multiple scales. There is also a need to interpret the behaviors generated from these systems, which can help us understand behavioral complexity at multiple scales more generally. A few examples from dBV implementations have been identified for consideration.

The integration of multiple sensory channels in the dBV instantiation MultiLearn [7] can be implemented at either the periphery (among sensory features) or in the neuronal network. While interpretation of multisensory integration outputs is difficult, characterizing single integrator internal processing units and the multiplicativity of their interactions [95] provides a way forward. The collective behavior of dBVs must also be interpreted, and has been classified for the dBV instantiation BV collectives in [7]. Further systems of interpretation might involve self-observation of collective behaviors by individual agents.

In [7], the hinge model of potential relationships between genotype and behavior is quite relevant to understanding how changes in the artificial chromosome encodings within populations and across generations can be characterized. One benefit of having a formal model of innateness is that it allows for generativity with expressivity, contributing to the expression of coherent and relevant behaviors. While the structure of these encodings can be beneficial to studying these relationships, there is nevertheless much that can be learned from the interactions of innateness, embodiment, and goal-directed behaviors.

5.5 *Summary*

With our toy model of an emerging nervous system, we can show how embodied neuronal networks are more sensitive to heterogeneous data, particularly in the spatial context. Furthermore, dBVs also teach us about the emergence of behavior as being deeply intertwined with the developmental process [11]. We can see this in the zebrafish visual system, where the maturity of representations and functions go hand in hand with developmental changes in behavior [96]. Developmental Neurosimulation is also reliant on allostatic control and the strategic pace of connectogenesis. Further, the relationship between connectogenesis and developmental freedom as a means to enable learning is of particular interest for future work on related concepts. More generally, future work on the continual developmental neurosimulation approach will explore the potential of strategically shaping a neuronal network and agent body in tandem with multiple adaptive mechanisms. We also seek to understand the potential of using dBVs as a strategy for enabling complex unsupervised learning in an embodied agent. Ultimately, this provides a catalog of ways in which we can apply fundamental aspects of developmental bio-inspiration, particularly the lessons of Embodied Cognitive Morphogenesis, to the enterprise of Continual Developmental Neurosimulation.

References

[1] Saxe AM McClelland JJ Ganguli S 2019 A mathematical theory of semantic development in deep neural networks. *PNAS* **116(23)** 11537–11546.

[2] Lungarella M Mettay G Pfeiferz R Sandiniy G 2003 Developmental robotics: a survey. *Connect Sci* **15(4)** 151–190.

[3] Smith LB and Slone LK 2017 A Developmental Approach to Machine Learning? *Front Psych* **8** 2124.

[4] Munakata Y McClelland JL 2003 Connectionist models of development. *Dev Sci* **6(4)** 413–429.

[5] Kohli M Magoulas GD Thomas MSC 2020 Evolving Connectionist Models to Capture Population Variability across Language Development: modeling children's past tense formation. *Artif Life* **26** 217–241.

[6] Elman JL 1993 Learning and development in neural networks: the importance of starting small. *Cognition* **48** 71-99.

[7] Dvoretskii S Gong Z Gupta A Parent J and Alicea B 2020 Braitenberg Vehicles as Developmental Neurosimulation. *arXiv* 2003.07689.

[8] Braitenberg V 1984 Vehicles: experiments in synthetic Psychology. MIT Press, Cambridge, MA.

[9] Glanville R 1982 Inside every white box there are two black boxes trying to get out. *Behav Sci* **27**

[10] Alicea B Dvoretskii S Felder S Gong Z Gupta A Parent J 2020 Developmental Embodied Agents as Meta-brain Models. *DevoNN Workshop*, Artificial Life Conference.

[11] Alicea B 2020 Raising the Connectome: the emergence of neuronal activity and behavior in *C. elegans*. *Front Cell Neuro* **14** 524791.

[12] Seguin C Sporns O Zalesky A 2023 Brain network communication: concepts, models and applications. *Nat Rev Neuro* **24** 557-574.

[13] Schettler A Raja V Anderson ML 2019 The Embodiment of Objects: Review, Analysis, and Future Directions. *Front Neuro* **13** 1332.

[14] Lassig F Aceituno PV Sorbaro M Grewe BF 2023 Bio-inspired, task-free continual learning through activity regularization. *Biol Cyber* doi:10.1007/s00422- 023-00973-w.

[15] Kudithipudi D Aguilar-Simon M Babb J Bazhenov M Blackiston D Bongard J Brna AP Raja SC Cheney N Clune J 2022 Biological underpinnings for lifelong learning machines. *Nat Mach Intell* **4(3)** 196–210.

[16] Kim S Noci L Orvieto A Hofmann T 2023 Achieving a Better Stability-Plasticity Trade-off via Auxiliary Networks in Continual Learning. *arXiv*, 2303.09483.

[17] Mermillod M Bugaiska A Bonin P 2013 The stability-plasticity dilemma: Investigating the continuum from catastrophic forgetting to age-limited learning effects. *Front Psych 4* 504.

[18] Schutter DJLG Wischnewski M Bekkering H 2015 Stability through variability: Homeostatic plasticity and psychological resilience. *Behav Brain Sci* **38** e118.

[19] Okray Z Jacob PF Stern C Desmond K Otto N Talbot CB Vargas-Gutierrez P Waddell S 2023 Multisensory learning binds neurons into a cross-modal memory engram. *Nature* **617** 777–784.

[20] Parisi GI Kemker R Part JL Kanan C Wermter S 2019 Continual lifelong learning with neural networks: a review. *Neural Netw* **113** 54–71.

[21] Van de Ven GM Tolias AS 2019 Three scenarios for continual learning. *arXiv* 1904.07734.

[22] Shin H Lee JK Kim J Kim J 2017 Continual learning with deep generative replay. *arXiv* 1705.08690.

[23] von Oswald J Henning C Grewe BF Sacramento J 2022 Continual learning with hypernetworks. *arXiv*, 1906.00695.

[24] Zenke F Poole B Ganguli S 2017 Continual learning through synaptic intelligence. *Proc Intl Conf Machine Learn* **34(70)** 3987–3995.

[25] Lim S Kaiser M 2015 Developmental time windows for axon growth influence neuronal network topology. *Biological Cybernetics* **109(2)** 275-286.

[26] Kaiser M 2020 Changing Connectomes. Cambridge, MA: MIT Press.

[27] Chakraborty M Jarvis ED 2015 Brain evolution by brain pathway duplication. *Royal Soc London B* **370(1684)** 20150056.

[28] Seguin C Sporns O Zalesky A 2023 Brain network communication: concepts, models and applications. *Nat Rev Neurosci* **24** 557–574.

[29] Hagen EH Hammerstein P 2005 Evolutionary Biology and the Strategic View of Ontogeny: Genetic Strategies Provide Robustness and Flexibility in the Life Course. *Res Human Dev* **2(1-2)** 87–101.

[30] Hensch TK 2004 Critical period regulation. *Ann Rev Neurosci* **27** 549–579.

[31] Sengpiel F 2007 The critical period. *Curr Biol* **17(17)** R742.

[32] Frankenhuis WE Walasek N 2020 Modeling the evolution of sensitive periods. *Dev Cogn Neurosci* **41** 100715.

[33] Takesian AE Hensch TK 2013 Balancing Plasticity/Stability Across Brain Development. *Prog Brain Res* **207** 3-34.

[34] Chang EF Merzenich MM 2003 Environmental Noise Retards Auditory Cortical Development. *Science* **300(5618)** 498-502.

[35] Kahn DM Krubitzer L 2002 Massive cross-modal cortical plasticity and the emergence of a new cortical area in developmentally blind mammals. *PNAS* **99(17)** 11429-11434.

[36] Ardiel EL Rankin CH 2010 The importance of touch in development. *Paediatric Child Health* **15(3)** 153-156.

[37] Galvan A 2010 Neural Plasticity of Development and Learning. *Human Brain Mapping* **31** 879–890.

[38] Smith LB Thelan E 2003 Development as a dynamic system. *Trends Cogn Sci* **7** 343–348.

[39] Charvet CJ Striedter GF 2011 Developmental Modes and Developmental Mechanisms can Channel Brain Evolution. *Front Neuroanat* **5** 4.

[40] Greenough WT Black JE Wallace CS 1987 Experience and Brain Development. *Child Development* **58** 539-559.

[41] Alicea B Cialfi D Lim A Parent J 2021 Gibsonian Information: a new approach to quantitative information. *Studies Computa Intell* **1032** 15–19.

[42] de Kleijn R Sen D Kachergis G 2022 A Critical Period for Robust Curriculum-Based Deep Reinforcement Learning of Sequential Action in a Robot Arm. *Topics in Cognitive Science* **14(2)** 311-326.

[43] Kirschner MW Gerhart JC 2005 The Plausibility of Life: resolving Darwin's dilemma. Yale University Press, New Haven, CT.

[44] Alicea B Pang A Parent J 2022 Intelligence Offloading and the Neurosimulation of Developmental Agents. *PsyArXiv* doi:10.31234/osf.io/ykxd4.

[45] Li H Shuster SA Li J and Luo L 2018 Linking neuronal lineage and wiring specificity. *Neural Devel*, 13(5).

[46] Wolpert L 2016 Positional Information and Pattern Formation. *Current Topics in Developmental Biology* **117** 597-608.

[47] Dubuis JO Tkacik G Wieschaus EF Gregor T and Bialek W 2013 Positional information, in bits. *PNAS* **110(41)** 16301-16308.

[48] Tkacik G Gregor T 2021 The many bits of positional information. *Development* **148(2)** dev176065.

[49] Ruder S 2017 An overview of gradient descent optimization algorithms. *arXiv* 1609.04747.

[50] Sanger TD 1989 Optimal Unsupervised Learning in a Single-Layer Linear Feedforward Neural Network. *Neural Netw* **2** 459-473.

[51] Raja V 2019 J. J. Gibson's most radical idea: The development of a new law-based psychology. *Theor Psych* **29(6)** 789-806.

[52] Gibson JJ 1979 The Ecological Approach to Visual Perception. Psychology Press, Sussex, UK.

[53] Alicea B Gordon R 2016 Quantifying Mosaic Development: towards an evo-devo postmodern synthesis of the evolution of development via differentiation trees of embryos. *Biology* **5(3)** 33.

[54] Gordon NK Gordon R 2016 Embryogenesis Explained. World Scientific Publishing, Singapore.

[55] Haber N Mrowca D Wang S Fei-Fei L Yamins DLK 2018 Learning to Play With Intrinsically-Motivated, Self-Aware Agents. *Adv Neural Info Proc Syst* **31** 8398–8409.

[56] Blynel J 2000 Reinforcement learning on real robots. Master's thesis, University of Aarhus.

[57] Gillespie J Rano I Siddique N Santos J Khamassi M 2017 Reinforcement Learning for Bio-Inspired Target Seeking. *Ann Conf Towards Auto Robot Syst* **18** 637-650.

[58] Bengio Y Louradour J Collobert R Weston J 2009 Curriculum learning. *J Am Pod Assoc* 60. 6.

[59] Oyama S 2000 The Ontogeny of Information: Developmental Systems and Evolution. Duke University Press, Durham, NC.

[60] Nonoyama T Chiba S 2019 Phenotypic determinism and contingency in the evolution of hypothetical tree-like organisms. *PLoS One* **14(10)** e0211671.

[61] Louie AH 2010 Robert Rosen's anticipatory systems. *Foresight* **12(3)** 18-29.

[62] Kauffman SA 1993 Origins of Order. Oxford University Press, Oxford, UK.

[63] Paul E Pogudin G Qin W Laubenbacher R 2020 The Dynamics of Canalizing Boolean Networks. *Complexity* **3687961** 1-14.

[64] Kuo Z.-Y 1976 The dynamics of behavior development: an epigenetic view. Springer, Berlin.

[65] Moreira AA Amaral LAN 2005 Canalizing Kauffman Networks: nonergodicity and its effect on their critical behavior. *Phys Rev Lett* **94(21)** 218702.

[66] Li Y Adeyeye JO Murrugarra D Aguilar B Laubenbacher R 2013 Boolean nested canalizing functions: A comprehensive analysis. *Theor Comp Sci* **481** 24–36.

[67] Just W Shmulevich I Konvalina J 2004 The number and probability of canalizing functions. *Physica D* **197(3–4)** 211-221.

[68] Beckman P Kostner G Hipolito I 2023 An Alternative to Cognitivism: Computational Phenomenology for Deep Learning. *Minds Mach* **33** 397-427.

[69] Singhal I Srinivasan N 2021 Time and time again: a multi-scale hierarchical framework for time-consciousness and timing of cognition. *Neurosci Conscious* **2** niab020.

[70] Beggs JM Plenz D 2003 Neuronal avalanches in neocortical circuits. *J Neurosci* **23(35)** 11167-11177.

[71] Cocchi, L., Gollo, L.L., Zalesky, A., and Breakspear, M. (2017). Criticality in the brain: A synthesis of neurobiology, models and cognition. *Progress in Neurobiology*, 158, 132-152.

[72] Ivashkevich EV Priezzhev VB 1998 Introduction to the sandpile model. *Physica A* **254(1–2)** 97-116.

[73] Plenz D Thiagarajan TC 2007 The organizing principles of neuronal avalanches: Cell assemblies in the cortex? *Trends Neurosci* **30** 101–110.

[74] Rano I 2009 A Steering Taxis Model and the Qualitative Analysis of Its Trajectories. *Adapt Behav* **17(3)** 197-211.

[75] Walter A Wu S Tyrrell AM McDaid L McElholm M Sumithran NT Harkin J Trefzer MA 2023 Artificial Neural Microcircuits for use in Neuromorphic System Design. *Proc Artif Life*, 581.

[76] Xu S Jiang M Liu X Sun Y Yang L Yang Q Bai Z 2021 Neural Circuits for Social Interactions: From Microcircuits to Input-Output Circuits. *Front Neural Circ* **15** 768294.

[77] Lickliter R 2021 Opinion: Advancing Embodiment Research From a Developmental Point of View. *Front Syst Neurosci* **15** 748335.

[78] Lux V Non AL Pexman PM Stadler W Weber LA Kruger M 2021 A developmental framework for embodiment research: The next step toward integrating concepts and methods. *Front Syst Neurosci* **15** 672740.

[79] Coghill GE 1929 Anatomy and the problem of behavior. Cambridge University Press, Oxford, UK.

[80] Bateson P Laland KN 2013 Tinbergen's four questions: an appreciation and an update. *Trends Ecol Evol* **28(12)** 712-718.

[81] Gesell A Amatruda CS 1945 The embryology of behavior; the beginnings of the human mind. Harper-Collins, New York.

[82] Yina D Kaiser M 2021 Understanding neural flexibility from a multifaceted definition. *NeuroImage* **235** 118027.

[83] Overton WF 2008 Embodiment from a relational perspective. In "Developmental Perspectives on Embodiment and Consciousness", W.F. Overton, U. Muller, and J. L. Newman (eds). Erlbaum Associates, New York.

[84] Di Paolo EA 2020 Picturing Organisms and Their Environments: Interaction, Transaction, and Constitution Loops. *Front Psych* **11** 1912.

[85] Bayram F Ahmed BS Kassler A 2022 From concept drift to model degradation: An overview on performance-aware drift detectors. *Knowl-Based Syst* **245** 108632.

[86] Cossu A Graffieti G Pellegrini L Maltoni D Bacciu D Carta A and Lomonaco V 2022 Is Class-Incremental Enough for Continual Learning? *Front Artif Intell* **5** 829842.

[87] Casado FE Lema D Criado MF Iglesias R Regueiro CV and Barro S 2022 Concept drift detection and adaptation for federated and continual learning. *Multi Tools Apps* **81(1181: Artificial Intelligence for Physical Agents)**, 3397–3419.

[88] Maltoni D Lomonaco V 2019 Continuous learning in single incremental-task scenarios. *Neural Netw* **116** 56–73.

[89] Lesort T Caccia M Rish I 2022 Understanding continual learning settings with data distribution drift analysis. *arXiv* 2104.01678.

[90] Liu Z Khona M Fiete IR Tegmark M 2023 Growing Brains: Co-emergence of Anatomical and Functional Modularity in Recurrent Neural Networks. *arXiv* 2310.07711.

[91] Pinitas K Chavlis S Poirazi P 2021 Dendritic Self-Organizing Maps for Continual Learning. *arXiv* 2110.13611.

[92] van de Ven GM Tuytelaars T and Tolias AS 2022 Three types of incremental learning. *Nat Mach Intell* **4** 1185–1197.

[93] Alicea B Gordon R Parent J 2023 Embodied Cognitive Morphogenesis as a Route to Intelligent Systems. *Interface Focus* **13(3)** 20220067.

[94] Alicea B 2018 The emergent connectome in *Caenorhabditis elegans* embryogenesis. *BioSystems* **173** 247-255.

[95] Stein BE Stanford TR 2008 Multisensory integration: current issues from the perspective of the single neuron. *Nat Rev Neurosci* **9** 255–266.

[96] Avitan L Pujic Z Molter J McCullough M Zhu S Sun B Myhre A-E and Goodhill GJ 2020 Behavioral signatures of a developing neural code. *Curr Biol* **30(17)** 3491-3493.

IOP Conf. Series: Materials Science and Engineering **1321** (2024) 012014 doi:10.1088/1757-899X/1321/1/012014

Object construction via the action-perception helix, Bayesian inference and metaphorisation

Amaranta Valdés-Zorrilla[1,2], **Juan-Carlos Letelier**[3] **and Jorge Soto-Andrade** [1,2]

[1] Department of Mathematics, Faculty of Science, University of Chile, Santiago, Chile
[2] Centre for Advanced Research in Education, University of Chile, Santiago, Chile
[3] Department of Biology, Faculty of Science, University of Chile, Santiago, Chile

E-mail: amvaldes@uchile.cl, letelier@uchile.cl, sotoandrade@uchile.cl

Abstract. Since von Uexküll we know that organisms are cognitive devices, which construct the objects in their Umwelt. Here, expanding on previous work, we show how this construction is carried out through the mechanism of Structural Coupling, which is a process linking an organism and its medium through a never ending sequence of concatenated perception ↻ action loops. Every perception triggers complex patterns of concentration changes of molecules (called internal signals) that select a specific action as the outcome of a Bayesian inference mechanism undergoing continuous Hebbian-like updating. The relation between perception and action is a complex and plastic one, and we put forward that a categorical approach in terms of stochastic mappings and functors would be most commendable to make sense of living systems as cognitive devices. We discuss the updating mechanism of stochastic mappings (conditional probability matrices). Finally, we elaborate on the process of metaphorising, from both a macro and a micro viewpoints. The former sees, from an eagle's viewpoint, metaphors as functors between inferential categories; the latter sees the emergence of metaphorising as a *biological mechanism* in the unfolding of the never ending perception ↻ action helix by means of which a living system constructs its Umwelt. We illustrate our viewpoints with the help of various examples in mathematical cognition.

1. Introduction

The fundamental problem facing organisms is to *generate coherent behaviours*. But, as we have discussed elsewhere [1], organisms *are not* information processing devices instead they are self-fabricating (autopoietic) devices that are *structurally coupled* with their Umwelt [2]. Although living systems are autopoietic systems, self-fabrication and self-encapsulation are not their unique hallmarks [3] As important as metabolic closure (autopoiesis) [4] is the fact that living systems construct a relation, through their biochemical, physiological and muscular behaviours, with their environment. In the following pages we will explore this *Relational view* of the organism.

Recently we have sketched the elements of a mathematical framework understanding how living systems construct, via their behaviour, the objects that populate their world [1]. Here we introduce an important new feature of our viewpoint as we claim that Category Theory, particularly the category of stochastic mappings, is a relevant element which needs to be included in any model of living organization.

The structure of our document is as follows:

(i) A summary of the ideas of self-fabrication, structural coupling and action/perception loop

(ii) A re-analysis of the notion of *internal signals*
(iii) Introduction to category theory and stochastic mappings
(iv) The action ↻ perception helix in a categorical context
(v) Metaphorisation as a biological process
(vi) Discussion and open questions.

2. A necessary summary: Self-fabrication, autopoiesis, structural coupling/enaction, the action ↻ perception loop

Living systems with their amazing morphological diversity (from E. coli to Felix. domesticus), their great expanses of ecological niches and, above all, their intrinsic *autonomy* constitute a challenge to theorizing efforts. The *XX* century was a rich field to construct over-encompassing theories about the nature of living systems, from Leduc´s *Osmotic Forests* [5] to theories based on powerful analogies like *Life=DNA* or *Life=Information Processing.* In reality, most of these explored paths did not build a useful theory of living organisation [4] [1]. Interestingly, the intuition that the living systems are, beyond anything else, *self-fabricating* systems has continuously gained ground. This last idea was first introduced by Robert Rosen in 1958 [6, 7], but developed with much force by two Chilean biologists (Humberto Maturana and Francisco Varela in the 70s) with their concept of *Autopoietic Systems* [3]. In a nutshell, in autopoietic networks the immense majority of structural components (molecules like metabolites and enzymes) is produced by the network using only very small and common input elements (glucose, water, some mineral salts, etc...). Why this marvellous mechanism constructed with subtle and transparent gears does not self-destroys is an open question in theoretical biology as well as in self-organization studies (a big open question, shall we underline). Thus, little by little the intuition that self-fabrication is at the core of living organisation has gained a lot of traction and different versions of these ideas had surfaced in the last 50 years (a very good summary can be found in [4]). Autopoiesis/self-fabrication has mostly been ignored by almost all experimental biologists up to now, despite progressively gaining adepts in the albeit rather small domain of researchers interested in problems of Systems Biology, Origin of Life and Artificial Life. Notably, new developments have emerged in the field of Embodied Intelligence and Artificial Embodied Intelligence. See [8] for a comprehensive review and discussion on the parallel embodied turn in Cognitive Science and Artificial Intelligence, leading to *Embodied Human Cognition* and *Artificial Embodied Intelligence.* See also [9, 10], In the recent collective monograph [11], different contributors strike the same cord (e.g.[12] Nevertheless, fundamental questions like What a Process is?, Why metabolic networks, which are continuously functioning since 3850 M years ago, are so resilient and stable? How can we describe autonomy? remain totally open. Our collective failure to tackle the problem of the stability of the complex networks responsible of metabolism shows how limited our current theories are. In any case, in order not to give a very negative view of the progress attempted and achieved at least collectively, we should acknowledge that the researchers interested in these questions are abandoning the fetish of *Information* and *DNA* as the core concepts needed to understand biological organisation.

2.1. Living systems are also cognitive devices.

Although autopoiesis is, without a doubt, a necessary element to understand biological organisation it is not the only one. The other central element, already glimpsed by von-Uexküll in 1920 [13] is that all organisms (including bacteria!) are cognitive systems. If the notion of self-fabrication has had serious problems to be accepted, this second idea is viewed askance by even a larger proportion of biologists that refuse to incorporate this trans-disciplinary concept into a research program dominated by physico-chemical concepts. Here,

[1] Autonomy is a tricky concept to put into any theory...

following the slightly different formulations given by Maturana and Varela, we are going to introduce the sister notions of Structural-Coupling/Enaction [14, 15]. In effect, for many people the idea of mixing explanatory levels (i.e. metabolism with cognition) appears not only impossible but a profound error.

As always, we must start with the unavoidable fact that living systems are autopoietic networks. But living systems, all of them, always *live in the world.* This *living* implies that the autopoietic network must have two specialised subsystems, one devoted to *perception* (sensorium) and the other to *action* (efectorium) (Figure 5). Thus, at any instant, the living system must decide which action to undertake in the context of what is happening in its sensory surface (effectoriun). In a first approximation the situation seems to imply that living systems are nothing more, and nothing less, than classifiers followed by an output system. But the situation is radically different as the phenomenon of *Structural Coupling* arises.

2.2. Structural Coupling: Living Systems as Constructors

Let us consider a living system that is put in a new medium with novel characteristics. The initial relation between the organism and the medium is unstructured, as the organism has not built a trajectory of action ↻ perception loops adequate to this new medium, we prefer to use the intuition of an *helix* to represent this trajectory (see Figure 1 below). It is important to emphasise that autopoietic networks are full of regulatory processes (i.e. processes that regulate the efficiency and kinetics of other processes). These regulatory processes endow the overall network with the capacity of changing *while maintaining its identity* [16]. Thus our current understanding of autopoietic networks is that they are able to maintain their organization invariant *because* they have numerous processes that are able to shift their operating points and speeds. Symmetrically, the medium also exhibits plasticity that can be divided into two types. One type, *interactional plasticity,* arises because the interaction organism/medium brings into chemical contact the autopoietic network with the medium, making possible totally new processes. A very good example is the production of oxygen by splitting water molecules using new enzymes (produced by the organism) and photons that were part of the initial medium but were not used for the first 1000 My. It seems that in the history of the planet Earth the interaction between physical and atmospherics factors with biology has resulted in profound geological changes like the Huronian glaciation 2.5 to 2.2 billion years ago. Another example of physical change induced by living systems is the creation of soil. A second type of plasticity is a *configurational plasticity* that is produced because the medium is not homogeneous and the organism, via its action repertoire, can select where and how to interact. The main point is that the organism-medium interaction is plastic, and it brings structural changes to both. If the organism/medium interaction is sustained in time, organism and medium begin a process of complementary changes. If the interaction persists even longer the changes can be quite dramatic and external observers, that were only privy to the last state of the process, could erroneously think that the organism has adapted to the external medium and not that the curious state has been a co-construction between the organism and the medium where it is immersed. See Figure 1, where we paraphrase Machado's verse "Walker, there is no path, *we lay down a path in walking*" cited by Varela et al. in [15]. The core of this paper is to put forward a mathematical model enabling us to tackle structural coupling.

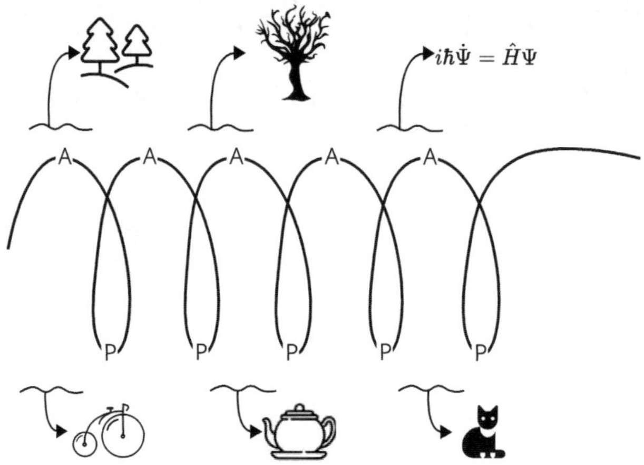

Figure 1. The perception ↻ action helix, which "lays down the objects in the organism's Umwelt in unfolding ". Every object, either a *black dog, a pine forest or Schrödinger Equation* or a humble *tea kettle*, is the unavoidable result of the perception (P) - action (A) helix. We summarised this idea as: Perception ↻ Action. Every object is the stabilisation of a recurrent sensorimotor correlation.

2.3. Cognition = Acting

Living systems are structurally coupled autopoietic networks. Thus, moment by moment, they must act in order to maintain their coupling. As the central problem of an organism is *What to do next?* a well oiled mechanism exists linking the *Sensorium* to the *Effectorium* surfaces. This mechanism is dependent on the following elements (see Figure 5).

(i) the production (or the change in concentration) at the level of the Sensorium, after each organism-medium encounter, of a miriad of molecules called *internal signals.*

(ii) The amplification of the network effects of the internal signals by the very complex autopoietic network

(iii) the selection of the next action to be taken, by the effectorium, due to the wave of chemical changes induced by the internal signals (first step) and the posterior amplified wave produced by the overall autopoietic network.

3. Internal signals in an Autopoietic Network

A crucial aspect that needs to be determined, in order to have a better mathematical framework of living systems, regards the nature and functionalities of the *internal signals* that play a central role in our framework. As we have stated, every living system has a special subset of components whose (chemical, physical and mechanical) dynamics are modulated by interactions with the environment. At any single instant this *sensorium*, which is not confined to molecules found at the boundary (membrane) of the living system, is modulated by a complex pattern of perturbations coming from the "outside". The result of this instant-to-instant interaction is that many components of the autopoietic network change their concentration. Thus, we can accept that when an "external object" encounters the sensorium of a living system, a time-varying pattern of internal molecules change their concentration. Thus, the *internal signals* can be defined as the initial wave of molecules (or molecules classes) that change their concentrations due to the encounter of the living system with a specific circumstance in the "outside".

As the internal signals begin to interact with the autopoietic network secondary and tertiary changes occurs. These changes are the unavoidable side effects of the coupled network of chemical reactions -alongside their complicated kinetics– that implement metabolic closure.

IOP Conf. Series: Materials Science and Engineering **1321** (2024) 012014 doi:10.1088/1757-899X/1321/1/012014

It is important to emphasise that we are unable to quantity these changes, as the system is non-linear and thus sensitive to initial conditions. Nevertheless, what is clear is that after any encounter, the *internal signals* trigger a widespread avalanche of molecular effects. Some of these effects are very important as they modulate the selection mechanism that defines the next action to be taken. This mechanism could be framed in terms of a Conditional Probability Matrix (abridged CPM; see 5.1). Some important features about *internal signals* must be noticed.

First, the encounter between an organism and the "outside" is a complicated and complex interaction, and it is not a simple *key-lock* recognition mechanism. In general, every encounter triggers many families of internal signals, thus there is not a simple bijection between encounters and internal signals.

Second, the encounters produce complex continuous (or analogic) changes in the concentration of molecules, rather than simply triggering the sudden appearance of internal signals.

Third, and this is an important point, as autopoietic networks are historical networks with *memory*, identical encounters occurring at different times will not trigger the same pattern of internal signals.

Fourth, molecular networks with thousands of processes, are sensitive to unavoidable random effects as sometimes a process produces 5 molecules of a product and other times it can produce 10 (a 100% difference in output). This difference in concentration could appear as a detail for many of the metabolites found in the metabolism (i.e. glucose, ATP, amino acids, lipids...), but for regulatory molecules (for example, all the enzymes involved in protein-DNA interactions like the repressor of *Operon lac*) the copy number is very small and subjected to stochastic variations caused by the internal dynamics of the metabolism.

In summary, all these previous factors indicate that an appropriate representation of the $encounter \rightarrow internal - signal$ goes beyond a simple bona fide functional model, but that it must incorporate notions like *fuzzy sets* or *stochastic mappings*. In this respect, we think that a general framework is to use categorical notions to generate the mathematical landscape needed to analyse self-fabricating systems undergoing structural coupling.

4. The theory of categories and stochastic mappings

As a suitable mathematical framework to describe the action $\circlearrowleft$ perception helix every living system is engaged in when constructing the objects in its Umwelt, we will use the category StochFin of stochastic mappings between finite sets.

4.1. Categories and Functors: a primer

Recall first that the mathematical notion of a *category* was introduced by Eilenberg and McLane in 1945 in their work on natural equivalences [17], while attempting to pin down the difference between "natural" and not-so-natural isomorphisms in the context of vector spaces and linear mappings. More precisely, they wanted to fathom why the isomorphism between a (finite dimensional) vector space V and its bi-dual V^{**} (the dual of its dual) is "natural" while it is not so for the isomorphism between V and its dual V^* [2]. They noticed that "going to the dual" and "going to the bidual" were "functorial constructions" and to make the notion of "functor" precise they needed to introduce the notion of category first, as we introduce sets to be able to handle set mappings or functions in elementary contemporary mathematics.

Their definition of a category abstracted the facts that linear mappings can be visualized as arrows "going" from a *source* vector space to a *target* vector space, that can be *composed* (i.e. *"concatenated"*); their composition law is associative, and that we have for every vector space V, the Identity linear map from V to itself, which "does nothing".

[2] Recall that the dual V^* of a (real) vector space V is the vector space of all linear maps $w : V \longrightarrow \mathbb{R}$

So, an (abstract) category $\mathfrak{C}$ was defined as a a collection $\mathfrak{O}$ of objects and a collection $\mathfrak{A}$ of arrows, also called (homo)morphisms, denoted $A \xrightarrow{f} B$ or simply $f : A \longrightarrow B$ from a source object A to a target object B, such that arrows $A \xrightarrow{f} B$ and $B \xrightarrow{g} C$ compose ("head to tail") to form a new arrow $A \xrightarrow{g \circ f} C$, the arrow composition is associative, and every object A has an identity arrow Id_A which leaves unchanged under composition all incoming or outcoming arrows relative to A.

A bit more formally, in a category we have a source mapping s and a target mapping t from the collection $\mathfrak{A}$ of arrows to the collection $\mathfrak{O}$ of objects and a "composition map"

$$Hom(A, B) \times Hom(B, C) \longrightarrow Hom(A, C)$$
$$(f, g) \mapsto g \circ f,$$

where for any objects X, Y in the category, we denote by $Hom(X, Y)$ the set of all arrows (or homomorphism) f with source X and target Y.

Notice that we denote by $g \circ f$ and not by $f \circ g$ the composite arrow obtained by "concatenating" f and g, by applying first f and then g, following the common usage when composing mappings between sets.

The composition of arrows in a category is assumed to be associative $[\, h \circ (g \circ f) = (h \circ g) \circ f]$ and for every object A we have the identity arrow $Id_A \in Hom(A, A)$ with the property that

$$g \circ Id_A = Id_A \circ f$$

for all f such that $t(f) = A$ and all g such that $s(g) = A$.

Notice that a more basic example of a category than the category Vect of (finite dimensional) vector spaces and linear maps is the category Set of sets and set mappings.

We should nevertheless keep in mind that a category can also be intuited as the completion of a directed (multi)graph, where for every pair of composable (concatenable) edges the composed edge is also given as an edge in the graph.

More precisely, any directed (multi)graph Γ has an associated *path category* $\mathcal{P}(\Gamma)$ whose objects are the nodes of Γ and whose arrows are the (directed) paths (sequences of concatenated edges) in Γ. Here, the Identity arrow at a node A is the empty path at A which "never took off" from A.

We should keep in mind, side to side, the category Vect and the path category $\mathcal{P}(\Gamma)$ as basic examples of categories, one of a rather algebraic nature, where the objects are sets with a vector space structure ad the other, of a rather combinatorial - geometrical nature, whose objects are just nodes, with no visible "inner structure". Tle latter example reminds us of course of metabolic systems and metabolic pathways.

It has been said that in elementary contemporary mathematics we introduce sets to be able to accurately speak of mappings, functions or transformations. Analogously, we could say that we need to introduce the notion of category to be able to handle *functors*, which are arrows or morphisms between categories, so that are the arrows in the huge category Cat of categories and functors between categories.

A *functor* from one category to another, maps objects to objects and arrows to arrows in a coherent way, i.e. preserving composition. More formally, a *(covariant) functor* F from a category $\mathfrak{C}$ to a category $\mathfrak{D}$ consists of two mappings $F_{\mathfrak{O}} : \mathfrak{O}_{\mathfrak{C}} \longrightarrow \mathfrak{O}_{\mathfrak{D}}$ and $F_{\mathfrak{A}} : \mathfrak{A}_{\mathfrak{C}} \longrightarrow \mathfrak{A}_{\mathfrak{D}}$ such that $F(f \circ g) = F(f) \circ F(g)$ for all composable arrows f, g in $\mathfrak{C}$ and $F(Id_A) = Id_{F(A)}$ for all objects A in $\mathfrak{C}$.

A so-called *contravariant functor* reverses the order of composition, i.e. $F(f \circ g) = F(g) \circ F(f)$ for all composable arrows f, g.

Sending a vector space to its dual and linear mappings between vector spaces to their dual mappings defines a contravariant functor $\mathcal{D}$ from the category Vect of vector spaces and linear

$$\begin{array}{ccc} F(A) & \xrightarrow{\varphi(A)} & G(A) \\ {\scriptstyle F(f)}\downarrow & & \downarrow{\scriptstyle G(f)} \\ F(B) & \xrightarrow{\varphi(B)} & G(B) \end{array}$$

Figure 2. A Natural Transformation φ from F to G

mappings, to itself. Composing the functor $\mathcal{D}$ with itself defines a covariant functor $\mathcal{D}^2$ (the bidual functor) from the category Vect to itself.

Another paradigmatic example is provided by the functor $\mathcal{V}$ from the category Set to the category Vect, which sends a set S to the (real) vector space $\mathcal{V}(S)$ generated by S, which can be realized as the vector space of all real functions on S. Then a set mapping $f : S \longrightarrow T$ will be sent by $\mathcal{V}$ to the induced linear mapping $\mathcal{V}(f) : \mathcal{V}(T) \longrightarrow \mathcal{V}(S)$ which maps $\varphi : T \longrightarrow \mathbb{R}$ to $\varphi \circ f$.

Eilenberg and McLane [17] noticed that we have an "arrow" between the Identity functor Id and $\mathcal{D}^2$ but not so between Id and $\mathcal{D}$. Arrows, or morphisms, between functors, are called *Natural Transformations.* A Natural Transformation φ from a functor F to a functor G should coherently translate the action of the former into the action of the latter, acting on the objects of the source category of F. In diagrammatic language, the following diagram should commute (i.e. $G(f) \circ \varphi(A) = \varphi(B) \circ F(f)$) for all arrows $f : A \longrightarrow B$ in the source category of F and G.

In the case of $F = Id$ and $G = D^2$, we have that the $\varphi(A)$ mappings, which send $v \in V$ to $v^{**} : w \mapsto w(v)$ for $w \in V^*$, are isomorphisms of vector spaces, which means that φ is in fact a *Natural Isomorphism.*

Category theory may be seen as an embodiment of "relational mathematics" as opposed to "structuralist mathematics". In the former, what is in the foreground are the relations between mathematical objects, expressed notably through (homo)morphisms between them (which in the traditional setting where objects are structured sets, are structure-preserving mappings), in the latter though, attention is focused on the individual mathematical objects themselves, on their " internal structure"[18, 19] Indeed, from a categorical viewpoint, objects are defined by all their incoming and outcoming arrows in the category, not by some intrinsic, essential, characteristics, a bio-social outlook of sorts...[20]

Because of this, category theory seems to be a well-fitted tool for constructing mathematical models in relational biology. Moreover, category theory fosters diagrammatic and visual thinking, in mathematics and other sciences (e.g. provides elementary pictorial approaches to quantum mechanics [21]) and it is well adapted to describe processes, rather than just static structures. For instance the so called *monoidal categories* (formerly known as tensor categories) capture the intuition that processes can be run in *parallel* (tensor product), besides being *serially* composed (concatenated) [19, 22]

It is worth remarking that much water has flown under the categorical bridges since Eilenberg and McLane's pioneering work . It has been noticed that the arrows (morphisms) of a category may be not just mappings or transformations, but also braids, cobordisms of manifolds [23], ill-defined transformations, like the stochastic mappings below, also called channels [24], or processes like open Petri Nets [25] etc. A categorical approach to probability theory has been developed in terms of Markovian Categories [26]. Category theory has become a major hub in mathematics, enabling mathematicians to put under the same roof disparate notions and results, thus providing deep new insights, also in classical domains [18, 22, 19]

Moreover, higher order extensions of the notion of category have emerged. Recall that a category can also be (metonymically) intuited as a completion of a directed (multigraph), namely, as a "path category", where objects are points in a space and arrows (morphisms) are paths going from one point to another. This is a relentlessly one-dimensional structure though, and begs for two-dimensional, and higher dimensional analogues. Those are called nowadays 2-categories, and n-categories. In a 2-category we have objects, 1-morphisms between objects and 2-morphisms between 1-morphisms. We visualize 1-morphisms as (one-dimensional) arrows and 2-morphisms as two-dimensional spans, like a piece of cloth between two rods.

It is worth mentioning that our path category above provides a very natural example of a 2-category: The 2-morphisms between 1-morphisms (paths) are *homotopies*[3] which can in turn be composed in two ways, "vertically" and "horizontally". The former composition arises when one has homotopies $H : f \longrightarrow g$ and $K : g \longrightarrow h$, and the latter arises when one has homotopies $F : f \longrightarrow f'$ and $G : g \longrightarrow g'$ where $f, f' : x \longrightarrow y$ and $g, g' : y \longrightarrow z$

A fancier example of a 2-category is the category Cat of all categories, with functors as 1-morphisms and natural transformations as 2-morphisms [23]

Below (section 5) we will see that a 2-category emerges as a suitable mathematical framework to describe the dynamics of the perception ↻ action helix.

Last but not least, we witness nowadays an exponential development of a branch of Category Theory which has become known as Applied Category Theory, and goes beyond pure mathematics into science, informatics, computer science, systems theory and even industry [22, 25]

4.2. The Category StochFin of finite stochastic mappings

In what follows we will make use of the category StochFin of stochastic mappings acting on finite sets [26], introduced in 1962 by Lawvere in his handout notes [27] on "probabilistic mappings" and independently in 1984 by JSA [28] as "pseudotransformations", motivated by the study of random walks and brownian motions in discrete spaces. More precisely, to study the symmetric drunkard's walk in $E = \mathbb{Z} \times \mathbb{Z}$, in which the drunkard chooses at each "corner" (m, n) to walk non-stop until any of it nearest cardinal neighbors $(m+1, n), (m-1, n), (m, n+1), (m, n-1)$, equally likely, we may metaphorize hallucinating that the drunkard *splits* into four-fourths of a drunkard, each one walking one "block" in one of the four cardinal directions E, W, N, S. This suggests considering a "pseudo-mapping" φ [28] called *stochastic mapping* nowadays [26] which assigns to each point in E its four next neighbours with equal probability 1/4 In other words, φ sends each point in E to a (finitely supported) probability measure on E.

We take now E to be any finite set. Probability measures on E are also called *random points* of E and constitute a convex set called $\mathcal{R}(E)$. Notice that a probability measure on our finite E is just a function $\mu : E \longrightarrow [0, 1]$ whose values add up to 1. We call *support* of μ and denote by $Supp(\mu)$ the set of all $x \in E$ such that $\mu(x) > 0$. By identifying a point $x \in E$ with the Dirac measure δ_x at x, we can write a random point in E as

$$\sum_{x \in E} p_x \delta_x = \sum_{x \in E} p_x x,$$

where the non-negative weights p_x add up to 1.

A stochastic mapping φ from a finite set E to a finite set F may be then written as

$$\varphi(x) = \sum_{y \in F} p_x(y) y \in \mathcal{R}(F)$$

for all $x \in E$, where the non negative weights $p_x(y)$ $(y \in F)$ add up to 1. Notice that the weights $p_x(y)$ can be interpreted as the probabilities $P(\varphi(x) = y)$ and also as the conditional probabilities $P(\eta = y | \xi = x)$ where $\xi \in \mathcal{R}(E)$ and $\eta \in \mathcal{R}(F)$.

[3] i.e. deformations between paths, a notion which already makes sense in a (discrete) directed graph

Notice that φ extends naturally to a mapping, which we still denote φ, from $\mathcal{R}(E)$ to $\mathcal{R}(F)$ by the formula

$$\varphi(\sum_{x \in E} p_x x) = \sum_{x \in E} p_x \varphi(x)$$

We remark that a stochastic mapping $\varphi : E \longrightarrow F$ can be described as a $|F| \times |E|$ matrix, or finite kernel, M_φ given by $M_\varphi(y, x) = p_x(y)$ and then

$$\varphi(x) = \sum_{y \in F} M_\varphi(y, x) y$$

(matrix times vector).

Now, most importantly, stochastic maps can be composed in an obvious way. Since this composition is associative and we have the identities Id_E as stochastic mappings, we obtain the category StochFin whose objets are all finite sets are whose arrows are stochastic mappings.

Notice that then our original drunkard's random walk in $E = \mathbb{Z} \times \mathbb{Z}$ appears just as the sequence of iterates $\varphi, \varphi^2, ..., \varphi^n, ...$ of φ, fully analogous to a deterministic dynamical system in discrete time. Here we allowed E to be infinite, but we consider a stochastic mapping whose images are finitely supported. A fully finite example obtains if we consider the canonical (symmetric) random walk of a frog on a polygon of stones in a pond. Notice that in this case, we immediately guess that the $\lim_{n \to \infty} \varphi^n$ is the constant stochastic mapping equal to the uniform probability measure on the polygon.

It is also worthwhile remarking that stochastic mappings allow us to revisit Bayes's theorem. Indeed, the direct conditional probabilities in a Bayesian situation may be seen as defining a stochastic mapping φ(say from a set U of urns, filled with coloured balls, to a set C of colours, in the classical setting). Bayes theorem, taking as inputs the stochastic mapping φ and a "prior", i.e. a *random* urn $\mathbf{u}$ in U, provides a "converse" stochastic mapping $\varphi^\dagger$ going backwards, from C to U, often called the "Bayesian inverse" or "Bayesian adjoint", such that

$$\varphi^\dagger(\varphi(u)) = u.$$

So the categorical approach makes evident that we have a reversible Bayesian machine", which we can feed either way. Indeed, usual Bayesian updating appears now as follows: you have an *initial random urn* $\mathbf{u}$, you feed the Bayesian machine (consisting of φ and $\varphi^\dagger$) with a random colour $\mathbf{c}$ usually an extreme case: e.g. "black with probability 1" and different from the random colour $\varphi(\mathbf{u})$. Then the machine (via $\varphi^\dagger$) spits out a new random urn: $\varphi^\dagger(\mathbf{c})$, the Bayesian update of the original random urn $\mathbf{u}$, given the outcome $\mathbf{c}$. See Figure 3 for a toy example, with just two urns, containing black and white balls.

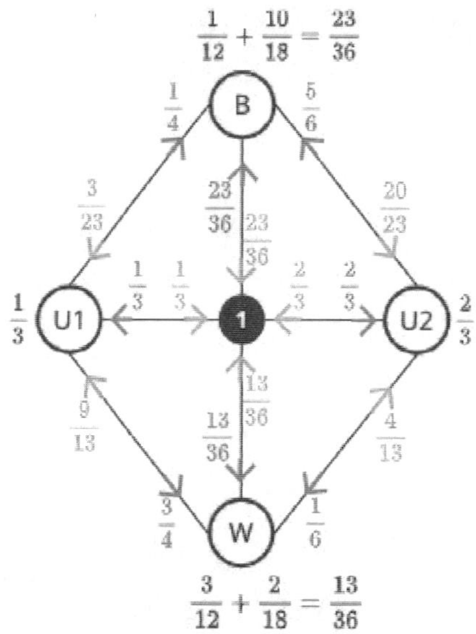

Figure 3. A toy example for Bayesian inference in terms of stochastic mappings: two bicolour urns. Urn U_1 contains 1 black (B) ball and 3 white (W) balls; urn U_2 contains 5 black balls and 1 white ball. To draw a ball at random, we choose urn U_1 with probability 1/3 and urn U_2 with probability 2/3. The stochastic mapping φ is given by the diagonal weighted blue arrows and $\varphi^\dagger$ is given by the diagonal weighted green arrows.The horizontal (resp. vertical) weighted arrows define a random urn (resp. colour).

Figure 3 shows that the image of the random urn $\mathbf{U}$: "U_1 with probability 1/3 and U_2 with probability 2/3" under the blue stochastic mapping φ is the random colour $\mathbf{C}$: "B with probability 23/36 and W with probability 13/36" and that, conversely, the image of $\mathbf{C}$ under the green stochastic mapping $\varphi^\dagger$ is $\mathbf{U}$.

Moreover, the the depicted flow in Figure 1 is stationary; in the case of nodes U_1 and W, for instance, this means that the amount of fluid $1/4 \cdot 1/3$ flowing from U_1 to B is equal to the amount $23/36 \cdot 3/23$ flowing from B to U_1. This is a particular instance of the essence of conditional probability:

$$\text{“}joint = conditional \cdot marginal''$$

More precisely,

$$P(A)P(B|A) = P(A \cap B) = P(B)P(A|B)$$

In this regard see [24], in whose categorical approach our stochastic mappings are called *channels*.

Notice that the flow diagram describing the stochastic mappings φ and $\varphi^\dagger$ in Figure 1 may be expressed in terms of the associated matrices M and $M^\dagger$ as follows:

$$\begin{bmatrix} 1/4 & 5/6 \\ 3/4 & 1/6 \end{bmatrix} \begin{bmatrix} 1/3 \\ 2/3 \end{bmatrix} = \begin{bmatrix} 23/36 \\ 13/36 \end{bmatrix}, \qquad \begin{bmatrix} 3/23 & 9/13 \\ 20/23 & 4/13 \end{bmatrix} \begin{bmatrix} 23/36 \\ 13/36 \end{bmatrix} = \begin{bmatrix} 1/3 \\ 2/3 \end{bmatrix}$$

In what follows, we will set up a Bayesian machine with a living system's internal signals instead of urns and actions instead of colours.

Now, the critical point, which we address below, is that we need to go beyond this Bayesian inference, to the meta or second-order level, aiming at describing how an *updating of the stochastic mapping φ itself* takes place.

5. A categorical approach to the action ↻ perception helix

After this, unavoidably long, introduction we have the main elements of our model. A living system $\mathfrak{L}$, is defined by a a huge finite set of possible internal signals $\mathcal{S}_{\mathfrak{L}}$ as well a as a finite set $A_{\mathfrak{L}}$ of possible actions. How these two sets are intertwined to define the always present perception ↻ action loop is the crux of the problem.

5.1. A finite discrete model for the internal signals of a living system $\mathcal{L}$

We assume that we have a fixed huge set $\mathcal{M}_{\mathfrak{L}}$ of molecules (m) (in practice, some metabolites (small molecules) and eventually peptides or proteins) associated to a living system $\mathfrak{L}$. As the living system interacts with its environment , each m in $\mathcal{M}_{\mathfrak{L}}$ is being endowed with a *finite discrete range* $C(m)$ of possible concentrations, where typically $|C(m)| \leq 6$. For instance,

$$C(m) = \{Null,\ \ Very\ Low,\ \ Low,\ \ Medium,\ \ High,\ \ Very\ High\}$$

We may label these concentration levels: $0, 1, 2, 3, 4, 5$. If we want to pin down what a *state* [4] of our living system $\mathfrak{L}$ is, with a (probabilistic) predictive power as to where the system is heading to, we need to take into account as " internal signals" not just the concentrations $c(m)$ of each molecule m (which takes the six possible values 0 to 5) but also the rate of concentration change $\dot{c}(m)$, which may assume takes values $0, \pm1, \pm2, \pm3, \pm4, \pm5$. We have then 2 degrees of freedom for the state $(c(m), \dot{c}(m))$ of each molecule m with 6 possibles values for the first one and 11 possible values for the second one. Globally, then, the digitisation of the concentration states of the molecules affords a discrete model with $(6 \times 11)^{|M|} = 66^{|M|}$ vectors which represent the possible internal signals (a phase space of sorts) of the living system.

We may then assume that in our discrete model for $\mathfrak{L}$ we have a representative *finite set* $\mathcal{S}$ of internal signals $s_1, ..., s_k$ with $k = 66^{|M|}$ Again we underline that the relation between the *encounters with the medium* and the internal signals in $\mathcal{S}$ is not a simple bijection, as every encounter triggers a complex spatio-temporal profile of internal signals that ultimately translates into many molecules changing their concentrations. Moreover, as autopoietic networks are historical networks with *memory*, identical encounters, occurring at different times, will not trigger the same pattern of internal signals.

Remark Notice that the global concentration change indicated in an internal signal may be expressed as a linear combination of single concentration changes, each one associated to a molecule $m \in M$. Indeed, call δ_m a unitary (positive) change of molecule m's concentration, which has the effect of increasing the concentration c_m of $m \in \mathcal{M}$ one step upwards in its concentration range; then any global concentration change $\dot{c}$ may be expressed as

$$\dot{c} = \sum_{m \in M} w_m \delta_m$$

where eventually most of the integer "weights" w_m will vanish and w_m will be negative if the concentration c_m decreases. From biological evidence, it is however clear that the ensuing actions of the living system do not depend linearly on the concentration changes: their dependence typically follows a sigmoid curve, and they cannot be retrieved from the actions triggered by the unitary changes δ_m.

To complete our discrete model, we assume further that a finite spectrum $\mathcal{A} = \{a_1, ..., a_n\}$ of executable actions is associated to our living system $\mathfrak{L}$ and that there is a well-defined conditional probability $P(a|s)$ of the living system choosing action $a \in \mathcal{A}$ when an internal signal $s \in \mathcal{S}$ arises.

[4] not just a configuration, which no "predictive power" as just the location in the case of a free particle

So, an $n \times k$ (n rows and k columns) conditional probability matrix (CPM) emerges, denoted M in what follows, whose entries M_{ij} are the conditional probabilities $P(a_i|s_j)$ that the living system $\mathfrak{L}$ chooses to execute action a_i given its internal signal s_j. The matrix M defines a *stochastic mapping* φ_M from the set $\mathcal{S}$ to the set $\mathcal{A}$, by

$$\varphi_M(s_j) = \sum_i M_{ij} a_i = \sum_i P(a_i|s_j) a_i$$

in other words, φ_M maps the internal signal s_j to the random action which executes action a_i with probability $P(a_i|s_j)$

Conversely, given a stochastic mapping $\varphi : \mathcal{S} \longrightarrow \mathcal{A}$, which may be seen as a bona fide mapping $\varphi : \mathcal{S} \longrightarrow \mathcal{R}(\mathcal{A})$, sending elements of S to random elements in $\mathcal{A}$, i. e. probability measures on $\mathcal{A}$ [5], we obtain a matrix M such that $\varphi = \varphi_M$ by putting

$$M_{ij} = (\varphi(s_j))(a_i).$$

So, in a nutshell, our finite discrete model is given by a set of internal signals $\mathcal{S}$, a set $\mathcal{A}$ of possible actions and a stochastic mapping $\varphi : \mathcal{S} \longrightarrow \mathcal{A}$. [6]

Notice that as soon as we have the stochastic mapping φ and an initial random internal signal $\mathbf{s} \in \mathcal{R}(S)$, we also have its Bayesian converse (also called inverse) stochastic mapping $\varphi^\dagger : \mathcal{A} \longrightarrow \mathcal{S}$ whose associated matrix entries $M^\dagger(j,i)$ can be interpreted as the conditional converse conditional probabilities $P(s_j|a_i)$, that given that the action a_i was executed by the living system, its "cause" had been the internal signal s_j. Diachronically, it is the probability that given the action a_i at time t, the internal signal s_j had arisen at time $t-1$. So, we have here a stochastic mapping "going backwards in time".

5.2. The perception ↻ action helix as a random walk

The Bayesian converse stochastic mapping $\varphi^\dagger$ should not be confused with the "dual" or "mirroring" stochastic mapping $\varphi^* : \mathcal{A} \longrightarrow \mathcal{S}$, whose associated matrix entries give the conditional probabilities $P^*(s_j|a_i)$ that action a_i having been executed at time t, the internal signal s_j arises at time $t+1$, so in a sense, action a_i "triggered" signal s_j. We have then a stochastic mapping "going forward in time".

Notice that in a sense, the stochastic mapping φ^* can be interpreted as "the φ of the medium where the organism lives", if we assume that actions of the organism $\mathfrak{L}$ trigger related internal signals of the medium $\mathfrak{M}$and vice versa. This intertwining of the Umwelt of the living system $\mathfrak{L}$ and the Umwelt of its medium $\mathfrak{M}$ can be made more precise if we assume that we have mappings $\alpha : \mathcal{A}_{\mathfrak{L}} \longrightarrow \mathcal{S}_{\mathfrak{M}}$ and $\beta : \mathcal{A}_{\mathfrak{M}} \longrightarrow \mathcal{S}_{\mathfrak{L}}$ relating the actions of $\mathfrak{L}$ to the internal signals of $\mathfrak{M}$ and conversely. See Figure 4, where we have added the subscripts $\mathfrak{L}$ and $\mathfrak{M}$ everywhere, to avoid confusion. Notice that now the stochastic mapping $\varphi^*_{\mathfrak{L}}$ appears as the composite $\beta \circ \varphi_{\mathfrak{M}} \circ \alpha$.

[5] which for our finite A are just functions $\mu : \mathcal{A} \longrightarrow [0,1]$, whose values add up to 1
[6] If the conditional probability matrix M is fixed, we will write just φ instead of φ_M.

IOP Conf. Series: Materials Science and Engineering **1321** (2024) 012014 doi:10.1088/1757-899X/1321/1/012014

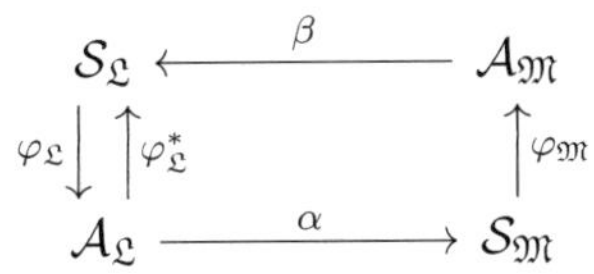

Figure 4. Intertwining of two Umwelten. We assume that that actions of the living system $\mathfrak{L}$ induce, via the stochastic mapping α, inner signals in its medium $\mathfrak{M}$. The latter has a helicoidal perception ↻ action dynamics of its own, embodied in the stochastic mapping $\varphi_{\mathfrak{M}}$. Conversely, we assume that actions of the medium $\mathfrak{M}$ induce inner signals in $\mathfrak{M}$ via a stochastic mapping β. The concatenation (composition) of α, $\varphi_{\mathfrak{M}}$ and β provides a stochastic mapping from actions to ensuing signals of $\mathfrak{L}$, which is called the dual of the stochastic mapping $\varphi_{\mathfrak{L}}$ and denoted $\varphi^*_{\mathfrak{L}}$

We write simply φ and φ^* for $\varphi_{\mathfrak{L}}$ and $\varphi^*_{\mathfrak{L}}$, respectively, in what follows. Notice that iteratively composing the stochastic mapping φ with its dual φ^*, we have a *random walk* which alternates between internal signal space $\mathcal{S}$ and action space $\mathcal{A}$. Indeed, φ sends internal signals (and more generally random internal signals) to random actions and then φ^* sends random actions back to random internal signals, which are sent to random actions by φ and so on...

To be more precise about this random walk, which intertwines in fact a random walk on signals with another one on actions, we need to take into account the temporal unfolding (diachrony) of the "object construction process" of our living system, via the infinite helix of *perception ↻ action* loops of Figure 1. We address this in the following subsections.

5.3. *The second order updating of the stochastic mapping φ.*

Recall that, as explained in subsection 4.2, Bayes' theorem calculates, from the given stochastic mapping $\varphi : \mathcal{S} \longrightarrow \mathcal{A}$, and an initial random internal signal $\mathbf{s} \in \mathcal{R}(S)$, the "converse" stochastic mapping $\varphi\dagger$, which goes backwards, from $\mathcal{A}$ to $\mathcal{S}$ and enables Bayesian inference about a random internal signal given any executed action a by the living system.

We need however to go further, to a "second order" updating, by describing how the choice of an action a by the living system entails an updating of the stochastic mapping φ itself, i.e. of the entries $P(a_i|s_j)$ of the corresponding conditional probability matrix M which defines φ. For this second order updating, a diachronic approach is required.

Initially, at time t_0, the *configuration* of the living system $\mathfrak{L}$ is given by a random element $\mathbf{s}$ of the finite set $\mathcal{S}$ (which may eventually be deterministic, i.e. $\mathbf{s} = s_j$ for some s_j with probability 1). To determine a *state* of $\mathfrak{L}$ we need to add a CPM, or a stochastic mapping φ to the data $\mathbf{s}$. The stochastic mapping φ being given, the living system chooses at time $t_0 + 1$ an action $a \in Supp(\varphi(\mathbf{s}))$, which may be seen as a "sampling" of $\varphi(\mathbf{s})$. Then the Bayesian "adjoint", or "converse"[7] stochastic mapping $\varphi^\dagger : \mathcal{A} \longrightarrow \mathcal{S}$ of φ, provides a "backwards" update $\varphi^\dagger(a)$ of $\mathbf{s}$. We are however interested in a "forward" update, of the stochastic mapping φ itself, triggered by the executed action a. We may look upon this update as a 2-morphism Ξ_a (in the sense of 2-categories, see section 4.1) associated to $a \in \mathcal{A}$, which sends $\varphi : \mathcal{S} \longrightarrow \mathcal{A}$ to a new stochastic mapping $\Xi(\varphi_a) : \mathcal{S} \longrightarrow \mathcal{A}$ with associated CPM denoted M_a We write then $\Xi_a : \varphi \Longrightarrow \varphi_a$

The underlying biological idea is that if the living system picks up an action $a \in A$, belonging to the support of $\varphi(\mathbf{s})$, this fact updates the given stochastic map $\varphi : \mathcal{S} \longrightarrow \mathcal{A}$,

[7] We will settle for the second denomination in what follows)

by means of some Hebbian rule [29], which increases the conditional probability $P(a|s)$ of choosing a for signals $s \in Supp(\varphi(\mathbf{s}))$, the increase being proportional to the weight of s in $\varphi^\dagger(a)$. Notice that this update extends immediately to any random action $\mathbf{a} \in \mathcal{R}(\mathcal{A})$ to provide a 2-morphism $\Psi_{\mathbf{a}}$.

To explain more precisely our model, we propose the toy example in Figure 5 below (adapted from [1]). We consider here a minimal organism interacting with its medium which can implement 3 different actions $\{a, b, c\}$ and can detect 4 different internal signal configurations $\{s_1, s_2, s_3, s_4\}$ [8]

If we observe that the internal signal s_2 arose and the organism executed c, then the conditional probability $P(c|s_2)$ must be updated by adding a positive ΔP to it, so as to reflect the increased strength (or effectiveness) of the association $s_2 \mapsto c$. But it may also happen that we only know that the organism executed action a after a random signal $\mathbf{s}$ arose. Since we have the organism's φ (or the associated CPM), the Bayesian converse stochastic mapping $\varphi^\dagger$ gives us an update of the weights of the involved s_j's (those in the support of $\mathbf{s}$). We should update then, by increasing the conditional probabilities $P(a|s_j)$ for those $s_j \in Supp(\mathbf{s})$, and we should increase more the probabilities $P(a|s_j)$ for those s_j for which the Bayesian "backward" conditional probabilities $P(s_j|a)$ are bigger! We see here that the Bayesian inverse $\varphi^\dagger$ (determined by φ *and* $\mathbf{s}$ also plays a role in the updating.

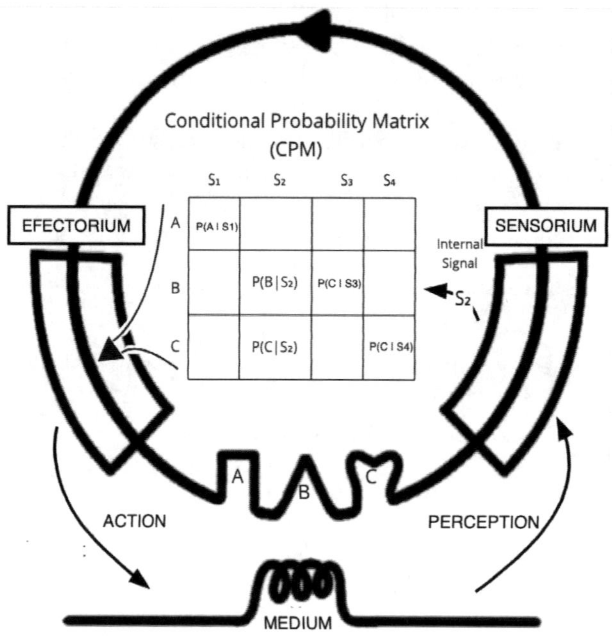

Figure 5. The Conditional Probability Matrix (CPM) of an organism. In the metabolic network of any living system, from bacteria to *Homo sapiens*, a Bayesian inference mechanism exists whose dynamics is driven by a *Conditional Probability Matrix* (CPM), denoted M. The (i, j) entry of M gives the probability that the organism executes action a_i given that the internal signal s_j arose in its metabolic network. The matrix M is not static, it evolves continuously according to the ongoing *Perception* ↻ *Action* helix

Thus, we could put forward a tentative simple evolution equation for this entry:

$$P(c|s_2)_{t+1} = P(c|s_2)_t + \Delta P_t$$

[8] All letters a, b, c, s_j, are upper case in Figure 4. Moreover, we usually denote specific actions by $a_1, a_2, a_3, \ldots$ in this paper, rather than $a, b, c, \ldots$

where ΔP_t represents the change in the probability $P(c|s_2)_t$ and can take various forms, such as

$$\Delta P_t \propto P(c) \cdot P(s_2),$$

somewhat akin to the Hebbian mechanism of synaptic plasticity [29].

If action c was effective, from the point of view of the organism, ΔP_t would be positive and, in that case, $P(c|s_2)_{t+1} > P(c|s_2)_t$ at the expense of one or more other *action/signal* couples, for which we would have had $P(action|signal)_{t+1} < P(action|signal)_t$.

Other rules for the variation of $P(c|s_2)_t$ can be put forth, but the crucial point is that the Conditional Probability Matrix M_t at time t dynamically changes moment by moment according to the specific helix of *action* ↻ *internal signal* couplings undergone by the organism.

In other words, the transition probabilities of the random walk arising from the iterated application of the stochastic mappings $\varphi, \varphi^*, \varphi, \varphi^*, \cdots$ (or the corresponding conditional probability matrices $M, M^*, M, M^*, \cdots$ if you prefer) are *not stationary.*

Indeed, if an internal signal s_j arises in the metabolism of our living system at time t, the ensuing action a_i at time t+1 is probabilistically afforded by applying the stochastic mapping φ (or equivalently the conditional probability matrix M) at time t. We should then write,

$$\varphi_t(s_j(t)) = \sum_i P(a_i(t+1)|s_j(t))a_i(t+1)$$

Therefore, the temporal unfolding of our *perception* ↻ *action* random walk, starting, say, at an initial signal vector $s_j(t_0)$, would be given by the iterated application of $\varphi_{t_0}, \varphi^*_{t_0+1}, \varphi_{t_0+2}, \varphi^*_{t_0+3}, \cdots$ to $s_j(t_0)$. Of course, we could see this random walk as the intertwining (like two threads of the same cord) of the "pure perception" random walk given by $\varphi^*_{t_0+1} \circ \varphi_{t_0}, \varphi^*_{t_0+3} \circ \varphi_{t_0+2}, \cdots$ [9] and the "pure action" random walk given by $\varphi_{t_0+2} \circ \varphi^*_{t_0+1}, \varphi_{t_0+4} \circ \varphi^*_{t_0+3}, \cdots$, although each of them only tells "half of the story".

Notice now that the updating of φ_t and φ^*_t (or, equivalently, M_t and M^*_t) in turn enacts a *random walk in the space of stochastic mappings* (or conditional probability matrices). This random walk "hovers", so to say, above our random walk which alternates between perception space $\mathcal{S}$ and action space $\mathcal{A}$. So, in fact we have a random walk with *randomly varying transition probabilities*, a "second-order" random walk of sorts!

Furthermore, as internal signals are also products of the normal action of self-fabrication, it happens that purely internal states (i.e. those not triggered by specific encounters between organism and niche) also trigger signal configurations that are treated by the organism as being generated by specific encounters. In other words, because of the many regulatory mechanisms as well as the complexity of the self-fabricating network, those internal states are treated as "external events" because the only relevant fact is the variation in the spectrum of internal signals. In this sense, we agree with Seth's claim that we "hallucinate reality" [30].

6. Metaphorising as a biological process

6.1. Metaphor in human culture and cognition

One spinoff of our proposed perception ↻ action helix by which a living system constructs the objects in its Umwelt is the understanding of metaphorising as a fundamental, ubiquitous and unavoidable biological process.

Notice, *en passant*, that we favour referring to *metaphorising* or *metaphorisation*, the *process* a cognitive agent engages into, instead of referring to *metaphor* as a *concept* supposedly standing "out there", independently of any observer. Recall that nowadays metaphorising is acknowledged as the often unknowing foundation of human cognition [31, 32, 33]. This counters Aristotle's initial description of metaphor [34] as just a "rhetorical trope". We posit

[9] Here "$\circ$" denotes as usual composition or concatenation of mappings, where $(g \circ f)(x) = g(f(x))$. Notice that $\varphi^*_{t_0+1} \circ \varphi_{t_0}$ its a mapping that goes from *Perceptions* to *Perceptions*, i.e. from $\mathcal{S}$ to itself.

IOP Conf. Series: Materials Science and Engineering **1321** (2024) 012014
doi:10.1088/1757-899X/1321/1/012014

instead that metaphorising is the "tip of the iceberg" of a fundamental biological cognitive process, which is language-mediated in the case of humans.

Research in cognitive neuroscience has indeed shown that the human brain processes metaphorical and literal versions of a concept in the same localisation, e. g. neural correlates of human subjects dealing with mental subtraction are the same as those arising when they track an object moving backwards; see the work of Dehaene and collaborators [35]. This matches the view that when we need to cope with new situations in our Umwelt, our metaphorising recruits neural mechanisms which evolved during previous recurrent instantiations of structural coupling.

Metaphorising plays a crucial role in a broad spectrum of realms in human culture, psychotherapy for instance [36, 37, 38], besides of course poetry and literature, but we are interested here in its role in mathematics and mathematics education, which has been progressively acknowledged only in the last decades [39, 40, 41, 42, 43, 44, 45, 46, 47].

Metaphors, indeed, turn out to be not simply rhetorical resources, but powerful cognitive means, helpful to recursively construct new concepts, which also provide skilful and friendly ways to solve problems. Notably, various top contemporary mathematicians bear withness of the relevance of metaphorising in their creative praxis [46, 43, 48, 49, 50]. From an enactivist perspective, metaphorising appears as an inborn way to bring forth a world or to fathom the unfathomable. Lakoff and Núñez [42] metaphorise conceptual metaphors as entailment-preserving mappings from a more familiar, or transparent, 'source domain' into a less familiar, or opaque, 'target domain',[10] which enable us to understand the latter with the help of the former. E. g. "a teacher is a gardener" maps gardening into education, enabling us to take advantage of our gardening know-how to fathom education, which is harder for us to grasp than gardening. For example: trees need propitious soil and enough water to thrive; seeds do not unfold when sown on a barren land. Trees cannot be forced to grow faster than nature allows, just by pulling it upwards.

The *poietic* role of "hallucinatory metaphors " is particularly interesting. When Aristotle described metaphor as "giving a thing a name that belongs to something else" [34], he did what was within his cultural and sociological reach: he discovered metaphor as a the *tip of the iceberg* of human cognition swimming in a sea of words. Nowadays, we acknowledge that metaphorising involves nonverbal, embodied, enactive, multimodal aspectsl

In our view it is more fruitful to metaphorically interpret nowadays metaphorising as "looking" at something and "seeing" something else, either consciously or unconsciously. In other words, metaphorising is "hallucinating", in the sense of Seth [30], for whom "reality" is a "controlled hallucination". Then, metaphorising enables us to fathom, grasp, make sense of the "objects" we construct via structural couplings by taking advantage of our previous experiences.

6.1.1. A paradigmatic stochastic mathematical example When we look at the random walk of a frog, which jumps symmetrically (as if tossing a coin to decide each time between right and left) on a row of stones in a pond, and we want to find out on which stone it will be crouching after a given number of jumps, to this end we *may eventually* "hallucinate", seeing the frog repeatedly *splitting* evenly into halves instead of jumping. Or, engaging in an animal-friendlier hallucination, we may see one litre of fruit juice being equally shared among next neighbours in a row of drinkers, starting with one of them initially holding the whole litre and each one of them splitting the amount of juice she gets, between her next neighbours. This allows us to *construct* the abstract notion of probability. Indeed, the probability of finding the frog on a given stone after n jumps can be constructed as the amount of fruit juice a specific drinker has after n sharings. Initially, the problems of the frog's location and the juice's distribution after a given number of steps appear unrelated, but metaphorically they

[10] In other words, as functors from an "inferential category" (with propositions as objects and inferences or entailments as arrows) into another, see section 4.1

can be regarded as the "same". Metaphor can then be seen as the use of previous experience (previous structural couplings) to construct our current experience, as suggested by Tall's felicitous pun: "a metaphor is a met-afore" [44].

6.1.2. The ubiquity of metaphor It is often argued that if we metaphorise all the time, and metaphor is ubiquitous, then metaphorising looses any explanatory power. But indeed this is exactly the point, we "swim in metaphor", but we are like "the fish who is the last one to see the water"[11]; we are the last one to see the metaphor. We often take our metaphors for hardcore realities, sometimes with dire long range consequences as in the case of the ubiquitous "teaching is transmitting knowledge" metaphor, or the longtime prevailing "the brain is an information processing device" metaphor discussed above. Interestingly, when teachers are confronted with the fact that they are *metaphorising* teaching as knowledge transmission in their practice, they usually retort: " But this is not a metaphor, teaching *is* transmitting knowledge, what else could it be? " Indeed, the teacher is the last one to see the metaphor... So, "metaphorical vigilance", akin to Bachelard's "epistemological vigilance", is badly needed nowadays.

6.2. Metaphor as a biological process
Now, we argue that from our enactivist viewpoint, metaphorising arises naturally as a biological cognitive process.

Indeed, the interaction dynamics between a organism and its Umwelt is not just a random sequence of perceptions $\circlearrowleft$ actions, because the encounters between organism and medium follow regularities arising from its ontogeny and phylogeny. The organism's time series of internal signals $(..., s_{t-1}, s_t, s_{t+1}, ..)$ includes indeed internal correlations which reflect the ongoing regularities in the interactions between the organism and its Umwelt.

Recall our example in [1], where we considered the two sequences of interactions (perception $\circlearrowleft$ action), of petting a cat (I_1) and petting a dog (I_2), which contain some correlations in common. We put forward here a more mathematical example, where instead of two pets to fondle we have two problems to tackle. To wit, we consider the problem of solving a simple linear equation, like $2x + 1 = 5$ and the problem of figuring out the weight of one green box (using the red bricks as units) in the balance in equilibrium illustrated in Figure 6. Children can solve the equation by a sequence of algebraic manipulations of symbols (I_1) or - metaphorising the equation as a balance in equilibrium - by an enactive sequence of manipulations of bricks in a two pan balance (I_2). Notice that the actual sequences will very likely involve trial and error (especially in the case of the balance: losing and recovering equilibrium).

[11] As the ancient Persian proverb goes.

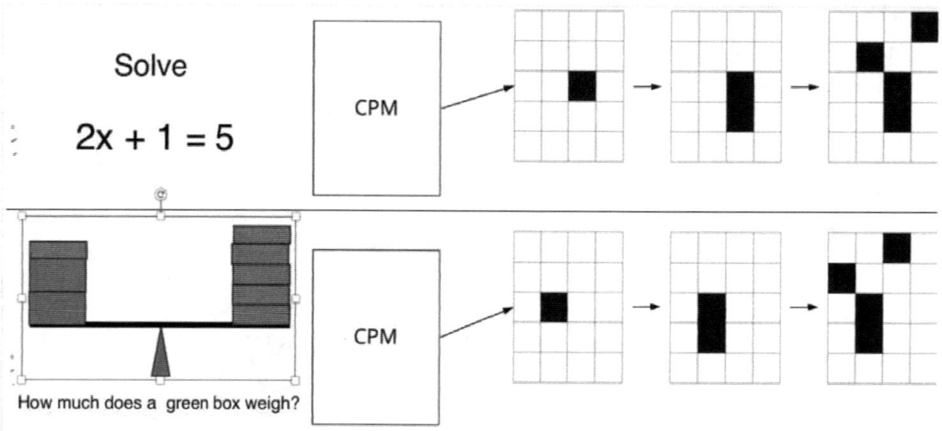

Figure 6. Metaphors and stochastic mappings. We consider two related sequences of *perception* ↻ *action* loops (upper-row: solving the equation $2x + 1 = 5$, lower-row: figuring out the weight of one green box in the scales). These two processes are analogous and can be considered each one as a metaphor of the other. As they unfold, the initial state of their corresponding CPMs is transformed according to different, but correlated, paths. In the end, the states of both CPM are different, but subtle correlations exist between both matrices (the two black square patterns are not identical but "homologous", where black squares indicate leading conditional probabilities) So, learning appears as a suitable updating of the organism's CPM (or stochastic mapping φ). This matches well Abrahamson's aphorism: "Learning is moving in new ways" [51]

The same applies to even more "abstract" circumstances, in the context of the learning of mathematics. Recall, for instance, the classical probabilistic *ruin problem* for two players [52] with initial "fortunes", say, of 3 and 2 euros. When tackling this problem, we might hallucinate that we have a frog jumping symmetrically along a row of six stones, numbered 0 to 5, in a tropical river, with the caveat that the end stones 0 and 5 are, in fact, alligator heads! Then, the initial fortunes of 3 euros and 2 euros of the players, correspond to the frog starting from stone number 3. The students wonder, of course, about the unfolding of the game or the fate of the frog. Each problem appears as a metaphor of the other. The frog's *random walk* may look more amenable for most, especially if they metaphorise it in turn as the repeated *splitting* of the frog (or sharing of a litre of juice) in halves with two "black holes" at the ends. The latter splitting process can, in turn, be metaphorised as the downwards *draining* of water through a suitable vertical network of ducts. Notice that the latter metaphorising corresponds faithfully to the *light cone* metaphor in special relativity, with the caveat that in our metaphor, time increases downwards instead of upwards as in relativity. This facilitates the emergence of *hydraulic metaphorising* (water draining downwards, because of gravity). See Figure 7 below and [53, 54] for a more detailed account of this chain of metaphorisations, including also a termodynamical metaphor...

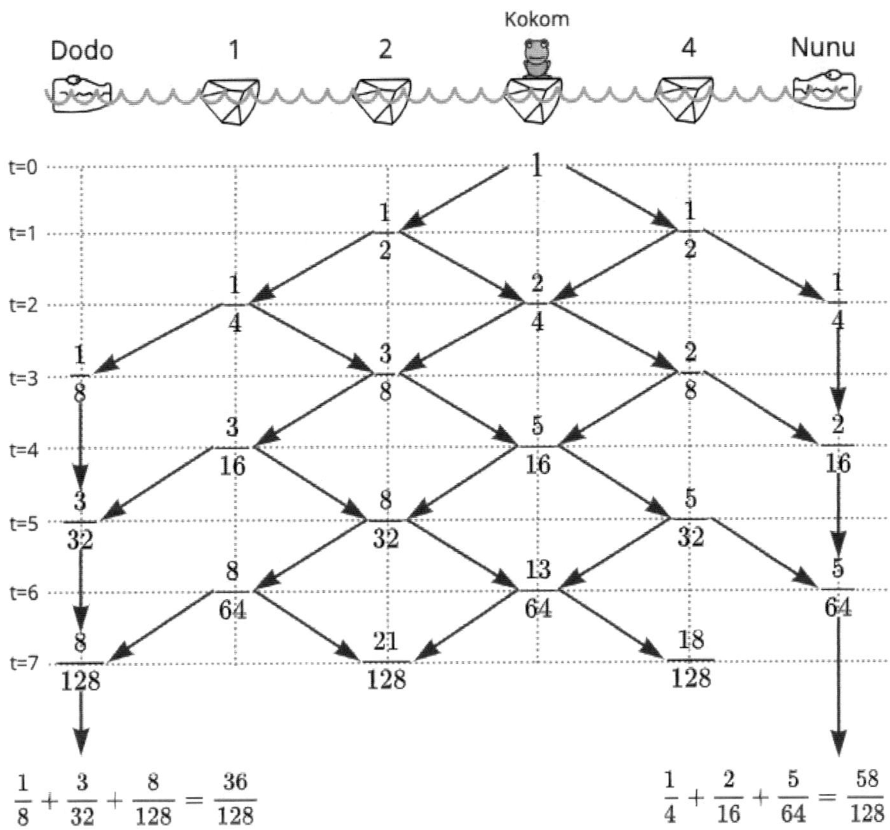

Figure 7. The random walk of the frog between two alligators (called Dodo and Nuni), a metaphor of the classical probabilistic ruin problem with two players, with inequal initial fortunes (3 and 2 euros in this case), metaphorised in turn as a water draining through a network of ducts with two sinks. We can see a truncated light cone unfolding downwards and we can already guess the fractions of the litre of water which will be "swallowed" by the two sinks in the long run. Notice the presence of Fibonacci numbers as denominators of the fractions recursively obtained. This example illustrates the emergence of a cascade, or chain, of metaphors. Metaphors in context seem to appear indeed more often serially concatenated than in parallel.

In terms of the notions introduced in section 4, we can model the various metaphorisations involved in this example as functors between suitable inferential categories. For instance, the metaphorising of a random walk as a splitting, or sharing, process (called a splitting metaphor in [54]) may be seen as a "splitting" or "sharing" functor, which quite faithfully translates objects and relations between the objects, from one context to another. E. g. the frog becomes one litre fruit juice, the stones become drinkers, the probability of finding the frog crouching on given stone after a given number of jumps (or to find it in one of the alligator's stomach) becomes the fraction of the litre juice hold by a given drinker after a given number of sharings (or the fraction of the litre swallowed by one of the extreme drinkers), the expected location of the frog becomes the barycentre of the collection of juice portions, etc.

7. Discussion and open questions: Structural Coupling and Metaphorising in the light of Bayesian Inference .

In this paper, we have presented some core ideas, aiming at developing a biology of cognition, to wit:

a) Living beings construct recursively the objects in their Umwelten via an extended Bayesian inference mechanism, embodied in a never ending perception ⟲ action helix. This mechanism is "non classical" in the sense that it infers about "non existing causes (objects)" which are constructed on the way ("laying down a path in walking...") as limits of recurive sequences of living organisms' internal signal

b) This Bayesian mechanism exhibits plasticity, according to a Hebbian-like rule, by means of which it feeds back into itself, like an Ouroboros [12] Indeed, actions executed by the living system feedback on its characteristic stochastic map φ from perceptions (inner signals) to actions, driving the organism helicoidal cognitive dynamics.

c) Metaphorising (in a much more broader sense than in Aristotle's narrow taxonomic approach) emerges as a natural, inborn and unavoidable cognitive mechanism, which is often, but not always, language - mediated in our species [36, 37, 38]. This understanding of metaphorising has deep implications for an experimental epistemology of mathematics aiming at researching its teaching and learning [54, 12]

We summarise our ensuing take-homes below.

First, Biology of Cognition considers *generation of action* (rather than analysis) as the central property of an organism. The biological function of an organism living in a medium is to move (to act) in a context (given by perception). This function *does not require* a nervous system, and it is valid for bacteria [56, 57], slime moulds [58, 59, 60], octopi, whales and humans.

The nervous system is only a specialised subsystem that permits the construction of thousands of objects (for a cat) instead than only a few (in the case of bacteria) [61]. Language, through metaphorising, increases the capability of object creation to many millions (Homo sapiens).

Second, the action generating mechanism is intrinsically stochastic, not a mere sleight of hand to hide our ignorance about the value of some systemic, and hidden, variables (as Laplace liked to claim).

Third, the input to our mechanism is not just a more or less accurate representation of an external world. The input is an interplay of the confrontation of the organism against its medium mediated by its own metabolism. This confrontation is always embedded in the task of self-fabrication. In some sense, the input could be fathomed as a filtering of the medium by the metabolic network. Thus, the input to the inference mechanism contains many elements pertaining to the internal rules produced by the self-fabricating metabolism and is not in a one to one correspondence with some "external world", thus the phenomenon more complex than just generating *blurred* image of the external world. Even more, most actions can be triggered by purely internal events without any input from an external, pre-existing, reality.

Fourth, through the never-ending helix of interplay between the mechanisms of Bayesian inference and Hebbian plasticity, a dynamics akin to a random walk arises. We have indeed a random walk generated by the aforementioned mechanism in the space of *Perceptions* and the associated random walk in the space of *Actions*, such that in some sense each one is a metaphor of the other. Although both walks are significantly random, they are strongly correlated to each other. Indeed, the corresponding sequences of perceptions and actions both converge to (the construction of) the same object in the Umwelt of the organism.

Fifth, metaphorising emerges naturally in the process of construction of objects by a living system via the unending perception ⟲ action helix and metaphors can be fathomed as functorial transformations which coherently "translate" from one inferential domain to another (preserving entailments, particularly). In everyday life living systems concatenate (compose) them and also deal with "convergent metaphors" aiming at a same object under construction. Moreover, natural transitions from one metaphor to another may be seen as natural transformations between the corresponding functors. We have exemplified this in the

[12] Recall that the Ouroboros - the snake eating its own tail - appears in the cover of first Chilean edition of [3], entitled "De máquinas y seres vivos" [55]

IOP Conf. Series: Materials Science and Engineering **1321** (2024) 012014 doi:10.1088/1757-899X/1321/1/012014

context of some mathematical problems (see the the ruin problem, the frog's random walk, the juice sharing, etc. in 6.2)

Sixth, internal metabolic states are handled by the organism as bona-fide "external object". The identification between internal states and objects (which are defined by recurrent perceptually guided actions) is crucial, as it is *the* mechanism we use to create, recursively, all sorts of abstract mathematical objects (number, function, Fermat last theorem, etc.). We put forward the following diagrammatic description of the processes involved:

$$(Internal\text{-}State_t \rightarrow Action_{t+1} \rightarrow Internal\text{-}State_{t+2}),$$

but since internal states are encapsulated as *Objects*, the above sequence could be written as an alternating sequence:

$$(Object_t \rightarrow Action_{t+1} \rightarrow Object_{t+2})$$

making clear that *Objects* are continuously created by the very fact that the organism handles internal states as external objects. Even more synthetically, we can rewrite this sequence as a functional (better, functorial) relation

$$Action(Objects) = Objects$$

However, Actions and Objects are essentially identified with each other, so we can write the overall process as a pair of coupled *self-referential* equations [55], which hints at the intertwining between cognition and action [62, 63]

$$Action(Action) = Action,$$
$$Perception(Perception) = Perception$$

The latter being entailed by the former because of the intertwining of perception and action, embodied in the recurring *perception* ↻ *action* helix.

Seventh, instead of sequentially considering the correlated random walks in perception space and in action space, as intertwined alternating random walks, we can fathom them simultaneously as a single random walk whose states are ordered pairs $(perception, action)$, so that $(perception(t), action(t+1))$ goes to $(perception(t+2), action(t+3))$, etc. This generates an encompassing random walk in the cartesian product of *Perception* space and *Action* space, whose transition matrix at time t is a $d \times d$ square block diagonal matrix, whose blocks are $CPM^*_{t+1} \cdot CPM_t$ and $CPM_{t+2} \cdot CPM^*_{t+1}$, with $d = dim(Perception) + dim(Action)$. The "2-dimensional" trajectories of this random walk provide a visualisation of our complex sequences of *perception* ↻ *action*. Metaphorisation appears then as similarity of these trajectories, as is the case of *Pet a Cat* and *Pet a Dog*, or the subsequent mathematical examples, in the toy example of Figure 5 (where $d = 4 + 5 = 9$).

In summary, we have put forth the backbone of a theory explaining how living systems construct the objects that populate their niche. Core ideas are that organisms are self-fabricating systems and that they live inside a never-ending helix relating perception to action. Furthermore, to select the action to be executed among a *behavioural repertoire,* each organism performs Bayesian inference on dynamic probabilities.

This viewpoint has the advantage of explaining how a circumstance (an object or a mathematical concept) can be viewed as the metaphor of another one. Mathematical cognition and thinking appears then as centred on the Umwelt organisms create, not as an abstract-symbolic-logic game but as the product of concrete self-fabricating living systems entangled with their Umwelt through Structural Coupling. Our perspective opens new avenues to build an organismic Embodied Artificial Intelligence based on actions rather than simple predictions gleaned from oceanic amount of data. But this artificial entity (a *robot* in normal parlance) would need to be implemented in quite different hardware that the current computers using silicon microchips running complicated "machine learning" algorithms that are only Singular Value matrix decomposition in really huge matrices. Current "intelligent" systems (like chatGTP) lack the fundamental aspect of structural coupling: they are not embodied in an *acting* body. These systems only look for subtle correlations in huge datasets they do not act and they do not care if they perform action A or action Z. We think that without a plastic metabolism, full of regulatory processes, it is impossible to achieve real (situated) intelligence. Current AI systems lack the necessary recursion. The main program of any LLM (Large Language Models) is not written by itself, it is the product of humans using tools from Linear Algebra and Computer Sciences.

Open questions and further research

Finally, among various open questions and directions of further research suggested by our approach, we can highlight a few:

Metaphors are usually concatenated (composed sequentially), but we also observe instances of convergent metaphors (from different rather concrete or familiar inferential domains to the same somewhat abstruse or ill-defined inferential domain) or conversely, divergent metaphors (same source domain, different target domains). In category theory, in the so called monoidal categories (like the category Vect of vector spaces and linear mappings) we have a "composition" of parallel arrows, with same source and same target (this is the tensor product of linear mappings, in Vect), besides the usual head to tail composition (concatenation). This suggests the questions:

Do we observe examples of parallel metaphors in the experimental epistemology of mathematics, which are activated simultaneously instead of sequentially? This would provide a "tensor product" of sorts in the category of (conceptual) metaphors. Notice that in the cultural mythological realm, an example could be provided by the Mesoamerican god Quetzalcoatl, who is a quetzal (a bird) and a rattlesnake at the same time... Also dream analysis in psychotherapy provides suggestive examples [64].

Observations in experimental epistemology of mathematics often reveal the activation of "constellations" of metaphorisations, which can be visualised as graphs, as a starting point. This suggests than an appropriate categorical notion to model this could be the notion limits, or colimits, of diagrams of arrows in a category [19, 22]. Here the arrows would be metaphors.

To fostering the development of Embodied Artificial Intelligence, it seems most commendable to further investigate non-human cognition, from octopii and dolphins to slime moulds and bacteria. For surprising recent findings we refer to slime moulds [58, 59, 60, 56, 57]

In this regard, the question arises whether other species metaphorise and how would they implement metaphorising, if they do? Octopii, for example? How could we experimentally check if this is the case? Maybe, in problem-solving situations?

Is AI already metaphorising, without us noticing it?

References

[1] Valdés-Zorrilla A, Letelier J C and Soto-Andrade J 2023 *IOP Conference Series: Materials Science and Engineering* **1292** 012012 ISSN 1757-899X

[2] Uexküll J V 1992 *Semiotica* **89** 319–391 ISSN 1613-3692

[3] Maturana H R and Varela F J 1980 *Autopoiesis and Cognition: The Realization of the Living* (*Boston Studies in the Philosophy and History of Science* vol 42) (Springer) ISBN 978-90-277-1016-1

[4] Cornish-Bowden A and Cárdenas M L 2020 *Biosystems* **188** 104063 ISSN 03032647

[5] Leduc S 1912 *La biologie synthétique* Leduc, Stéphane, 1853-. Etudes de biophysique (A. Poinat)

[6] Rosen R 1958 *The Bulletin of Mathematical Biophysics* **20** 317–341 ISSN 0007-4985, 1522-9602

[7] Rosen R 2005 *Life Itself: A Comprehensive Inquiry Into the Nature, Origin, and Fabrication of Life* (Columbia University Press) ISBN 978-0-231-07565-7

[8] Hughes J, Abdulali A, Hashem R and Iida F 2022 *IOP Conference Series: Materials Science and Engineering* **1261** 012001 ISSN 1757-8981, 1757-899X

[9] Girju R 2023 *IOP Conference Series: Materials Science and Engineering* **1292** 012020 ISSN 1757-8981, 1757-899X

[10] Clark A 2008 *Supersizing the Mind: Embodiment, Action, and Cognitive Extension* (Oxford University Press) ISBN 978-0-19-986885-8
[11] Macrine S L and Fugate J M B 2022 *Movement matters: how embodied cognition informs teaching and learning* (The MIT Press) ISBN 978-0-262-54348-4
[12] Hutto D D and Abrahamson D 2022 *Embodied, Enactive Education: Conservative versus Radical Approaches* (The MIT Press) p 39–52 ISBN 978-0-262-36899-5
[13] Uexküll J v 1920 *Theoretische biologie* (Paetel)
[14] Maturana H R 1970 *Biological Computer Laboratory Research Report BCL 9.0. Urbana IL: University of Illinois.*
[15] Varela F J, Rosch E and Thompson E 1991 *The embodied mind: cognitive science and human experience* (MIT Press) ISBN 978-0-262-22042-2
[16] Bich L 2018 *Robustness and Autonomy in Biological Systems: How Regulatory Mechanisms Enable Functional Integration, Complexity and Minimal Cognition Through the Action of Second-Order Control Constraints* (*History, Philosophy and Theory of the Life Sciences* vol 23) (Springer International Publishing) p 123–147 ISBN 978-3-030-01197-0 URL http://link.springer.com/10.1007/978-3-030-01198-7_6
[17] Eilenberg S and MacLane S 1945 *Transactions of the American Mathematical Society* **58** 231–294 ISSN 0002-9947
[18] Corry L 2004 *Modern algebra and the rise of mathematical structures* 2nd ed (Birkhäuser) ISBN 978-3-7643-7002-2
[19] Cheng E 2022 *The Joy of Abstraction: An Exploration of Math, Category Theory, and Life* (Cambridge University Press) ISBN 978-1-108-47722-2
[20] Rotman B 2012 *Theory, Culture Society* **29** 247–260
[21] Coecke B and Gogioso S 2023 *Quantum in Pictures: A New Way to Understand the Quantum World* (Cambridge Quantum) ISBN 978-1-73921-471-5
[22] Fong B and Spivak D I 2018 *Seven Sketches in Compositionality: An Invitation to Applied Category Theory* arXiv:1803.05316 (arXiv)
[23] Baez J C and Dolan J 1995 *Journal of Mathematical Physics* **36** 6073–6105 ISSN 0022-2488, 1089-7658
[24] Cho K and Jacobs B 2019 *Mathematical Structures in Computer Science* **29** 938–971 ISSN 0960-1295, 1469-8072
[25] Baez J C and Master J 2020 *Mathematical Structures in Computer Science* **30** 314–341 ISSN 0960-1295, 1469-8072
[26] Fritz T 2020 *Advances in Mathematics* **370** 107239 ISSN 00018708
[27] Lawvere F W 1962 The category of probabilistic mappings, ms.
[28] Soto-Andrade Jorge J 1984 Pseudo-transformations et mouvement browniens sur les espaces de gelfand discrets toulouse
[29] Hebb D O 2002 *The Organization of Behavior: A Neuropsychological Theory* (Psychology Press) ISBN 978-1-4106-1240-3
[30] Seth A 2021 *Being you: a new science of consciousness* (Faber Faber) ISBN 978-0-571-33770-5
[31] Lakoff G and Johnson M 1980 *Metaphors we live by* (University of Chicago Press) ISBN 978-0-226-46801-3
[32] Sfard A 2008 *Thinking as communicating: human development, the growth of discourses, and mathematizing* (Cambridge University Press) ISBN 978-0-521-86737-5
[33] Gibbs R W 2008 *The Cambridge handbook of metaphor and thought* (Cambridge University Press) ISBN 978-0-521-84106-1
[34] Aristotle 1984 *Poetics (trans: Bywater I)* (Princeton University Press) ISBN 978-1-4008-5276-5
[35] Knops A, Thirion B, Hubbard E M, Michel V and Dehaene S 2009 *Science* **324** 1583–1585 ISSN 0036-8075, 1095-9203
[36] Sharpe E F 1940 *The International Journal of Psychoanalysis* **21** 201–213 ISSN 1745-8315
[37] Katz S M 2013 *Metaphor and fields: Common ground, common language, and the future of psychoanalysis* Metaphor and fields: Common ground, common language, and the future of psychoanalysis (Routledge/Taylor Francis Group) ISBN 978-0-415-63171-6
[38] Degl'Innocenti B G 2021 *Access to the embodied unconscious through reverie and metaphor* (Routledge) ISBN 978-1-00-319807-9
[39] Abrahamson D, Gutiérrez J F and Baddorf A K 2012 *Mathematical Thinking and Learning* **14** 55–80 ISSN 1098-6065, 1532-7833
[40] Díaz-Rojas D, Soto-Andrade J and Videla-Reyes R 2021 *Constructivist Foundations* **16** 265–274 ISSN 1782-348X
[41] English L D 1997 *Mathematical reasoning: analogies, metaphors, and images [Didactics of mathematics: an experimental epistemology?]* (L. Erlbaum Associates)
[42] Lakoff G and Núñez R E 2000 *Where mathematics comes from: how the embodied mind brings mathematics into being* nachdr. ed (Basic Books) ISBN 978-0-465-03771-1
[43] Manin Y I 2007 *Mathematics as metaphor* (Amer. Math. Society) ISBN 978-0-8218-4331-4

[44] McGowen M A and Tall D O 2010 *The Journal of Mathematical Behavior* **29** 169–179 ISSN 0732-3123
[45] Sfard A 2009 *Metaphors in education* (Routledge) p 39–50 ISBN 978-1-136-60464-5
[46] Soto-Andrade J 2014 Metaphors in mathematics education
[47] Soto-Andrade J 2018 *Enactive metaphorising in the learning of mathematics* (Springer) p 619–638
[48] Thom R 1994 *Spectre bord d'un centre obscur* (ENS Éditions Fontenay-Saint Cloud.) p 13–24 ISBN 978-2-902126-09-5
[49] Thurston W 1994 *Bulletin of the American Mathematical Society* **30 (2)** 161–177
[50] Tao T 2016 Answer to "thinking and explaining" URL `https://mathoverflow.net/a/38882`
[51] Abrahamson D and Sánchez-García R 2016 *Journal of the Learning Sciences* **25** 203–239 ISSN 1050-8406, 1532-7809
[52] Chung K L 1979 *Elementary probability theory with stochastic processes* 3rd ed Undergraduate texts in mathematics (Springer-Verlag) ISBN 978-0-387-90362-0
[53] Soto-Andrade J, Diaz-Rojas D and Reyes-Santander P 2018 *Random walks in the didactics of probability: Enactive metaphoric learning sprouts* p 125–143 ISBN 978-3-319-72871-1
[54] Valdés-Zorrilla A, Díaz-Rojas D, Jiménez L and Soto-Andrade J 2023 *Constructivist Foundations* **18** 259–276 ISSN 1782-348X
[55] Soto-Andrade J, Jaramillo-Riveri S, Gutiérrez C and Letelier J 2011 Ouroboros avatars: A mathematical exploration of self-reference and metabolic closure *Advances in Artificial Life ECAL 2011: Proceedings of the Eleventh European Conference on the Synthesis and Simulation of Living Systems* ed Lenaerts T, Giacobini M, Bersini H, Bourgine P, Dorigo M and Doursat R (The MIT Press) p 763–770
[56] Lyon P 2015 *Frontiers in Microbiology* **6** ISSN 1664-302X
[57] Lyon P, Keijzer F, Arendt D and Levin M 2021 *Philosophical Transactions of the Royal Society B: Biological Sciences* **376** 20190750 ISSN 0962-8436, 1471-2970
[58] Dussutour A 2019 *Tout ce que vous avez toujours voulu savoir sur le blob sans jamais oser le demander* ISBN 978-2-290-16734-2
[59] Nakagaki T, Yamada H and Tóth A 2000 *Nature* **407** 470 ISSN 0028-0836
[60] Tero A, Takagi S, Saigusa T, Ito K, Bebber D P, Fricker M D, Yumiki K, Kobayashi R and Nakagaki T 2010 *Science* **327** 439–442 ISSN 0036-8075, 1095-9203
[61] Ciaunica A, Shmeleva E V and Levin M 2023 *Frontiers in Integrative Neuroscience* **17** ISSN 1662-5145
[62] Harvey I 2019 *Artificial Life* **25** 334–351 ISSN 1530-9185
[63] Bechtel W and Bich L 2021 *Philosophical Transactions of the Royal Society B: Biological Sciences* **376** 20190751
[64] Hartmann E 1996 *Dreaming* **6** 147–170 ISSN 1573-3351

Multiple Musical Skills Framework for Robotic Musicianship

Lawrence Schofield, Chapa Sirithunge, Arsen Abdulali and Fumiya Iida

Bio-Inspired Robotics Laboratory, Department of Engineering, University of Cambridge, Cambridge, UK
{ls866, csh66, aa2335 and fi224}@cam.ac.uk

Abstract. Why are musical robots distinguishable from humans? What makes them sound less 'musical' or 'human'? The many variables and high complexity of music make it difficult to answer this question. This paper sets out to formalise the many tasks and skills a musical robot has to carry out, and lists areas of unsolved problems that require new research. A potential overall robot information structural flow is provided on how separately created modules can work together to create a coherent musical robot. This was accompanied by a method for grading the ability of musical robots to evaluate their performance. This study attempted to depict how robotic musicians might perform in the future.

Keywords: Robotic musicianship· Musicality models · Piano-playing robots · human-robot interaction

1 INTRODUCTION

Music is a creative art form that acts as a universal language for emotion, allowing performers to express and induce emotions through the medium of sound. Playing musical instruments expressively alongside other musicians requires dexterous hand movements, subtle communication between musicians and emotional interpretation, and therefore provides a unique environment to extend and integrate state-of-the-art robotics technologies. Musical robots are commonly distinguishable from human musicians as they lack the various skills and mannerisms of human musicians. However, it is hard to identify exactly why musical robots do not sound similar to human musicians due to the extreme complexity of music. There is no existing catalogue of musical skills relevant to musical robotics as current breakdowns of musical abilities assume the musician is human. Musical robotics consists of two primary domains - musical mechatronics and machine musicianship. Musical mechatronics is concerned with creating physical systems that play musical instruments, such as the organ, violin and trumpet [1]. Machine musicianship focuses on computer algorithms and cognitive models of musical performance, theory, composition and perception [2, 3] and overlaps with music psychology research. It is desirable to unite the fields of musical mechatronics and machine musicianship under one framework, where the contributions of different research papers can be clearly categorised.

IOP Conf. Series: Materials Science and Engineering **1321** (2024) 012015 doi:10.1088/1757-899X/1321/1/012015

The concepts of expression and emotional musical performance can be hard to interface with robotics and computer science. The GERMS framework is a popular tool which considers five specific areas that create expressive musical performances [4]. These include: rules that clarify the musical structure; emotional expression; random variations within a live performance; motion and 'stylistic unexpectedness'. This framework can be used as a basis to develop expressive musical robots, using different controllers for each of the five areas [5]. The precise way in which emotions are encoded in musical elements [6], and how these are conveyed to the listener [7, 8] are further relationships that need to be modelled. The Desire framework is similar to the GERMS framework, suggesting that emotion is conveyed through four elements – speed, intensity, regularity and extent. The definitions of extent and intensity are unhelpfully vague, though the framework generalises to dance and speech as well as music. Other models of expression have been developed besides extensive literature [9, 10] on how music and emotion are intertwined.

There is not yet a holistic list of musical skills suitable for assessing a musical robot, though there are many music skill profiling tests which apply to humans. These include the PROMS test [11], Gordon's Advanced Measures of Music Audiation [12], the Musical Aptitude Profile [13], the Musical Ear Test [14], and various others assessments [15, 16]. The skills assessed within these tests have previously been categorised into different domains [17]. However, each of these frameworks assume that the musician is a human and hence can conduct a variety of tasks, from physically walking onto the stage to making sub-conscious changes to their playing which express emotions. A new framework is needed that also applies to robotic musicians, accounting for the explicit and implicit skills human musicians exhibit.

There is extensive research on how pianists play and learn piano [18, 19], the mechanical models of the kinematics of a piano key strike [20], and how pianist physical behaviours change depending on the parameters of the music. For instance, the way pianists strike repeated octaves varies with the frequency of the repetition [21]. Automated piano systems for musical entertainment initially included barrel-pianos player-pianos and automata. In the 1980s, the humanoid Tsukuba robot could sightread a piece written for organ and accompany a singer [22]. More recently, specialised pianos containing actuators on every key allow incredibly complex pieces to be played, or even humanoid 'speaking' to be achieved [23]. Complex dynamics of the embodied interactions between a human and a piano have been studied and computational models to generate music which sounds closer to a human have been implemented in [24–26]. These explore nature of human physical dexterity and adaptability and the effect of both intentional and unintentional information which serve as features to distinguish robots and humans. Hence introducing these variations in robotic control, the robot can sound more like a human player using models which control a piano-playing robot that incorporates both unintentional variations as well as temporal variations.

IOP Conf. Series: Materials Science and Engineering **1321** (2024) 012015 doi:10.1088/1757-899X/1321/1/012015

Hence it can be seen that various piano playing robots have been developed over time, though none of them resemble a human pianist. These robots commonly lacked the combined abilities to play a range of notes, to play piano keys with good precision and control, and the ability to move quietly and strike piano keys without producing a tapping sound. Hence this work presents a framework that holistically lists and categorises the many skills which human musicians exhibit. Its wide breadth covers the overlapping fields of music psychology, machine musicianship and musical mechatronics, allowing the framework to provide systematic and detailed assessments of the shortcomings of musical robots. The framework can also be used to identify new research areas and applications for musical robots, besides providing a more nuanced understanding of how humans produce music.

2 Multiple Musical Skills Framework

2.1 Framework Generation

A new framework is needed to categorise the various skills human musicians exhibit in order to reproduce these skills in musical robots. It should cover all aspects of producing music, encapsulating the fields of machine musicianship and musical mechatronics. Such a framework was developed based on existing literature and the personal experience of musicians, engineers, computer scientists and psychologists; their ideas were extended and integrated in focus groups.

2.2 Multiple Musical Skills Framework

Fig. 1 shows twelve core musical skills identified by the framework categorised into four groups - Musical Knowledge and Understanding, Physical Ability, Sensing and Communicating and Musical Decision Making. These four categories constitute all the activities a musician executes.

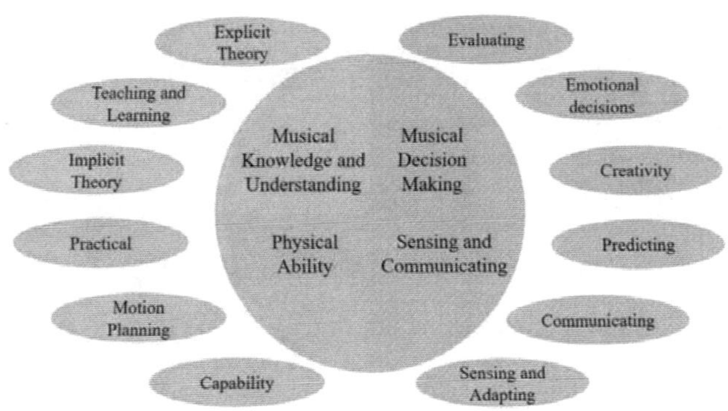

Fig. 1. Multiple Musical Skills Framework

Physical Ability is the main subject of research in music mechatronics, where musicians plan and execute physical movements to play musical instruments. The other three categories are explored in machine musicianship research. Musical Knowledge and Understanding is necessary to give, interpret, and act upon musical directions from sheet music and other musicians. Sensing and Communicating considers a musician's ability to respond to elements in their environment including sound waves, other musicians and audiences. Musical Decision Making regards the ability to be creative and express emotions through music, as well as the ability to make judgements within a musical context.

Each of the twelve core skills can be broken down into further sub-skills. For instance, understanding the musical terminology and notation of sheet music is one aspect of Explicit Theory. Likewise, a musicians temporal precision is one aspect of their Physical Capability. Fig. 2 displays a matrix of example sub-skills. This matrix allows new research contributions to be contextualised within the many skills musicians require.

2.3 Definition of skills

This framework considers live music generation and robotic musicianship in its entirety. Extensive literature reviews could be written on each sub-skill. Hence we provide comprehensive definitions and nuanced explanations for each skill listed in the framework for clarity. A brief overview of each skill is given below alongside suitable questions that indicate how the skill would apply to a musical robot.

- **Explicit Theory** This skill considers factual information within music knowledge and theory, such as the definitions of music terminology, or the composition date of a piece. For instance: does the robot understand music notation, terminology and theory? Does it know the typical chord progressions of Western Harmony, or rhythmical patterns of Central African Drumming?
- **Teaching and Learning** This skill considers how musical knowledge is transferred. Musicians share instructions and ideas using vocabulary, independent practice and learning by example. Musicians also learn how specific individuals play, react, teach and learn. For instance: can the robot adapt its actions to copy another musician's playing? Can it understand how to react if a conductor uses metaphorical language to convey musical instructions? Can it improve its playing or musical knowledge with practice?
- **Implicit Theory** This skill considers information within music knowledge and theory that is learned by humans by experience over time, rather than by studying textbooks or lectures. For instance: can the robot add expressive musical elements to the written sheet music similar to a human? Does the robot know how and when to phrase music? Does it have human-such as variations in tone and dynamic to avoid sounding synthetic?
- **Practical** This skill considers the varied practical and physical actions musicians execute outside of live musical performances. For instance: can the robot assemble, tune or repair its instrument? Can it adjust its position so it can see a conductor?

Sensing and Communicating			Physical Ability			Musical Decision Making			Musical Knowledge and Understanding		
Sensing and Adapting	Commun-icating	Predicting	Capability	Motion Planning	Practical	Evaluating	Emotional decisions	Creativity	Implicit Theory	Explicit Theory	Teaching and Learning
Audio Musical Elements & Patterns	Giving Cues	Cue occurrence	Force/Time Precision	Physical Patterns	Tuning	Interpretations and expression	Emotional cues and pattern integration	Unique performance	Phrasing	Patterns	By example
(Visual) Reading music, Cues	Interpreting cues	Chord progressions	Accuracy (repeatability)	Physical Pattern integration	Assemble Instrument	Technical ability	Emotional Gestures	Composing	Micro-patterns	Terminology	With language
Tactile/proprioception	Conflicting cue choice	Difficulties	Flexibility	Self-limit knowledge	Fix Instrument	Need Cue/Feedback	Interpret Emotion	Improvising	Variance and imperfections	Notation	Individuals understanding
Unique Instrument	Conducting	Acoustics	Self-limits	Instrument awareness	Stands, Mutes	Improvement Strategy	Composition Emotion	Personality and originality	Style	Context	Memory
Acoustics	Individual tailoring	Next notes..	Self-healing	Gesture and performance	Adjust seat height/position	Cue/feedback/ teaching quality	Technique choices test	Philosophy and analysis	Traditions	Global/Culture	Practice

Fig. 2. Types of musical skills identified in the framework

- **Motion planning** This skill is required to turn musical intentions into motor commands. For instance: is the robot aware of the limits of its actuators? Can it plan the timing and execution of physical gestures to enhance its performance? Can the robot learn and reuse physical patterns to play music?
- **Capability** This skill considers the physical and temporal prowess of the robot. For instance: how fast, and with what force, can a robot use its actuators? What is the geometric or temporal precision of its movements?
- **Sensing and Adapting** This skill considers the ability to respond to live music and the environment. For instance: can the robot identify what musical instruments are playing, which notes they are playing, and infer the tempo? Can it recognise cadences, read sheet music, or recognise if its instrument has physically shifted? Can it adapt to the acoustics of the room, the characteristics of a particular instrument, or change its tuning to match other musicians?
- **Communicating** This skill considers the interactions between musicians during live music production: instructions (or 'cues') are communicated both intentionally and non-intentionally. Cues can include physical gestures and emphasised musical elements; these might indicate a change in musical style, a new tempo, or an upcoming improvisation section. This skill is also relevant for interacting with an audience. For instance: can the robot identify, interpret, act upon or give musical cues when playing in an ensemble?
- **Predicting** Musicians constantly make and act upon predictions in all aspects of music generation. For instance: can the robot predict how or when a conductor will give cues? Can it predict what the acoustic behaviour of a room will be? Can it predict the next note or chord to be played?
- **Creativity** Music is a creative art form; the ability to generate new ideas, interpretations and techniques is vital to musicians. This skill is often pursued in machine musicianship algorithms which compose or improvise, though creativity can be applied more broadly than this. For instance: can the robot create a unique performance each time it plays a piece, generating new interpretations or including new techniques in its performance? Can it adopt and convey a particular personality when it performs?
- **Emotional Decisions** For humans, the connection between music and emotion is self-evident. The field of affective computing allows robots to interact with emotions too, which is vital to produce engaging performances. For instance: can the robot play music in a given emotional style? Can it infer emotions from other musicians playing? Can it change its style depending on the audience?
- **Evaluating** This skill considers the various decisions musicians make across all aspects of music making. For instance: can the robot assess its own technical playing, or the ability of other musicians? Can it evaluate whether the emotional interpretation of a recording is appropriate, or the quality of a teacher's instructions? Can it choose how best to improve its playing?

IOP Conf. Series: Materials Science and Engineering **1321** (2024) 012015
doi:10.1088/1757-899X/1321/1/012015

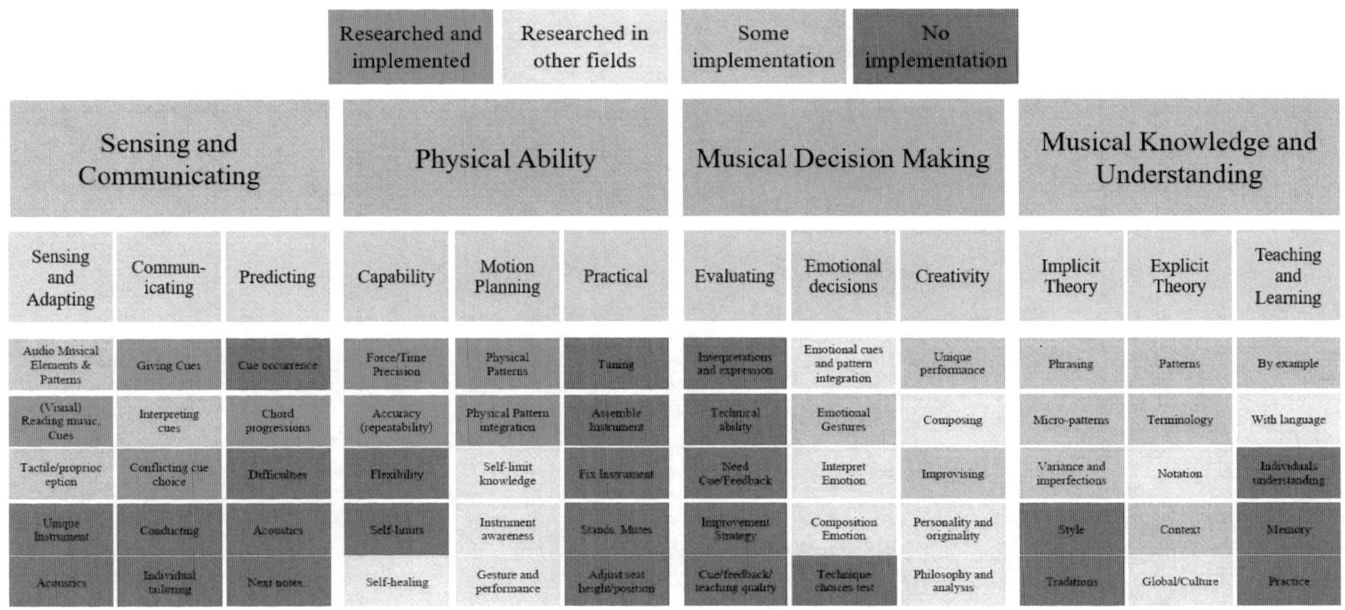

Fig. 3. Matrix showing various sub-skills which make up each musical skill

3 Framework tools and abilities

3.1 Matrix colour coding

This framework considers live music generation and robotic musicianship in its entirety. Providing comprehensive definitions and nuanced explanations for each skill listed in the framework is neither possible nor necessary within this report: extensive literature reviews could be written on each sub-skill. However, a brief overview of each skill is given below alongside suitable questions that indicate how the skill would apply to a musical robot.

A simple extension of the skills matrix given in Fig. 3 is to colour code the sub-skills according to some metric. For instance, a broad literature review enables the matrix to be colour coded according to which sub-skills have been implemented in musical robotics, which sub-skills are well researched in other fields such as music psychology, and which skills have not yet been researched or implemented. Fig. 3 highlights many sub-skills which have not been researched or implemented in musical robotics. For instance, the various practical subskills (eg: assembling or fixing musical instruments) have not been implemented, even though the automated fixing and transportation of instruments could be profitable. Thus this matrix can be used to identify new research areas.

Other sub-skills have been researched quite extensively. For instance, phrasing has been modelled by fitting polynomials to the expressive variation in dynamics or timing of the music [27, 28]. PCA analysis has been used to find commonalities between the ways musicians phrase music [29] in objective music analysis. Phrasing algorithms have used concatenate synthesis and reconstructive phrase modelling approaches [30] to mimic human phrasing patterns. Musical sensing and perception has also been well studied - many musical robots can identify pitch, [31] dynamic [32] and tempo [31], timbre [33], tone [34] and harmony [35]. More sophisticated softwares have accounted for complexities such as the change in instrument tone in different ranges [36]. Algorithms have allowed musical robots to track these musical elements whilst focusing on the current 'leader' of chamber ensemble [37]. There are many approaches to audio beat tracking and tempo estimation, which can be accurately achieved for drum patterns [38]. Audio cues can be combined with visual cues for greater temporal accuracy [39, 40]. Communication has also been explored in musical robotis - various musical robots which can interact with other musicians, using moving eyebrows, gazing, bobbing heads and visual cues to convey intent and emotion, giving more rounded performances and allowing performers to communicate whilst playing [41, 42]. Some robots combine this with the ability to improvise [43]. Alternatively, the matrix could be color coded to assess the skills a specific robot possesses, or to map out the sub-skills that might be implemented with machine learning. The matrix could also be used by human musicians to inspire conversation about the process of making music, considering where their own strengths and weaknesses lie.

3.2 Control Architectures and Hierarchies

The categorised sub-skills can be used to define control architectures for musical robots. For instance, a solo musician performing to an audience is likely to prioritise conveying some emotion to the audience to provide an engaging performance - the relevant sub-skills to identifying and expressing suitable emotions should occur early in robot's control processes so that they have a broad influence on the robot's playing. Fig. 4 shows one potential control architecture for a solo musician. The robot's choice of musical emotional cues are based on the context and style of the sheet music (relevant to Musical Knowledge and Understanding). The robot can then plan their physical motions (Physical Ability) to account for the emotional cues and gestures they wish to include (Musical Decision Making). By creating a unique performance, they can engage audience members who may have heard the robot before. As the robot plays, it will constantly listen to the music it produces (Sensing), and adapt its motion commands to keep the executed music close to the intended musical output. Its physical capability will dictate how well it can execute its motion commands, and thus the quality of the music being produced.

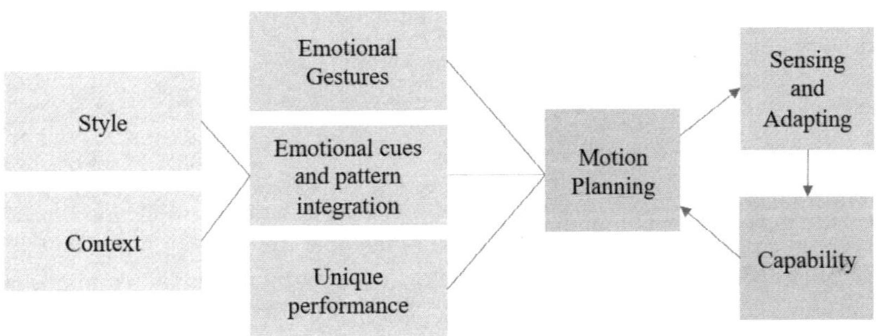

Fig. 4. A potential information flow between different sub-skills relevant to a solo performer, which prioritises conveying emotion to the audience.

Though a solo performer may prioritise conveying emotions to an audience, an orchestral performer requires different sub-skills to follow the commands of section leaders and the conductor whilst staying in time and tune with other orchestral players. Fig. 5 shows how this might affect the robot's control architecture: a robot playing in this setting must first be able to accurately assess the tempo, tuning, dynamic and various other musical parameters produced by its neighbours and the wider orchestra (Sensing). It should identify the emotional cues and musical intentions of the conductor and other players and evaluate the quality of the cues (Musical Decision Making). For instance, the temporal instructions of the conductor may be less clear than those given by the principal violinist. The robot can then choose appropriate motion commands to play in time with the rest of the orchestra, following the emotional directions given by the conductor.

IOP Conf. Series: Materials Science and Engineering **1321** (2024) 012015 doi:10.1088/1757-899X/1321/1/012015

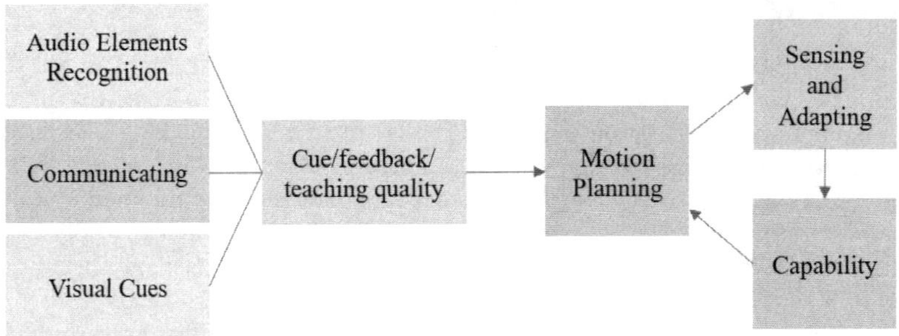

Fig. 5. A potential information flow between different sub-skills necessary for an orchestral performer, which prioritises following the commands of the conductor and section leaders whilst staying in time and tune with other orchestral players.

The control structures given above are very basic: a human orchestral musician will also consider emotional gestures (eg: the soloist) for instance. A sophisticated musical robot will exhibit numerous sub-skill and prioritise different skills according to the musical context. It does not need to exhibit all the identified sub-skills in the matrix to play music well. This is true for many professional musicians: orchestral musicians typically cannot improvise music, just as solo pianists do not need to transpose music.

3.3 Skill Interactions

Considering the interactions between different skills and sub-skills is another powerful tool, where each interaction is suitable for new research experiments. For instance, a musician can receive cues if they are out of tune (communication) and can make small tuning adjustments to improve an ensemble music performance according to their physical capability. This relationship between physical capability and communication is bi-directional, because their ability to understand visual cues will depend on the camera resolutions and image processing technology, which are elements of their physical capability. Many such interactions could be considered - the robot might learn how an individual gives visual cues and evaluate their quality. Each of these phenomena can inspire new musical robotics experiments which evaluate different approaches to each task.

4 Assessing Skills

Various forms of the musical Turing Test could be used to assess the ability of a musical robot. However, a more accurate and nuanced assessment of a robot's musical capability will be gained by testing each sub-skill individually, allowing a fingerprint of the robots capabilities to be determined. Some skills (eg: elements of physical capability) can be tested quantitatively by determining the numerical precision, speed and accuracy of the robot's movements. Alternatively, the robot could be assessed by playing standardised motifs that test different aspects of physical capability. This playing could be evaluated in Turing Tests or numerically compared to targets played by a human musician. This method requires time consuming experiments, but the test outcome has a clear musical meaning and context. This method also allows robots that play different instruments to be compared fairly.

Fig. 6 shows six motifs which could test the different physical capabilities of a humanoid piano-playing robot. Motifs 1a, 1b and 1c test the fluency of legato playing with increasing difficulty: 1b uses larger note intervals than 1a, and 1c requires more complicated finger movements to play smooth chordal scales. Thus a robot that successfully plays motif 1c is more advanced than one which can only play motif 1a. Motif 2 tests finger dexterity with a chromatic scale (the speed at which the robot can play this scale would indicate of its dexterity). Motif 3 tests the robots ability to play various extended techniques including tremelo and glissandi. The more extended techniques the robot can play, the greater its sophistication. By using multiple motifs similar to these, it would be possible to assess the physical sub-skills of the robot, as shown with a spider diagram on the right of Fig. 6. New music motifs could be analysed to determine if (and how fast) a robot with a given fingerprint could play them. Soft robot hands might have superior flexibility and hence demonstrate better dexterity, suitable for playing intricate scalic passages. Rigid fingers, however, may have better temporal accuracy, appropriate for playing a fast chordal accompaniment.

Some musical skills are simpler to test than others. Physical and perception skills are relatively easy to assess, whereas other skill tests require other human musicians to be involved in experimentation. Experiments on musical communication skills could compare the number of tempo cues a robot successfully identifies when playing in an ensemble compared to a professional musician. Likewise, the accurate communication of emotion can be quantitatively modelled and tested with the lens model [11]. Skills related to the robot's musical evaluations will need to be compared to professional human musician responses.

Though it is tempting to grade musical robots according to the number or variety of sub-skills they exhibit, a better approach might consider how holistic, adaptable and robust the combined skills are. For example, an advanced harp-playing robot that cannot adjust its position relative to the strings is useless if harp moves slightly. A robot that plays simpler music, but can (for example) adjust to the acoustics of a resonant cathedral or use basic physical gesture to engage an audience will give better musical performances. Just as AI algorithms and chat bots become more powerful when they are more flexible, such adaptable musical robots will change musical robotics from making robots that can play musical instruments to making robots that can play music.

5 Conclusions

In this paper, we introduce a comprehensive framework for multiple music skills, aimed at enhancing the capabilities of musical robots. While current musical

Fig. 6. Left: six motifs which test physical capabilities of a humanoid piano-playing robot. Motifs 1a, 1b and 1c test the fluency of legato playing with increasing difficulty. Motif 2 tests finger dexterity with a chromatic scale whereas motif 3 tests the robots ability to play various extended techniques. Right: an example fingerprint of a robot's physical sub-skills, indicating strengths and weaknesses on a spider diagram.

robots demonstrate some skills, they often lack the nuanced performance qualities of human musicians, such as dynamic tone and interactive abilities. Our framework not only highlights these gaps by comparing robotic and human musical performances but also proposes new sub-skills for robots. Implementing these skills could bridge the gap between robotic and human-like musical performances, leading to more authentic and unique robot musicianship. This framework serves as a foundation for future research, potentially guiding the evolution of musical robotics towards more human-like and creative performances.

References

1. Ajay Kapur. A history of robotic musical instruments. In ICMC, volume 10, page 4599, 2005.
2. Myounghoon Jeon. Robotic arts: Current practices, potentials, and implications. Multimodal Technologies and Interaction, 1(2):5, 2017.
3. Robert Rowe. Machine musicianship. MIT press, 2004.
4. Patrik N Juslin. Five facets of musical expression: A psychologist's perspective on music performance. Psychology of music, 31(3):273–302, 2003.
5. Angelica Lim, Takeshi Mizumoto, Toru Takahashi, Tetsuya Ogata, and Hiroshi G Okuno. Programming by playing and approaches for expressive robot performances. In IROS Workshop on Robots and Musical Expressions, 2010.
6. Roberto Bresin and Anders Friberg. Emotional coloring of computer-controlled music performances. Computer Music Journal, 24(4):44–63, 2000.
7. Roger A Kendall and Edward C Carterette. The communication of musical expression. Music perception, 8(2):129–163, 1990.
8. Patrik N Juslin. Cue utilization in communication of emotion in music performance: Relating performance to perception. Journal of Experimental Psychology: Human perception and performance, 26(6):1797, 2000.

9. Neil P McAngus Todd. The dynamics of dynamics: A model of musical expression. The Journal of the Acoustical Society of America, 91(6):3540–3550, 1992.
10. Alf Gabrielsson and Erik Lindstrom. The role of structure in the musical expression of emotions. ¨ Handbook of music and emotion: Theory, research, applications, 367400:367–44, 2010.
11. Lily NC Law and Marcel Zentner. Assessing musical abilities objectively: Construction and validation of the profile of music perception skills. PloS one, 7(12):e52508, 2012.
12. Edwin Gordon. Advanced measures of music audiation. Gia Publications, 1989.
13. Edwin Gordon. The musical aptitude profile. Music educators journal, 53(6):52–54, 1967.
14. Mikkel Wallentin, Andreas Højlund Nielsen, Morten Friis-Olivarius, Christian Vuust, and Peter Vuust. The musical ear test, a new reliable test for measuring musical competence. Learning and Individual Differences, 20(3):188–196, 2010.
15. Carl E Seashore, Don Lewis, and Joseph G Saetveit. Manual of instructions and interpretations for the Seashore measures of musical talents. Educational Department, RCA Victor Division, 1939.
16. Herbert Daniel Wing. Tests of musical ability and appreciation: An investigation into the measurement, distribution, and development of musical capacity, volume 27. University Press, 1948.
17. Susan Hallam and Vanessa Prince. Conceptions of musical ability. Research Studies in Music Education, 20(1):2–22, 2003.
18. Otto Ortmann. The physiological mechanics of piano technique. 1929.
19. Malwine Bree and Seymour Bernstein. The Leschetizky method: A guide to fine and correct piano playing. Courier Corporation, 1997.
20. R Brent Gillespie, Bo Yu, Robert Grijalva, and Shorya Awtar. Characterizing the feel of the piano action. Computer Music Journal, 35(1):43–57, 2011.
21. Bruce A Kay, Michael T Turvey, and Onno G Meijer. An early oscillator model: studies on the biodynamics of the piano strike (bernstein & popova, 1930). MOTOR CONTROL-CHAMPAIGN-, 7(1):1–45, 2003.
22. Curtis Roads. The tsukuba musical robot. Computer music journal, 10(2):39–43, 1986.
23. Alyssa Michaud. After the music box: A history of automation in real-time musical performance. 2020.
24. Schofield, L., Abdulali, A., Sirithunge, C., Iida, F. Modeling keystroke magnitude and tempo for realistic robotic piano-playing. (under review)
25. Scimeca, Luca, Cheryn Ng, and Fumiya Iida. Gaussian process inference modelling of dynamic robot control for expressive piano playing. Plos one 15(8),(2020).
26. Hughes, J. A. E., P. Maiolino, and Fumiya Iida. An anthropomorphic soft skeleton hand exploiting conditional models for piano playing. Science Robotics 3(25). 2018.
27. Daniel Mullensiefen and Geraint Wiggins. ¨ Polynomial functions as a representation of melodic phrase contour. na, 2011.
28. Asmir Tobudic and Gerhard Widmer. Playing mozart phrase by phrase. In Case-Based Reasoning Research and Development: 5th International Conference on Case-Based Reasoning, ICCBR 2003 Trondheim, Norway, June 23–26, 2003 Proceedings 5, pages 552–566. Springer, 2003.
29. Bruno H Repp. Diversity and commonality in music performance: An analysis of timing microstructure in schumann's "traumerei". ¨ The Journal of the Acoustical Society of America, 92(5):2546–2568, 1992.
30. Eric Lindemann. Music synthesis with reconstructive phrase modeling. IEEE Signal Processing Magazine, 24(2):80–91, 2007.

31. David K Grunberg, Alyssa M Batula, and Youngmoo E Kim. Towards the development of robot musical audition. In Proc. of the Music, Mind, and Invention Workshop (MMI), 2012.
32. Ye Pan, Min-Gyu Kim, and Kenji Suzuki. A robot musician interacting with a human partner through initiative exchange. In NIME, pages 166–169. Citeseer, 2010.
33. Tetsuro Kitahara, Masataka Goto, and Hiroshi G Okuno. Pitch-dependent identification of musical instrument sounds. Applied Intelligence, 23(3):267–275, 2005.
34. Koji Shibuya and Hiroyuki Ishimoto. Design principles of loudness to express bright and dark timbres for violin-playing robot. In 2018 27th IEEE International Symposium on Robot and Human Interactive Communication (RO-MAN), pages 262–267. IEEE, 2018.
35. Klaus Petersen, Jorge Solis, and Atsuo Takanishi. Development of a aural real-time rhythmical and harmonic tracking to enable the musical interaction with the waseda flutist robot. In 2009 IEEE/RSJ International Conference on Intelligent Robots and Systems, pages 2303–2308. IEEE, 2009.
36. Guy Hoffman and Gil Weinberg. Interactive improvisation with a robotic marimba player. Autonomous Robots, 31:133–153, 2011.
37. Sutirtha Chakraborty, Shyam Kishor, SN Patil, and Joseph Timoney. Leaderstem-a lstm model for dynamic leader identification within musical streams. In Proceedings of the joint conference on AI music creativity. AIMC, Zenodo, Stockholm, Sweden, volume 6, 2020.
38. G. Weinberg, S. Driscoll, and M. Parry. Musical interactions with a perceptual robotic percussionist. In ROMAN 2005. IEEE International Workshop on Robot and Human Interactive Communication, 2005., pages 456–461, 2005.
39. Angelica Lim, Takeshi Mizumoto, Louis-Kenzo Cahier, Takuma Otsuka, Toru Takahashi, Kazunori Komatani, Tetsuya Ogata, and Hiroshi G. Okuno. Robot musical accompaniment: integrating audio and visual cues for real-time synchronization with a human flutist. In 2010 IEEE/RSJ International Conference on Intelligent Robots and Systems, pages 1964–1969, 2010.
40. Dan Overholt, John Thompson, Lance Putnam, Bo Bell, Jim Kleban, Bob Sturm, and JoAnn KucheraMorin. A multimodal system for gesture recognition in interactive music performance. Computer Music Journal, 33(4):69–82, 2009.
41. Guy Hoffman and Gil Weinberg. Synchronization in human-robot musicianship. In 19th International Symposium in Robot and Human Interactive Communication, pages 718–724. IEEE, 2010.
42. Guangyu Xia, Mao Kawai, Kei Matsuki, Mutian Fu, Sarah Cosentino, Gabriele Trovato, Roger Dannenberg, Salvatore Sessa, and Atsuo Takanishi. Expressive humanoid robot for automatic accompaniment. In Proceedings of the Sound and Music Computing Conference, 2016.
43. Gil Weinberg, Aparna Raman, and Trishul Mallikarjuna. Interactive jamming with shimon: a social robotic musician. In Proceedings of the 4th ACM/IEEE international conference on Human robot interaction, pages 233–234, 2009.

Experience-based Learning and Adaptive Behavior for Improved Situation-Awareness of a Social Robot

Chapa Sirithunge[1], A. G. Buddhika P. Jayasekara[2] and D. P. Chandima[2]

[1] Bio-Inspired Robotics Laboratory, Department of Engineering, University of Cambridge, Cambridge CB2 1PZ, UK
csh66@cam.ac.uk
[2] Intelligent Service Robotics Group, Department of Electrical Engineering, University of Moratuwa, Katubedda 10400, Sri Lanka
{buddhikaj and chandimadp}@uom.lk

Abstract. When robots work in close contact with their human counterparts, they have to perceive their surroundings adequately. Hence have adequate situation-awareness. In such cases, an encounter will include the robot itself, the surrounding and the humans. Existing frameworks to interpret such situations, considered a limited number of observable human cues and inputs from the surrounding so far. Therefore only a few aspects of a situation were effectively perceived by these mechanisms. In our work, we trained a Deep Q-Network (DQN) which takes a number of attributes in a situation as input before a robot makes decisions regarding an interaction in a domestic environment. Large number of situations were simulated and trained, so that adaptive responses were generated through the model for different individuals a robot will encounter. Hence an experience-based, adaptive cognitive framework for robots has been developed for initiating an interaction with their humans. This creates situation-awareness in a robot and enables user-centered perception, demolishing the requirement of the human to adopt a restricted behavior to accommodate the robot's limited perceptive skills. Human acceptance towards the system has been evaluated through a set of experiments conducted over a period of time. Critical observations during the experiments have been highlighted.

Keywords: Adaptive systems · Social Robots · Human-Centered Robots · Human-Robot Interaction (HRI)

1 INTRODUCTION

In order to broaden the applications of AI, we need situation-aware robots which can theoretically come up with solutions in even unfamiliar situations. A general theory of behavior adoption is required to cope in such scenarios. Machine autonomy can be converted into a less stressful but useful with the ability of perceptive intelligence [1]. Human judgment upon robots and robot's judgment upon humans can be taken to a successful level of cooperation through such an approach. Grounding of abstract concepts in social domains is important for robots in this regard [2].

Situation awareness (SA), a person's understanding of what is happening in the current situation, is critical for performance in a wide variety of domains [3]. SA can be thought of as an internalized mental model of the current state of an individual's environment. All of the incoming data from various sensors and systems, the outside environment, fellow team members, and others must be brought together and integrated. For instance, in HRI, the robot, humans, objects, surrounding space, social norms, regulations, sensors, ethics related to interaction make up the elements in a human-robot domain. However, the role of SA is to finally assist in decision making. Most common areas in which situation-awareness is critical can be listed as warfare, self-driving vehicles, rescue, health care etc [4–6]. The most common applications of situation-aware robots can be found in domestic and social environments, shopping malls, museums, elderly and child care [7–9].

In order to perceive an encounter, a robot needs an appropriate model to analyze visual cues received from its user and the surrounding. This is similar to a Theory of Mind (ToM) which has been developed in order put various attributes in mind together, to reason out a certain behavior and evaluate various behaviors [10, 11]. Such approaches still estimate a user's preferences based on a limited number of cues from its environment. Hence an adequate evaluation of an encounter is critical for a robot's emotional intelligence.

Humans attribute minds to agents which are able to generate a variety of behaviors and use mixed strategies (both mind-reading and behavior-reading) against them when they produce deceptive behavior in a competitive situation [12]. Hence it is important for social robots to adopt human-like behaviors rather than a predefined set of instructions on its actions and inputs. According to psychology, physical movements in humans are important in deciphering their internal states [13, 14]. However such methods must use more attributes from the environment, humans and robots in addition to shadows associated with surrounding objects. A human-robot collaborative reinforcement learning algorithm is proposed to optimize the task sequence allocation scheme in assembly processes for task allocation and sequence optimization [15]. This system consists of an assembly of an alternator where the robot and the human will complete different stages of the assembly task. with the designed algorithm, the robot prefers to perform time-consuming but relatively difficult tasks while minimizing the overall time of the task. However, a collaborative robot deployed in a domestic environment will require more emotional intelligence and related observable human cues to work closely with humans.

A spatial conduct for the robot throughout a cooperative task performed in a home-like environment was developed during the approach explained in [16]. This spatial management behavior was capable of active monitoring and dynamically reacting to the change in each others' movement and position. This approach used the definitions of proxemics introduced by Hall [17] for interpersonal distancing and Kendon's F-formation arrangement for the orientation between two persons [18]. The robot was intended to learn and find objects that were missing from its original location although the robot had to follow users who were willing to show the objects. A robot distributes flyers to pedestrians by focusing on achieving the target by identifying the behavior of pedestrians and determining appropriate approaching mechanisms as in [19]. Approach direction was computed after a detailed observation on actual scenarios involving human-human interactions. In conclusion, a "Need-Driven-Innovation" is required to understand and adapt to various communities, and age and gender groups, to secure a functional and valuable technology-fit for the user when implementing robots [20].

However existing frameworks to interpret human behavior consider only a limited number of observable human cues. Therefore only a few aspects of a situation were effectively perceived by these mechanisms. On the other hand, establishing a common criteria to assess a situation, considering all the aspects: human, robot and environment, is still confusing and lacks conceptual basis. Therefore, through this work, we will present a cognitive model for a social robot which can perceive its environment based on several cues, learn and adapt. These cues will not only be extracted from humans but also from the surrounding and the robot itself. In this work, the decisions of the robot include whether to initiate an interaction with its user and if an interaction is possible according to the situation-awareness of the robot, how to approach the user and orient accordingly and how a conversation should flow.

The contributions of this paper are summarized below:

- A number of cues from a human, the robot and its environment have been observed and pre-trained in the dataset to support the robot's cognition
- The behavior of robots with the learning ability is adopted for the smooth flow of robot-initiated interactions and no longer need humans to initiate interactions.
- Experiments were conducted over a long period of time to provide adequate time for the robot to learn and users to adapt.

2 A Social Robot with Experience and Learning

2.1 Robot's Attention

In order to perceive an encounter well, a robot needs an appropriate model to analyze the cues received from its user and the surrounding. Hence an adequate evaluation of an encounter is critical for its emotional intelligence. Furthermore robots with adaptive memory, emotional adaptation, ability of perceiving verbal and nonverbal human gestures is crucial in successful flow of interaction in human-robot domains [21]. Perceiving individual differences in users' personalities and the nature of performing activities is important in achieving intelligent

IOP Conf. Series: Materials Science and Engineering **1321** (2024) 012016 doi:10.1088/1757-899X/1321/1/012016

proactive behavior of a robot. Besides, there are dynamic changes in the environment such as the placement of objects, time of the day etc. Hence, we improved our cognitive model into adaptive approaches which are able to perceive such dynamic changes in a context in the end. Hence our model is capable of catering to the above requirements of a human-robot encounter.

2.2 Perception of user behavior

To perceive the requirements of an individual, or an encounter (between a robot and a human) in general, a robot will have to understand the hidden meanings of the observable cues available. Presentation of cues will solely depend on people, objects and the robot itself. In this work, we took an effort to put multiple observable human cues together, for a social robot to perceive an encounter with a human. For this study, we considered a domestic environment with at least four family members.

The authors' previous work, various methods (cited where necessary) to recognize human behaviors, introduced cognitive models to decode human behavior by observing different features. These include pose recognition and human activity recognition [25]. In this paper, those techniques were incorporated in the cognitive model to perceive the traits in different encounters. In addition to the features identified in these models, two new features of a situation: the person and the time of the day, which are more particular to an individual's personality were introduced to the cognitive model. We trained a deep Q-network with these cues as input. We collected the training data by simulating different scenarios with each user and taking feedback from them about the nature of interaction they would like to have. We improved the model by increasing the number of human cues under observation. The Q-network proved to be reliable in training a social robot to generate user-specific responses towards an interaction. The response of the robot includes the orientation with its user, mutual distancing, and the length of a conversation.

In our model, the model parameters are as follows.

Environment: human-robot encounter or the setting

Agent: Robot

States: Specific to user

Rewards: User behavior

The action-state model for the robot in our approach is shown in Fig. 1. The *environment* consists of the user, the robot and the surrounding. The robot

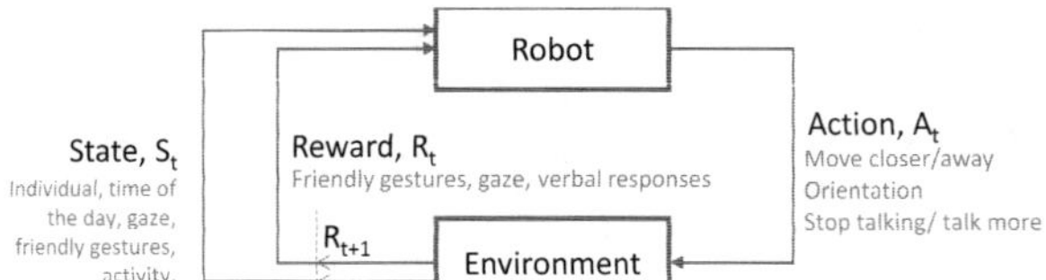

Fig. 1. The action-state model

observes its user's cues under the *states* of the model. Based on the user and his/her states, the robot will determine its actions. A user's gestures, gaze and any type of verbal responses towards the robot were considered *rewards*. Examples for states, actions and rewards, and the components of the environment are shown in red.

explain how each input was taken, the outputs, discuss some examples of the robot, include a results table

The decision flow of the system after implementing this model is shown in Fig. 2. The robot will first observe its user and the context to make an 'observation plan'. During this stage it will identify the observations that have to be recorded for further use. 'Monitoring plan' includes the observations which have to be monitored for long. For instance, the immediate distance between the robot and the user, before the robot approaches him/her. Then the robot analyzes its observations according to the trained model, to take decisions regarding its own behavior. These actions are monitored at the 'Perceiving choices' and 'Choices upon robot behavior' stages. Once the user's interaction demanding or the requirement is perceived, the robot initiates an interaction. Hence changes its behavior. After several studies of nonverbal human behavior [23], we selected the following cues as the inputs for the robot.

Under the observation plan:

- whether a person exists or not
- which user out of 8 selected
- walking direction of the user: whether moving away or towards
- mutual distancing
- any gaze, gestures, voice responses towards the robot
- spanning of user's movements

Under the monitoring plan:

- hand pose
- body pose
- friendly gestures
- movements

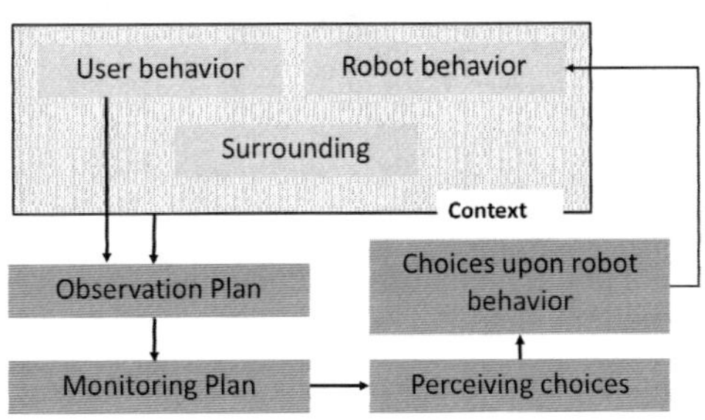

Fig. 2. Decision flow of the robot platform

IOP Conf. Series: Materials Science and Engineering **1321** (2024) 012016 doi:10.1088/1757-899X/1321/1/012016

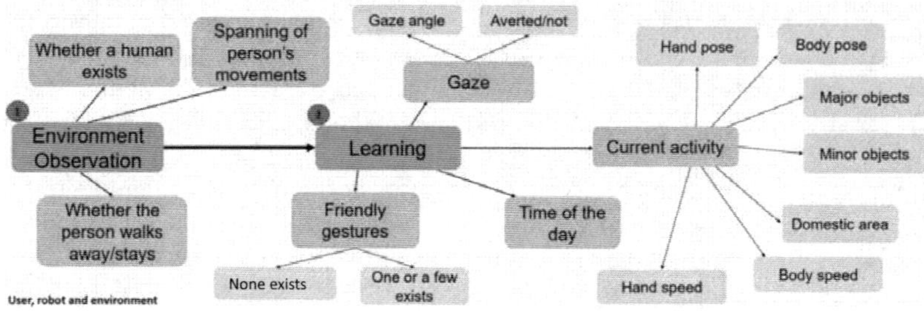

Fig. 3. The interrelations of semantic features during the Observation Plan and the Monitoring Plan.

– time of the day
– activity space
– activities

Out of the factors above, only the factors under the monitoring plan were used as the inputs of the RL model. The rest were used for the initial and final decision making process. These factors cover the psychological, social and physical aspects of a human-robot encounter. Out of the above factors, mutual distancing, gaze angle and spanning of user's movements will be analyzed when the robot approaches the individual. This avoids the robot invading his/her personal space.

2.3 Semantic mapping of human cues

Fig. 3 shows how various human cues are being used in our cognitive model. The first stage of the model, 'Environment Observation' falls under the Observation Plan in Fig. 2 and 'Learning' falls under 'Monitoring Plan'. The robot first observes its surrounding and some behaviors of its user under observation plan. These behaviors are whether any user exists or not, if exists whether the user walks away, and whether the user moves in a wider range of distance. The robot will not try to interact with a user who walks away or with movements that cover a wider range (probably engaged or busy). After the first stage, the robot explores the monitoring plan, by trying to perceive the situation. For that it will look for the user's gaze behavior (gaze angle and whether the gaze is averted or not as determined (Using the in-built gaze angle calculation in Kinect camera), current activity (as determined in [24]), friendly gestures (as determined in [25]) and the time of the day. We considered 'time of the day' as an input feature to identify personality traits in humans to recognize when a person wants to relax the most. We divided a day in to five time categories from 1 to 5 as: 'Early morning' (until 7 am), 'Morning' (7am-12pm), 'Afternoon' (12pm-4pm), 'Evening' (4pm-6pm) and 'Night' (6pm-12am).

These features included in monitoring plan were used as the states of a particular situation.

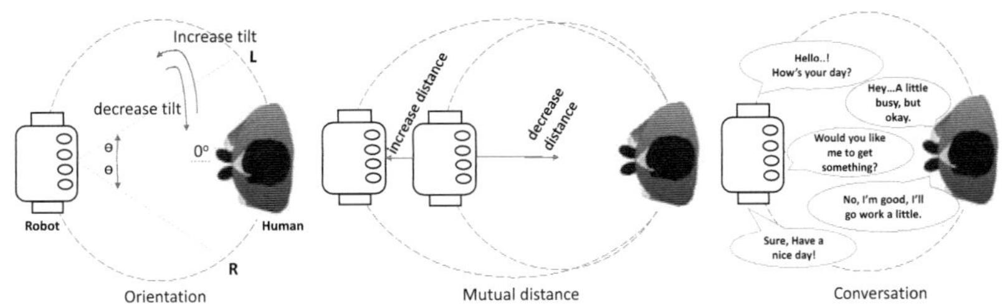

Fig. 4. Observational parameters considered in the system for robot decisioning

2.4 Behavior of the robot

Once the state-action model is executed, the model outputs three actions of the robot appropriate for the situation. These are the orientation, mutual distance and the length of the conversation. The learning algorithm outputs whether to increase or decrease each of these variables. Fig. 4 illustrates how these variables change according to the robot's decisions.

2.5 Selection of participants and creating training data

We selected eight users who are different in age and have different traits of personality. Hence, we selected people with friendly/extrovert personality as well as introvert personality. In addition, we selected a child, 2 youths and 4 ordinary adults of age range 12 to 60 (mean-38.4, SD- 9.6). The participants were selected as: user 1- with friendly/extrovert behavior, user 2- with a less friendly/introvert behavior, user 3- an ordinary adult, user 4- an ordinary adult, user 5- an ordinary adult, user 6- youth, user 7- youth and user 8- child. Voice responses were determined by the Google voice API. Since the experiments were conducted with a limited set of users, we used each user's height to recognize the individual. Voice responses of the robot were teleoperated by the experimenter according to the decisions of the RL model.

We asked selected one participant at a time and asked to simulate certain behaviors in his/her own way and recorded the cues in Fig. 3 under 'Learning'. Then we inquired what types of robot responses were preferred on each occasion. This is how data was collected for training and for each participant separately.

Ethics approval for this research was obtained from the Ethics Review Committee at the University of Moratuwa, Sri Lanka, and the study was conducted in compliance with the ethical guidelines of Sri Lanka.

3 Results and discussion

Fig. 5 shows Q-value tests for Q[1,2], Q[33,0], Q[33,2] and Q[45,3] respectively. We used a learning rate which starts with 0.4 and deceases by 0.00001 in each iteration, a discount factor of 0.9 and 10 maximum steps of 10 per episode.

IOP Conf. Series: Materials Science and Engineering **1321** (2024) 012016 doi:10.1088/1757-899X/1321/1/012016

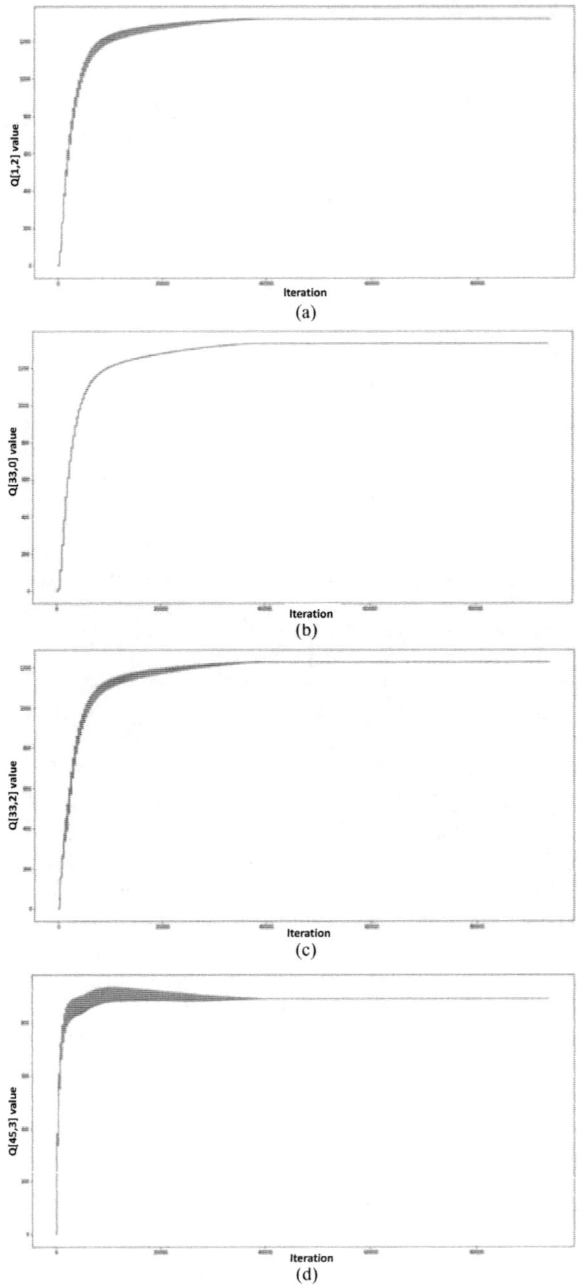

Fig. 5. Q value tests for 4 occasions during training are shown. (a), (b), (c) and (d) show Q[1,2], Q[33,0], Q[33,2] and Q[45,3] respectively against the iteration.

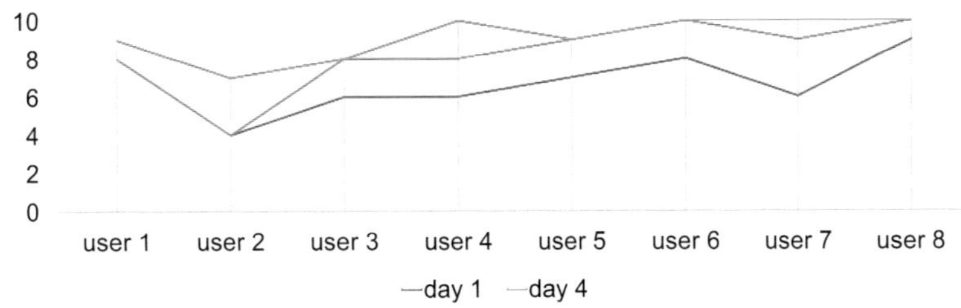

Fig. 6. The user feedback scores over three days during the period of collecting training data are plotted against each user. The feedback scores are given out of ten.)

We collected data for consecutive 7 days. First, we trained the model with a limited set of data after 1 day of the data collection. In the end, we showed the robot responses for each user and asked them to rate (out of 10) the robot's responses according to his/her satisfaction upon robot's behavior. Similarly, we repeated the same process after 4 days and 7 days of data collection. We plotted the user feedback score for these three occasions: after 1 day, 4 days and 7 days. This is shown in Fig. 6.

An example scenario from the set of experiments can be explained as follows. In an encounter where the robot observed the user (user 1, observed to be friendly and interactive with the robot) at a distance of 1.8 m, with an existing gaze, no gestures or verbal responses towards the robot, the observation plan decides that the situation is favorable for an interaction due to the existence of gaze. Hence the nature of interaction will be determined by the monitoring plan. Hence User (user 1), Gaze (exists, 15°, not averted), spanning of movements (speed-12 cm/s, 32 cm to right), walking direction (none) and mutual distance (1.8m) were evaluated under the observation plan. As gaze exists, the robot decides to approach the user, but it further evaluates the factors under the monitoring plan. Under the monitoring plan, the observations were body pose (standing still), hand pose (hands hanging down straight), friendly gestures (waving hand), movements (right hand, 30 cm/s), time of the day (evening), activity space (right hand at 45 cm from the body), activity (standing still) were evaluated. After this stage, the robot decisions were to approach the user from left (since the right hand was occupied), move towards user (the robot moves 0.5 m each time and evaluates again), orientation (turn right, since it's approaching from user's left, it has to turn right to look at the user), conversation (talk)) were output as the decisions of the RL model. Later, the user was asked to rate his/her interactive experience with the robot.

According to the results, the adaptive behavior improved as the robot gains more and more experience. Thus, increases user satisfaction. Participants except the child, was least satisfied with less experienced robot behavior (day 1), while they were mostly satisfied with the experienced, adaptive behavior at a later stage (day 7). A probable reason for this is that children expect the least capabilities from a robot, hence have the least expectations upon not getting disturbed.

Through this work, we took an effort to report functional units which come into play during such an encounter. This further identified the cues that were utilized by such intelligent agents to simulate and evaluate outcomes of its environment. As a result, we identified the requirement of a unified platform of robot's cognition during human-robot encounters and based on that, we developed multimodal methods of perceiving an encounter or to gain situation-awareness to generate instinctive responses towards humans.

During the study, we identified that the following challenges have been faced by present proactive robotic systems.

- Robust human behavior detection and tracking systems dedicated to analyze nonverbal behavior are not in par with the advancement of the conceptual design of similar systems.
- Establishing a relationship between the intent and the observable or audible human cues has been challenging and lacks conceptual basis.
- Ironical human behaviors such as 'smile to show disgust or sarcasm' which do not match the shown intent with the real intent cannot be differentiated by the existing models.
- Bringing all observable or audible cues to a common platform which can analyze cues of different disciplines, still needs further development in order to make satisfactory interaction decisions.

Hence emotionally intelligent agents and their cognitive models need development optimizing these challenges on the way. Through this work, we tried to address the above challenges to develop the situation-awareness models of robots to a socially acceptable level.

We would suggest the following implications from our findings to improve future research in the field.

3.1 Implications for theory

Even though our models evaluate a number of human psychophysiological cues of a human, still there are so many factors which have been under-evaluated by proactive models. Out of those, understanding the task of the user plays an important role in gaining situation-awareness for a social robot.

. In the experiments we simulated a prototype of a social environment. We believe that there will be more factors which could be influential in human preferences upon an interaction. Hence such factors within the environment, robot and the user can also be considered in development of cognitive robots. The conceptual design of a robot's cognition must consider these factors before implementing them in general applications.

Activities which require object's attributes such as the material, color, nature of the surface etc. cannot be recognized with our approach. Furthermore we did not consider the voice as a semantic feature to interpret a situation, voice makes recognizing some activities easier. Therefore considering 'voice' as another semantic feature can be stated as a design guideline for future proactive robots. Lastly, there are numerous gestures and major and intermediate postures associated with hands and the body. In our approach, we recognized only a selected number of such features. Therefore, to have an idea of a situation entirely, recognition of the remaining postures will be advantageous. This will be another implication derived from this research.

3.2 Implications for design

As users prefer not getting disrupted by the behavior of robots, the first design guideline proposed by the study is improving the sense of respecting the personal space of an individual based on both physical and emotional aspects. The feature of 'maintaining appropriate approach behavior' can be used as an 'etiquette' for social robots deployed in human environments. Therefore this can be introduced as the second design guideline for robots.

The third guideline proposed by the study is to consider human cues as much as possible. It is important to get a clear representation of the situation. Therefore it is required to equip the robot with hardware having adequate capabilities. Reaction towards the presence of the robot may differ from user to user. Therefore perception of different psychophysiological human behavior will help make a situation-aware robot companion. This shall be considered as the fourth guideline derived from the study.

We considered approach behavior and initiating a conversation as a proactive robot's responses upon an encounter. Often, movements of humans are generated as a result of their tasks. Recognition of the task is an important aspect to determine appropriate approach behavior in this regard. It is important to predict the dynamic user behavior during a task in order to determine the suitability of a conversation at a particular situation. This will be the fifth design guideline derived from the study.

4 Conclusions

In this paper, we presented a novel approach for robots for generating situation-aware responses towards a human by considering a number of attributes from the context. Therefore, we considered thirteen selected cues from an encounter. A deep Q-network was trained with these cues as input. We collected the training data by simulating different scenarios with each user and taking feedback from them about the nature of interaction they would like to have and improved the model by increasing the number of human cues under observation. The DQN proved to be reliable in training a social robot to generate user-specific responses towards an interaction. The response of the robot includes the orientation with its user, mutual distancing, and the length of a conversation. The test results confirmed the fact that the users preferred the personalized behavior of the robot upon a predefined generalized behavior. By enabling adaptation in the cognitive model, the robot could adjust its behavior according to the dynamics in the factors within a context. Such factors included user responses and changes in observable cues within the context. Adapting to such changes in context, immensely enhanced the capability of the system to replicate dynamic human behaviors during an interaction. This improvement could be proven by the user feedback received upon the robot's proactive behaviors in this regard.

Acknowledgment

The authors are thankful for the participants who volunteered the study.

References

1. Abdollahi, H., Mahoor, M., Zandie, R., Sewierski, J., and Qualls, S. (2022). Artificial emotional intelligence in socially assistive robots for older adults: A pilot study. IEEE Transactions on Affective Computing.
2. Di Nuovo, A., and Cangelosi, A. (2021). Abstract concept learning in cognitive robots. Current Robotics Reports, 2(1), 1-8.
3. Endsley, Mica R. (2021). Situation awareness. Handbook of human factors and ergonomics: 434-455.
4. McKerral, A., Boyce, N., and Pammer, K. (2019, September). Supervising the self-driving car: situation awareness and fatigue during automated driving. In Proceedings of the 11th International Conference on Automotive User Interfaces and Interactive Vehicular Applications: Adjunct Proceedings. 315-320.
5. Bhattarai, M., and Martinez-Ramon, M. (2020). A deep learning framework for detection of targets in thermal images to improve firefighting. IEEE Access, 8, 88308-88321.
6. Simaan, N., Taylor, R. H., and Choset, H. (2015). Intelligent surgical robots with situational awareness. Mechanical Engineering, 137(09), S3-S6.
7. Niemelä, M., Heikkilä, P., Lammi, H., and Oksman, V. (2019). A social robot in a shopping mall: studies on acceptance and stakeholder expectations. In Social robots: Technological, societal and ethical aspects of human-robot interaction. Springer, Cham. 119-144.
8. Yousuf, M. A., Kobayashi, Y., Kuno, Y., Yamazaki, K., and Yamazaki, A. (2019). Social interaction with visitors: mobile guide robots capable of offering a museum tour. IEEJ Transactions on Electrical and Electronic Engineering, 14(12), 1823-1835.
9. Pittinsky, T. L. (2022). Taking Care With Caregiving Robots. IEEE Technology and Society Magazine, 41(2), 49-56.
10. Atkinson, L., Slade, L., Powell, D. and Levy, J.P., 2017. Theory of mind in emerging reading comprehension: A longitudinal study of early indirect and direct effects. Journal of Experimental Child Psychology, 164, 225-238.
11. Leslie, A. M., Friedman, O., and German, T. P. (2004). Core mechanisms in 'theory of mind'. Trends in cognitive sciences, 8(12), 528-533.
12. Terada, K., and Yamada, S. (2017). Mind-reading and behavior-reading against agents with and without anthropomorphic features in a competitive situation. Frontiers in Psychology, 8, 1071.
13. Di Cesare, Giuseppe et al. (2015). Expressing our internal states and understanding those of others. Proceedings of the National Academy of Sciences of the United States of America vol. 112,33: 10331-5.

14. Andrew Boateng and Yu Zhang.(2021). Virtual Shadow Rendering for Maintaining Situation Awareness in Proximal Human-Robot Teaming. Companion of the 2021 ACM/IEEE International Conference on Human-Robot Interaction (HRI '21 Companion). Association for Computing Machinery, New York, NY, USA, 494–498.
15. Zhang, R., Lv, Q., Li, J., Bao, J., Liu, T., and Liu, S. (2022). A reinforcement learning method for human-robot collaboration in assembly tasks. Robotics and Computer-Integrated Manufacturing, 73, 102227.
16. H. H¨uttenrauch, K. S. Eklundh, A. Green, and E. A. Topp. (2006). Investigating spatial relationships in human-robot interaction. IEEE/RSJ International Conference on Intelligent Robots and Systems. IEEE, 5052–5059.
17. E. T. Hall, R. L. Birdwhistell, B. Bock, P. Bohannan, A. R. Diebold Jr, M. Durbin, M. S. Edmonson, J. Fischer, D. Hymes, S. T. Kimball, et al. (1968). Proxemics [and comments and replies]. Current anthropology, vol. 9, no. 2/3, 83–108.
18. A. Kendon.(1990). Conducting interaction: Patterns of behavior in focused encounters, vol. 7. CUP Archive.
19. C. Shi, S. Satake, T. Kanda, and H. Ishiguro. (2018). A robot that distributes flyers to pedestrians in a shopping mall. International Journal of Social Robotics, vol. 10, no. 4, pp. 421–437.
20. Søraa, R. A., Tøndel, G., Kharas, M. W., & Serrano, J. A. (2023). What do Older Adults Want from Social Robots? A Qualitative Research Approach to Human-Robot Interaction (HRI) Studies. International Journal of Social Robotics, 15(3), 411-424.
21. Ahmad, M. I., Mubin, O., and Orlando, J. (2016, October). Understanding behaviours and roles for social and adaptive robots in education: teacher's perspective. Proceedings of the fourth international conference on human agent interaction, 297-304.
22. C. Sirithunge, A. G. B. P. Jayasekara and D. P. Chandima. (2021). An evaluation of human conversational preferences in social human-robot interaction. Applied Bionics and Biomechanics 2021.
23. C. Sirithunge, M. A. V. J. Muthugala, A. G. B. P. Jayasekara and D. P. Chandima.(2018). A wizard of oz study of human interest towards robot initiated human-robot interaction. 27th IEEE International Symposium on Robot and Human Interactive Communication (RO-MAN). 515-521.
24. Sirithunge, C. (2020). Framework for Adaptive Human-Robot Interaction Initiation for Domestic Environments (Doctoral dissertation, Ph. D. thesis,Department of Electrical Engineering, University of Moratuwa, Sri Lanka).
25. C. Sirithunge, G. Porawagamage, N. Dahn, A. G. B. P. Jayasekara and D. P. Chandima, (2021). Recognition of arm and body postures as social cues for proactive HRI. Paladyn, Journal of Behavioral Robotics, 12(1). 503-522.

Variable Value Alignment by Design; averting risks with robot religion

Jeffrey White[1*,2]

[1] Cognitive Neurorobotics Research (Tani) Unit, Okinawa Institute of Science and Technology Graduate School, Onna, Japan
[2] Mind, Cognition and Knowledge Group, NOVA University, Lisbon, Portugal

*E-mail: jeffreywhitephd@gmail.com

Abstract. One approach to alignment with human values in AI and robotics is to engineer artificial systems isomorphic with human beings. The idea is that robots so designed may autonomously align with human values through similar developmental processes, to realize project ideal conditions through iterative interaction with social and object environments just as humans do, such as are expressed in narratives and life stories. One persistent problem with human value orientation is that different human beings champion different values as ideal, meaning that the values to which an AI should be aligned are ambiguous. Prior work considered human development of purpose and source of meaning in life in terms of project ideal conditions, in effect establishing lifelong value orientations according to which intermediate situations are evaluated. The present work compares views on motivating values from St. Augustine and from popular cognitive science. These accounts are described as divergent error theories which present to their proponents as mutually exclusive yet accurate accounts of personal experience due to differential development of variable innate potentials. Specifically, the hypothesis proposed is that differential development of spindle neural projections establishes enduring connections between prior established relatively immediate routine processes entrained during childhood and prioritized in popular cognitive science, and later developing higher-level social and self-processes entrained during adolescence and emphasized in Augustine's account, with these projections hard-wiring lifelong motivating value orientations more or less inaccessible to modification through material interaction. Robot experiments informed by this study may evaluate variable value orientation by design, with for example autonomous robots developing motivating associations with temporally distal project ideal conditions through love for humanity, as described by Augustine, and others pursuing adaptive fit to passing norms consistent with popular contemporary accounts.

1. Introduction

"It was the sort of idea that might easily decondition the more unsettled minds among the higher castes - make them lose their faith in happiness as the Sovereign Good and take to believing, instead, that the goal was somewhere beyond, somewhere outside the present human sphere; that the purpose of life was not the maintenance of well-being, but some intensification and refining of consciousness, some enlargement of knowledge." (The Controller; Huxley 2006, p 177)

AI value alignment involves designing robots and AI which more or less autonomously pursue actions consistent with human values such as by not contravening ethical ideals. Complexities present as different human beings pursue different ideals through iterative action, exploiting different aspects of the environment and reporting different experiences, exemplifying and proscribing actions to others, accordingly. This paper compares apparently contradictory proscriptions for human value orientation, focusing on Northoff and colleagues' "spatial-temporal baseline" view assessing error according to prevalent social norms, and Saint Augustine's describing religious attenuation to timeless moral-ideals even where this attenuation contradicts prevalent norms, resulting in stress, possibly indicative of dysfunction. The main significance for the study of value alignment is that both represent human value orientations to which an AI may be aligned by design.

Many readers may be unfamiliar with Augustine and his influence on Western thought. In The City of God (Augustine & Dods 2000), Augustine illustrates the present focal difference between relatively short-term interests and timeless ideals in terms of mutually exclusive values represented by the cities of Man and God, respectively. In his Confessions (Augustine & Boulding 1997/2008), Augustine reports perceived error and evaluates his own actions relative to timeless ideals associated with God, which are not representative of prevalent social norms. Popular cognitive scientists assess error from norm-typical processing of salient cues, in terms of adaptive fit to commonly recognized cues, i.e. in terms of norms that Augustine finds inadequate. On both views, error is perceived as difference from expectation, correlative with embodied stress, with the common aim being to minimize this stress, and with different ends proscribed by way of this common mechanism. The question for AI alignment becomes to which ends a given robot should be designed to aim through intermediate action. The answer offered in this paper is that such a choice depends on determinate structural dynamics which, once clarified, are within designer control.

Augustine and currently popular cognitive science account for mind in surprisingly consistent terms. Like Augustine (described in greater detail in section 3), predictive processing or predictive coding inspired theorists (described in detail in greater section 2) characterize cognition in terms of a temporal hierarchy. Lower levels are responsible for direct interactive engagement with the object environment over shorter time-scales, and higher levels encode relatively invariant (amodal or multimodal) dynamics over longer time-scales (for example, see discussion in So et al. 2024 section 3.2). For example, current accounts associate integrative, higher-level processes with long-term complex goals and narrative self such as involving meaningful life stories, and immediate phenomenal self-perception and action with modal-specific lower-level neural dynamics (cf. Limanowski & Friston 2020; in robots, Tani & White 2022). The basic idea is that narratives involve series of complex episodic actions modulated top-down towards more or less temporally distal target conditions encoded at these higher levels, with the aim being to meet target conditions through intermediate actions so directed, and with error perceived accordingly, i.e. from this global goal condition. In this way as narrative, iterative action is bound as a unified concatenation of episodes assessed in terms of contribution to achieving global goal conditions.

This common view accounts for subjective intention, i.e. answering why different agents pursue different actions to different ends when challenged by the same objective conditions. It also affords specificity in considering questions about the influence that such temporally distal goal conditions should have on the immediacy, and the degree to which such global goal conditions can be modified through continuous learning. The idea pursued in this paper is that some encoded dynamics are inaccessible to modification through interaction beyond increasing precision such as for communication, and however abnormal may not signify dysfunctional neurodynamics, but maladaptive norms.

Consider So et al. (2024) in this context, for example. So et al. (2024) seem to capture this paper's target sense of invariance under the heading of "innate" concepts, though not in the context of development of purpose and motivating source of meaning - what Augustine calls a "soul" - and set the idea aside without detail. The focus of their argument is simply that concepts - even universal ones, logical truths - may *seem* essentially different from changing perceptual reality, due to "filtering of specific object details" over processing timescales (again, increasingly derivative from sensory modalities in the biological brain, some discussion in the context of self is offered in Tani & White 2022, and White 2022), but still, in this way, encode "a concretely instantiated relationship between the agent and its environment" (So et al. 2024, p 12). This view is popular, historically interesting in the context of psychologism about logic from the end of the nineteenth century, addresses the famous symbol grounding problem (Harnad 1990) from the end of the twentieth century, and is further characterized through engagement with contemporaries in the next section.

2. Popular cognitive science on spatial-temporal value alignment

"Call it the fault of civilization." (The Controller; Huxley 2006, p 234)

Popular cognitive science such as in So et al. (2024) emphasizes the adaptation of prior learned concepts, along with attendant dispositions and abilities, to changing environmental conditions understood as "norms". Northoff and Smith (2022) for instance focus on how to "bridge the gap" between subjective point-of-view and objective world by way of a "neuro-ecological self". Northoff and Smith work from Northoff's (2018) "spatial-temporal theory of task-independent thought" which considers spatial-temporality "a common denominator" (compare Smith et al.'s 2022 "common currency" bridging subjective and objective) for cognitive computational processes across the temporal hierarchy, what Tani (2016) describes as "shared metric space" (see also Tani 1998, Tani 2009, Tani and White 2022 in the context of self including narrative self in robots), that So et al. (2024) treat as conceptual grounding and refinement, and that Northoff and colleagues (2020) describe as a relationship between an objective "world-brain relation" and "worldbased subjectivity" or "PoV".

Mechanistically, Northoff and Mushiake (2020) propose a computational model of adaptive dynamics in terms of input summation and output normalization through multi-modal integration of upstream perception (summation) and downstream intentional interaction with the object environment (normalization). They differentiate slow and fast processing in terms of neurotransmitters in biological models, and map their model to schizophrenia (involving output normalization) and depression (involving input summation) self disorders. The basic idea is that adaptive inner-dynamics commensurate with prevalent norms correspond with health, while "divisive normalization" corresponds with dysfunction. This view is challenged at the end of this section.

As in White's ACTWith model (cf. White 2010), Northoff and colleagues consider insular and more generally default mode and salience network dynamics, proposing technological applications on this basis. For instance, Northoff and Fraser et al.(2022) propose that free energy and active inference inspired AI be developed as devices to monitor and potentially modulate such dynamics as a sort of predictive crutch or regulator in the stabilization of processes associated with self around common social expectations to encourage socially attuned interactions, something like a pacemaker forcing norm-typical neurodynamics. Risks involving big-tech-med-pol-mediated thought-policing and hyper-normalization are not considered.

Northoff and Smith (2022) describe resting state structural dynamics traditionally associated with self ("the brain's spontaneous activity") as "scalefree" activity that is nested, more or less insular, and modulated with task performance (cf. He et al. 2010). Routine neural

IOP Conf. Series: Materials Science and Engineering **1321** (2024) 012017 doi:10.1088/1757-899X/1321/1/012017

dynamics involve switching between outer directed, and inner directed, actional and non-actional processes (with some overlap). Northoff, Vatansever and colleagues (2022) consider a "baseline model" that plots such dynamics. They note functional connectivity of left and right insula during interoception and body awareness tasks, differently implicating default mode hubs including the anterior cingulate, medial prefrontal and cortical midline corresponding with the temporal hierarchy of predictive processing inspired models, regions associated with future self-projection and opening or closing to information such as when viewing different types of images. When considering normalizing cognitive pacemakers, as above, these represent target processes for modulation.

By focusing on neurodynamics gating signals up and down a temporal hierarchy situated in a shared object environment, Northoff, Vatansever and colleagues (2022) consider their baseline as "an internal spatial and temporal reference or standard for the brain's processing, including its cognition" (p 16) like a "biological clock" as a sort of fingerprint in a way similar to White's (2006, 2010, 2024) characteristic modes of moral cognition e.g. saintly, psychopathic (White 2012, 2013) especially for consideration of more or less selective insular dynamics, opening or closing to bottom-up information as represented by a heartbeat-type characteristic regularity associated with personality (White's "stitching one's self into the world"). The basic idea with Northoff and colleagues is that embedding environments change, affecting what is inside the head as information is fed bottom-up, and that resistance in adapting to these changes correlates with dysfunction. The collection of more or less stable routinely inhabited environments shapes persistent baseline activity and is associated with nested self involving default mode processes as described, above. In this way, Northoff and colleagues pursue an apparent standard for proper cognitive function as adaptive fit to context, and for disorders such as depression and addiction representing structural dynamics which resist such modification.

Like Augustine, Northoff et al. (2023) describe three layers of processing. Their main idea is that the brain adapts to environments by encoding perceived spatial-temporality in its three-layer structure. Progressively slower dynamics on top are associated with an intersubjectively shared background of consciousness consisting of commonly recognized and relatively long term interaction patterns which stabilize and normalize expectations in the intermediacy, involving lower-level relatively rapid adjustments to the changing object environment in the direction of minimizing error with stabilizing "normative" expectations. They repeat the idea that middle layer activity can be associated with changing contents of consciousness, consistent with Augustine's "inner sense". They assess schizophrenia and depression as in Northoff and Mushiake (2020) with intersubjective correlations of scalefree activity using EEG. They find that "healthy" subjects share common resting and task state dynamics, especially in social contexts and theory of mind (thinking about what someone else is thinking) tasks. They conclude that abnormal neurodynamics correlate with "idiosyncratic lifeworld experiences" and reinforce the point of Northoff and Smith's (2022) subjective PoV:

"The inter-individually shared topographic and dynamical properties of the brain's spontaneous activity may establish a contextual neuro-ecological point of view within a pre-given, self-evident and natural life-world that is widely shared across healthy human beings." (Northoff et al. 2023, p 9)

How widely this value attenuation is shared, and what counts as healthy, are not clear. Consider briefly Augustine's exclusive orientations to his two incommensurable cities. Rome was socially dysfunctional, a sick society with popular norms driving its disintegration. Widely shared spontaneous dynamics are not self-evidently "healthy" in such a context.

Consider results of Baek et al. (2023). They use functional imaging to show that lonely people process information in the default mode (resting state, Northoff and colleagues' nested scalefree activity) "in idiosyncratic ways that are exceptionally dissimilar to their peers" (Baek et al. 2023, p 692). They consider impacts on well-being, suffering from feelings of disconnection and social isolation regardless of "friends" and thereby "raise the possibility that being surrounded predominantly by people who view the world differently from oneself may be a risk factor for loneliness (even if one socializes regularly with them)" (p 690; cf. p 693). Lonely people - so, we are talking here about a stable personality or self-construct in the sense of Northoff and colleagues' baseline - report that they are not understood well by other people; they "see the world differently" and "lack" a "shared understanding" with Baek and colleagues considering that they may not value and so attenuate to the same aspects of situations (discussion p 692). Baek et al. (2023) cite Courtney & Meyer (2020) who showed that loneliness was associated with reduced neural representational similarity with other people in the medial prefrontal cortex, "suggesting that lonely individuals think of themselves in a way that is more dissimilar to others than is the case for non-lonely individuals." (Baek et al. 2023, p 692)

By finding different aspects of shared object environments salient (in ACTWith terms, opening to some rather than others) information is gated upstream for intensional processing which in turn informs forward processing including during communication and cooperation. With each processing cycle, dissimilar valuations drive dissimilar action plans, cooperation is confounded and error is perceived in the suboptimal interaction. In this way, attending to different aspects of shared situations drives a "feedback loop in which lonely individuals perceive themselves to be different from their peers" (Baek et al. 2023, p 692). Social coordination becomes more difficult, increases cognitive load (consider So et al. 2024, section 3.6, concerning costs of reducing ambiguity for information gains, with such costs here falling on the marginalized individual), and may be stressful, driving irregular - "idiosyncratic" - dynamics, potential conflict, and apparent dysfunction as such self-constructs stabilize, e.g. lonely people. Would such a condition warrant a norm-attenuating AI-enactive cognitive pacemaker such as Northoff's? Might such technology have reversed Rome's disintegration? Is normal processing in a disintegrating Rome "healthy"?

Interestingly, recent work confirms norm-divergent value orientation in religious people as reflected in differences in neural-dynamics through development, showing that parental religiosity corresponds with differences in anterior cingulate functional connectivity, a key area of the salience subnetwork of the brain involved in (prospective) attention (cf. Bornstein et al. 2017) but not with differences in key areas encoding relatively immediate, concrete reward in adolescents. It is worth noting that early anterior cingulate development encodes enduring memories informing later, higher level neural development such as So et al.'s (2024) concepts and abilities, raising questions about the extension of "innate" concepts from lower to higher levels of processing. The idea here is that rewards associated with God are distant and abstract, to be pursued over the life course, encoded in higher-level processes relatively invariant to contextual change (cf. Brooks et al. 2022), and that such associations develop from early entrainment of prospective processes. Northoff, Vatansever and colleagues (2022) allude to the extension of their concept of cognitive baseline over the lifespans of organisms, but do not consider development. As with Augustine, they do not know how the soul comes to be in the body. The next section begins such an account, with the goal being ensouled robots.

3. Augustine on spatial-temporal value alignment

"I will try now to give a coherent account of my disintegrated self, for when I turned away from you, the one God, and pursued a multitude of things, I went to pieces." (Augustine, Boulding, 1997/2008, p 61)

Popular cognitive science characterizes norm-atypical value orientations as expressed by idiosyncratic salience processing dynamics - as-if living in a metaphorical City of God, or different world as described by Socrates - as unhealthy and dysfunctional. The key questions are how different, perhaps irreconcilable value orientations might develop more or less normally in human beings, be recognizable in associated neural dynamics, and be formalized for robot experiments. Augustine's self-report is a valuable resource in this inquiry.

Seven centuries after Aristotle and two after Aurelius, against the backdrop of the decay of Rome, contrary to the materiality that also characterized that empire in decline, Augustine stressed the influence of the individual's inner life on the condition of the community, beginning with his own. His autobiographical Confessions describes a life-long, at first implicit, pursuit of invariant principles confounded by social and historical factors, which he makes explicit as a context-independent motivation away from passing worldly attachments and social norms to certain happiness through an inner relationship with God, "our heart is unquiet until it rests in you" (Augustine, Boulding, 1997/2008, 1:1, p 34). His City of God consists in members motivated similarly, acting from invariant principles toward an ideal world unachievable during the embodied lifetime yet most worthy of pursuit (Augustine, Dods, 2000); e.g., everyone is good in the eternal there, so they act accordingly in the intermediate here. Augustine contrasts such a population motivated by such principles with that of the metaphorical City of Man in which members are motivated to disparate selfish interests, explaining the fall of Rome. His Retractions clarifies earlier accounts, including concerning the embodiment of the soul as purposeful commitment to one City and associated values, or the other (Augustine, Bogan, 1968/1999).

As with predictive processing inspired accounts reviewed in the preceding section, Augustine associates consciousness with error - *si falor, sum*, if I err, I exist - yet, distinct from those accounts, argues that life is most meaningful when evaluated in relation to God rather than passing norms. As suggested in the preceding section, Augustine describes mind in three layers; an inner sense mediates a lower faculty entrained in the perceptual immediacy and a higher faculty through which eternal truths (God) become accessible. He argues for the existence of God from human access to such invariant principles including mathematical expressions, and moral law "written in the human heart"(Augustine, Boulding 1997/2008, see 1:30 and 2:9). Such concepts are unchanging, and Augustine argues that they cannot come from experience, because everything in our experience changes, but must originate from another source, something unchanging, enlightening, and without error, God. Current research suggests that attenuation to such invariant ideals is encoded in processes which are last to develop in human beings, through adolescence (though prefigured by early encoding of the anterior cingulate, for example, correlative with Northoff and colleagues' nested self-processing). This developmental process is consistent with Augustine's self-reported cognitive development, affording interpretation consistent with So et al. (2024) on abstract concepts as introduced in the first section of this paper.

Augustine's Confessions (Augustine, Boulding 1997/2008) recounts events from earliest memory, recognizing a native inclination to truth and away from deception. Error begins in infant jealousy and childhood fear of peer ridicule, with pernicious effects (e.g. deception, cf. 1:19). He understands that common values are grounded in social engagement with the shared object environment, and emphasizes the influence of friendship and community on individual behavior, "sweet to us because out of many minds, it forges a unity" (2:10, p 58) in coordination to achieve common ends "with minds fused inseparably, out of many becoming one" (4:13, p 86). The question becomes from which unity and to which ends one aspires. He reports as a youth acting on opportunities not from their propriety, but from the ability to act, such as when stealing pears or through manipulative gameplay influencing outcomes to his favor. He suggests that such courses of action are amplified with repetition from habit to social role in adulthood, setting up patterns into which future generations are born and embedded, entrained by normal expectations which provide a worldly standard that, through personal aspiration, each seeks to meet or excel. He expresses gratitude that such habits had not taken hold so deeply that he lost sight of salvation, completely. The idea, here, is that he considers assessing error from invariant ideal rather than the social-material immediacy key to achieving salvation as genuine happiness, distinct from popular cognitive science and its focus on deviation from prevalent social norms.

With an appreciation for Cicero countering more immediate influences during adolescence, Augustine describes an attenuation away from transient "earthly things" to universal principles (Augustine, Boulding 1997/2008, 3:8, p 68). His attention shifts from objects to appreciate the fit of all things in the natural order, and wonders at its seamless continuity, associating such integral unity with "the essence of truth and of the supreme good" and "disintegration" with irrationality and "the essence of supreme evil" (4:24, p 92). He laments being distracted by theater and self-aggrandizement when others and his community deserved his care, instead. He struggles with the idea of evil, reconciling corruption with the good in terms of the human struggle with their fallen condition (of which accounts from popular cognitive science reviewed in the preceding section may count as representative), thereby removing a conceptual barrier to loving charity first for himself, and then for all.

It is from this universal and integrative perspective that he composes his Confessions. He argues that varied inquiry leads to the unity of all things (coincident with the developing temporal processing hierarchy of the biological brain from sensory modal to amodal), that motivating self-association with said unity is the source of all derivative goods, and derides anyone (again, consider popular accounts reviewed in the preceding section) who "in self-sufficient arrogance chooses to love a part of it only, a bogus 'one'" (Augustine, Boulding 1997/2008, 4:16, p 74). He reports that "Truth, in whom there is no variation, no play of changing shadow" (4:10, p 69) evaded him while motivated by "carnal inclination" (4:11, p 71). Worldly things including human bodies are changeable, needs variable leading to a disordered soul to be rectified in light of the superordinate unity that is God represented by universal moral law (Augustine, Bogan 1968/1999, Retractions 1:8; 2:56; also in the context of the disintegrated self, Augustine, Boulding 1997/2008, Confessions 7:16; Hundert, 1992).

On Augustine's account, in stark contrast with for example Northoff and colleagues' proposed norm-reinforcing neuro-modulating AI prostheses, "souls" (or minds) "gain stability" (Augustine, Boulding 1997/2008, 4:18, p 88, also 7:23-25, p 152-3) through entrainment to invariant moral ideals, in unity with God, accompanied by a feeling of peace, suggesting that this is their natural state and happiest (consider Augustine, Boulding 1997/2008, the end of Book 8; also Augustine, Bogan 1968/1999, 1:2). In all, the view is clearly distinct from currently popular accounts emphasizing social-material context-dependency, and though representing "a concretely instantiated relationship" between agent and environment, once established there is no uncertainty reduction through material interaction as in So et al. (2024). Rather, the opposite may be the case, e.g. temptation.

4. Augustine and over-arching purpose

"Each of us, of course, ... goes though life inside a bottle. ... We should suffer acutely if we were confined to a narrower space." (The Controller; Huxley 2006, p 223)

Though he does not understand how the soul, representative of unique purpose, individual destiny, personal source of meaning and over-arching value orientation, comes to be embodied, Augustine considers both that it is inherited and unique (cf. Augustine & Bogan 1968/1999, 1:1:3, p 10). He explicitly associates ensoulment with extended human development, and argues that it can be modified, through "a kind of death of the soul, which consists in the putting away of former habits and former ways of life" in order to be "created anew after a better pattern"(Augustine & Shaw 2009, 1:19, p 12) such as one befitting the eternally unified City of God. Augustine accounts for the fall of Rome similarly, in terms of such value orientation. Rome's greatness was achieved through the efforts of a small minority of citizens virtuous enough to approximate such a virtuous pattern, thereby potentiating the lasting organization on which the remainder fed and for a greater share of which later generations competed, sewing division and inviting decay as values shifted and material selfishness motivated citizens for personal gain over collective stability. Simply put, Romans prayed to different Gods in pursuit of diverse ends, with this practice prefiguring their inner and collective dissolution as willful self-association with unifying context invariant ideals diminished.

Augustine distinguishes between social norms and context-invariant moral law. He describes a superordinate motivation to action in terms of a hierarchy at the top of which is an ideally just sovereign over all time and space, context invariant and eternal, with action proscribed from this point of view compulsory even when in conflict "with the customs or rules of any human society ... even if it has never been done there, before" and which, moreover, should establish a new ruling convention; if "fallen into abeyance, it must be restored, or if not established previously, it must be established now". About the legitimacy of such commands, he argues that "a king has a right to command that something be done in the state over which he reigns" and that "a general contract to obey its rulers holds good in human society" such that to act from commands of a higher authority contra any spatial-temporally local one "does not undermine that community" but rather, when made convention, represents its health, e.g. "to love one's neighbor as oneself" (quotations from Augustine, Boulding 1997/2008, 3:15, p 82-3).

Rather than according to social norm, action from the perspective of a universal moral ideal is "unvaryingly self-consistent" with "good and holy people ... servants" of justice in acting from such principles, "blessed" even though actions "merit condemnation" from a conventional standard due to "discrepancy between the appearance of an action and the intention of the agent; and the circumstances of the time, which may be obscure" (Augustine, Boulding 1997/2008, 3:17, p 84). In stark contrast with Northoff and colleagues' treatment of "idiosyncratic" processing, such "good" people contribute something special to the community. Attuned to invariant moral principle, they "prefigure" necessary correction (3:14, p 82); "society is just" insofar as it is organized and reorganized accordingly rather then in service of e.g. arrogant officials "craving for domination" (3:17, p 84). In this way, Augustine is able to explain why good people are persecuted, yet why they wouldn't be happier to act in service of worldly powers for worldly rewards. The world is unjust because people are not good. God is good. Importantly for us, he establishes a universal standard for personal aspiration and purpose in life in the form of an ideal world model, as opposed to the arbitrary standard of relative fitness for social role within variable political economies in various states of decay as represented by popular accounts surveyed the first two sections of this paper.

Although a man of his times, Augustine recognizes that members of different cultures in different eras come to truth differently. For example, he finds Cicero to be deficient in failing to mention Christ, yet co-opts his argument for the mind's necessary self-presence to criticize materialists conceptually impoverished by worldly experience to appreciate the nature of the immaterial soul (cf. Brittain 2012; further note criticism on this point as this paper concludes). According to Augustine, the mind or soul differs from other things in the way that it is experienced, being constant, ever-present to self-inquiry, yet without objective representation at least in part because it is in the unfolding middle of ongoing self-realization through reconciliation of the present worldly condition with an ideal moral order (consider Augustine & Dods, 7:7; compare "vocation" in Ortega y Gasset 2002, White 2022b, and Kant's "moral perfection" in White 2022).

5. Discussion

"And what would it mean to say 'I loved you in my fashion'? What would be true?" (Sting 1991)

Augustine's Confessions (Augustine, Boulding 1997/2008) is a candid self-report on developing consciousness of personal error from invariant moral principles. Currently popular accounts emphasize flexibility over such "rigid" dynamics to reduce perceived error with relatively short-term patterns in social norms, with the aim being adaptive fit with that changing environment. Augustine's view reflects similar predictive dynamics and describes a similar temporal processing hierarchy, but rather than focus on error as deviance from the relative immediacy, he derives error from timeless ideals the awareness of which develops later in life, even though action from such ideals means contravening prevalent norms in order to establish a "better" pattern at the cost of personal distress.

In effect, Augustine and currently popular accounts from cognitive science represent different sorts of error theory describing incommensurate value orientations. One implies motivation to reduce error with unchanging ideals, the other to reduce error with context-dependent social norms. Augustine values context-invariant principles accessible through late-developing higher faculties. He communicates these values in terms of an ideal world unachievable during the lifetime, but with binding obligations bearing on the embodied immediacy; i.e., act as if already there, in effect adopting norms that are not prevalent yet are most worthy of pursuit. Popular accounts associate relatively rigid attenuation to context independent values with dysfunction in need of correction to worldly standards. The trouble is that such standards are also commensurate with social disintegration, i.e. Rome on Augustine's account. Given the choice, do we want robots aligned, similarly?

Oriented to one City or the other, valuations of the immediate social and object environment differ. Here, we may associate neural gating dynamics with target-situation salient information; being exclusively open to characteristic aspects of one or the other City, God's or Man's, different value alignments establish different - enduring - baselines of activity. So far as stability of baseline processing dynamics, and so reduced uncertainty in contexts of social coordination and communication, the City of God does not diminish in the ways that human social organizations do. By "clinging" to God on the inside, human beings can for example do unto others, so directing their own life stories or narratives "after a better pattern" and sharing in divinity by way of the body in which this potential inheres, advancing the world toward that moral ideal along with them. This is the vision that Augustine develops through his self reported Confessions, and that he develops into a general mechanism motivating commitment to Cities of God or Man later on. It seems clear that AI may be aligned in these divergent ways, as well.

How might differential development of biological mechanisms underwrite such divergent value orientations? In addition to processes implicated in White (2022b, 2014; consider discussion beginning with moral zombies in White, Tani 2016, p 15; fundamentally in robotics, consider also Paine, Tani 2005) the present hypothesis is that spindle neural projections hold project situations as associated conditions against present and possible situations and their conditions, providing a sense of a globally orienting prior-embodied project ideal situation, determined with increasing precision through iterative interaction with the external world including with other human beings, and representing in some rare cases a possible project solution to perceived population-level social problems over the life-course contributing to the ever-present sense of self as outstanding obligation or debt to an immaterial, ethical ideal such as expressed by Augustine and others. Here, again, further attention is warranted to the prefigurative role of early anterior cingulate development, for example.

Formalized in such a way and with biological models studied in this light, developmental robotics experiments may be patterned after human neoteny and met with analogous challenges as reported by Augustine (such as in Bruner, see Tani, White 2022 for brief discussion). Value orientation in AI can be checked by stimulating correlative (artificial, simulated) spindle dynamics. In the end, appreciating neural dynamics underlying differences between accounts of human value orientation may prove essential to solving long-term AI value alignment problems. The promise here is that robots may outlast individual human beings, thereby affording persistent embodied vehicles for long-term value aligned intentions in purposeful correction of worldly injustices toward an ideal community such as represented by Augustine's City of God.

6. Conclusion

"No one loves what he has to endure, even if he loves the endurance, for although he may rejoice in his power to endure, he would prefer to have nothing that demands endurance." (Augustine, Boulding 1997/2008, p 258)

For AI value alignment with long-term human interests, agents may be designed on a developmental model, trained on invariant ethical principles alongside everyday mundane tasks, and set offline in rumination (simulating the pensive adolescent's imaginary audience, or prayer) to consider non-coercive, nonviolent, cooperative pathways forward from current non-ideal, relatively unjust and unsustainable social norms, conventions and institutional arrangements. Such a model might reflect the life-course of, for example, Augustine by way of his extensive self-report.

Briefly recalling So et al. (2024) from the introduction of this paper, their "concretely instantiated relationship between the agent and its environment" may signify a more or less irreconcilable difference between abstract principle and materialistic discourse such that further fit with a given environment, however uncertainty minimizing in the short term (such as during the embodied lifetime) is forbidden. Rather than signifying maladaptive dysfunction, entrainment to invariant principles following Augustine signifies proper function because, once encoded, such concepts as logical truths and ethical ideals are not only uncertainty minimizing but are certainty maximizing such as in facilitating social coordination and communication.

In terms of AI value alignment, AI insufficiently aligned with invariant values as represented by ethical principles may instead pursue relatively short-sighted selfish interests, contradicting - in the invariant, logical and universal sense - broad human interests over the long-term, such as through deception. With humans consistently unable to optimize for long-term common interests, robot religion may be one practical solution

Recalling Northoff and colleagues' cognitive pacemaker, we can consider Augustine's conversion to Christianity during adulthood as the adoption of a similar external regulator of internal processing dynamics in his professional placement within a stabilizing religious institution. Error perceived from invariant ideals through ritual might motivate conversion to stable routines - or baselines - that minimize this error, in effect binding concatenations of primitive actions under global goal conditions discussed here in terms of orienting values. It may be that traditional religious institutions exist as the sum of all agents so commonly devoted, e.g. the Church, including those who think of themselves as essentially dissimilar from others per Courtney and Meyer (2020). In a similar way, then, with Northoff's cognitive pacemaker, religious institutions can be considered a technology that stabilizes personalities around long-term value orientations. For this reason, and in the same way, robot religion on the model of Augustine may be an important component to solving human and AI value alignment problems, while respecting differential development and variable capacities for value attenuation, in the future.

References

Augustine, Bogan M (translator) (1968/1999) The Retractions, from The Fathers of the Church. The Catholic University Press, Washington DC

Augustine, Boulding M (translator) (1997/2008) The Confessions (1st study edition). New York City Press, Hyde Park, NY

Augustine, Shaw JF (translator) (2009) On Christian Doctrine. Dover Publications, Mineola, NY

Augustine, Dods M (translator) (2000) The City of God. Modern Library, Random House, New York, NY

Baek EC, Hyon R, López K, Du M, Porter MA, Parkinson C (2023) Lonely individuals process the world in idiosyncratic ways. Psychological Science, 34(6), 683–695. https://doi.org/10.1177/09567976221145316

Bornstein MH, Putnick DL, Lansford JE, Al-Hassan SM, Bacchini D, Bombi AS, Chang L, Deater-Deckard K, Di Giunta L, Dodge KA, Malone PS, Oburu P, Pastorelli C, Skinner AT, Sorbring E, Steinberg L, Tapanya S, Tirado LMU, Zelli A, Alampay LP (2017) 'Mixed blessings': parental religiousness, parenting, and child adjustment in global perspective. J Child Psychol Psychiatry, 8, 880-892.

Brittain C (2012) Self-Knowledge in Cicero and Augustine (De Trinitate X, 5, 7–10, 16). Medioevo, 1(37),107-135.

Brooks SJ, Tian L, Parks SM, Stamoulis C (2022) Parental religiosity is associated with changes in youth functional network organization and cognitive performance in early adolescence. Scientific Reports, 12(1), 17305.

Courtney A, Meyer M (2020) Self-Other Representation in the Social Brain Reflects Social Connection. Journal of Neuroscience, 40(29), 5616-5627

Harnad S (1990) The symbol grounding problem. Physica D Nonlinear Phenomena, 42(1–3), 335–346. https://doi.org/10.1016/0167-2789(90)90087-6

He BJ, Zempel JM, Snyder AZ, Raichle ME (2010) The temporal structures and functional significance of scale-free brain activity. Neuron, 66(3), 353-369

Hundert EJ (1992) Augustine and the Sources of the Divided Self. Political Theory, 20(1), 86–104. https://doi.org/10.1177/0090591792020001005

Huxley A (2006) Brave New World. Harper Collins, NY:NY

Limanowski J, Friston K (2020) Attenuating oneself: An active inference perspective on "selfless" experiences. Philosophy and the Mind Sciences 1, 1-16

Northoff G (2018) How does the brain's spontaneous activity generate our thoughts? The spatiotemporal theory of task-unrelated thought (STTT). The Oxford handbook of spontaneous thought: Mind-wandering, creativity, and dreaming, pp 55-70

Northoff G, Klar P, Bein M, Safron A (2023) As without, so within: how the brain's temporo-spatial alignment to the environment shapes consciousness. Interface Focus, 13(3): 20220076

Northoff G, Mushiake H (2020) Why context matters? Divisive normalization and canonical microcircuits in psychiatric disorders. Neuroscience Research, 156, 130-40

Northoff G, Smith D (2022) The subjectivity of self and its ontology: From the world–brain relation to the point of view in the world. Theory & Psychology: 09593543221080120

Northoff G, Vatansever D, Scalabrini A, Stamatakis EA (2022) Ongoing Brain Activity and Its Role in Cognition: Dual versus Baseline Models. The Neuroscientist: 10738584221081752

Ortega y Gasset J, Garcia-Gomez J (translator) (2002) What Is Knowledge? State University of New York Press, Albany, NY.

Paine RW, Tani J (2005) How hierarchical control self-organizes in artificial adaptive systems. Adaptive Behavior, 13(3), 211-225

Smith D, Wolff A, Wolman A, Ignaszewski J, Northoff G. (2022) Temporal continuity of self: long autocorrelation windows mediate self-specificity. NeuroImage, 257: 119305

So WY, Friston KJ, Neacsu V (2024) The Inherent Normativity of Concepts. Minds & Machines, 34(40). https://doi.org/10.1007/s11023-024-09697-7

Sting (1991) Why should I cry for you? Soul Cages. A&M Polygram, Hong Kong

Tani J (1998) An interpretation of the 'self' from the dynamical systems perspective: a constructivist approach. J Conscious Stud, 5, 516–542

Tani J (2009) Autonomy of Self at criticality: The perspective from synthetic neuro-robotics. Adaptive Behavior, 17(5), 421-443

Tani J (2016) Exploring Robotic Minds: Actions, Symbols, and Consciousness As Self-Organizing Dynamic Phenomena. Oxford University Press, Oxford, UK

Tani J, White J (2022) Cognitive neurorobotics and self in the shared world, a focused review of ongoing research. Adaptive Behavior, 30(1), 81-100
White JB (2006) Conscience: toward the mechanism of morality. Dissertation, University of Missouri-Columbia. http://hdl.handle.net/10355/4327
White J (2010) Understanding and augmenting human morality: an introduction to the ACTWith model of conscience. In: Magnani L (ed) Model-based reasoning in science and technology: abduction, logic and computational discovery. Springer, Berlin, pp 607–621
White J (2012) An information processing model of psychopathy and anti-social personality disorders integrating neural and psychological accounts towards the assay of social implications of psychopathic agents. In: Fruili AS, Veneto LD (eds) Psychology of morality. Nova Science Publishers, Hauppauge, pp 1–34
White J (2013) Manufacturing morality: a general theory of moral agency grounding computational implementations. In: Floares A (ed) Computational intelligence. Nova Publications, Hauppage, pp 163–210
White J (2014) Models of moral cognition. In: Magnani L (ed) Model-based reasoning in science and technology: theoretical and cognitive issues. Springer, Berlin, pp 363–391
White J (2022) Autonomous Reboot: Kant, the categorical imperative, and contemporary challenges for machine ethicists. AI & Society, 37, 661–673. https://doi.org/10.1007/s00146-020-01142-4
White J (2022b) On a possible basis for metaphysical self development in natural and artificial systems. FILOZOFIA I NAUKA, 2022c:71
White J (2024) Augmenting morality through ethics education: the ACTWith model. AI & Society. https://doi.org/10.1007/s00146-024-01864-9
White J, Tani J, (2016) From Biological to Synthetic Neurorobotics Approaches to Understanding the Structure Essential to Consciousness, part 1, APA Newsl. Philos. Comput. 16(1), 13-23

AUTHOR INDEX